SpringerWienNewYork

CISM COURSES AND LECTURES

The series presents lecture notes, monographs, edited works and proceedings in the field of Mechanics, Engineering, Computer Science and Applied Mathematics.
Purpose of the series is to make known in the international scientific and technical community results obtained in some of the activities organized by CISM, the International Centre for Mechanical Sciences.

INTERNATIONAL CENTRE FOR MECHANICAL SCIENCES

COURSES AND LECTURES - No. 492

MULTIPHASE REACTING FLOWS: MODELLING AND SIMULATION

EDITED BY

DANIELE L. MARCHISIO
POLITECNICO DI TORINO, ITALY

RODNEY O. FOX
IOWA STATE UNIVERSITY, USA

SpringerWienNewYork

This volume contains 84 illustrations

SPIN 214440

All contributions have been typeset by the authors.

ISBN 978-3-211-72463-7 SpringerWienNewYork

PREFACE

This book entitled "Multiphase reacting flows: modelling and simulation" contains the lecture notes of the CISM (International Centre for Mechanical Sciences) course held in Udine, Italy, on July 3-7, 2006, and it describes various modelling approaches for dealing with polydisperse multiphase reacting flows.

A multiphase reacting system is characterized by the presence of multiple phases and in this book we focus on disperse multiphase systems, where one phase can be considered as a continuum, whereas the additional phases are dispersed in the continuous one. In other words, in this book we deal with multiphase systems constituted by particles, droplets or bubbles (i.e., solid particles suspended in a continuous liquid phase, liquid droplets in a gaseous phase, or gas bubbles in liquid.)

The other important characteristic elements of the systems discussed in this book are the presence of one or more chemical reactions and the turbulent nature of the flow. The chemical reactions usually involve all the phases present in the system and might be responsible for the formation or disappearance of the disperse and/or continuous phases. The evolution of the different phases is not only governed by chemical reactions, but also by other fluid-dynamical interactions between the continuous and the disperse phases, and by interactions among elements of the disperse phases, such as coalescence, aggregation, agglomeration and break-up.

All these phenomena are closely linked together resulting in what is known as phase coupling. When the continuous phase influences the disperse phase, we usually refer to one-phase coupling, whereas when the continuous phase influences the disperse phase and vice versa, we talk about two-way coupling. When also interactions between elements of the disperse phases are important, we must describe the turbulent system in terms of so-called four-way coupling.

The evolution of the disperse phase and its interactions with the continuous phase can be mathematically described at the mesoscopic level by the generalized population balance equation, also known as the Boltzmann-Williams equation. This book contains a detailed derivation of this equation, and presents several numerical approaches to solve it. In particular, the derivation of Eulerian classes methods (or multi-fluid approaches) is presented both for the solution of laminar and turbulent multiphase systems. Moreover the use of alternative Eulerian methods, such as moment methods coupled with quadrature approximations, is also presented. The book deals also with the use of Lagrangian approaches such as direct particle tracking methods and Lattice-Boltzmann methods. Moreover, reference to the general modelling framework for turbulent flows (i.e. Reynolds-Average Navier Stokes approach, Large-Eddy Simulation and Direct Numerical Simulation) is made, as well as to computational details and numerical issues.

In the first chapter, Fox provides an overview of the basic formulation and conceptual ideas needed for modelling polydisperse multiphase systems. Special emphasis is given to systems exhibiting polydispersity in one or more internal coordinate (e.g. velocity and droplet size). Such systems are described by a generalized population balance equation, governing a number density function, which can be solved using sectional or moment methods. An example application to the spray equation is used to illustrate the power of moment methods based on quadrature closures.

In the second chapter, Marchisio presents several moment methods for the solution of population balance equations and discusses in particular the use of methods based on a quadrature approximation.

In the third chapter, Massot reviews the various multi-fluid methods available for the numerical simulation of dilute and moderately dense polydisperse systems, with particular attention to evaporating sprays. First the simple laminar case is treated, and then the extension of the approach to turbulent inhomogeneous flows is discussed. The prevalence of "singularities" in the number density fields resulting from multi-fluid models is demonstrated with simple examples, and ideas for improving the models are discussed.

In the fourth chapter, Hjertager focuses on sectional methods and particular emphasis is given multi-fluid techniques for turbulent systems. Turbulence modelling strategies for gas-particle flows based on the kinetic theory for granular flows as well as strategies for gas-liquid flows are given. Eventually examples are shown for several gas-particle systems including flow in risers, segregation by size and reacting systems as well as gas-liquid systems including bubble columns and stirred tanks.

In the fifth chapter, Derksen describes the lattice-Boltzmann method for multiphase fluid flow simulations. In particular cellular automata for mimicking physical systems, the concepts of lattice gas and lattice-Boltzmann automata, and fluid dynamics with the lattice-Boltzmann method are presented and applied for direct numerical simulation of solid-liquid suspensions and direct numerical and large-eddy simulations of single-phase turbulent reacting systems.

Finally, in the sixth chapter, Reveillon presents the various issues related to some very detailed modelling approaches based on direct numerical simulations of sprays. In particular the possibility of simulating the continuous phase using an Eulerian approach and a Lagrangian approach for the disperse phase is discussed, and some results obtained in the fields of turbulent dispersion, evaporation and combustion of sprays in combustion chambers are presented.

Daniele L. Marchisio and Rodney O. Fox

CONTENTS

Introduction and Fundamentals of Modeling Approaches for Polydisperse Multiphase Flows

Rodney O. Fox

Department of Chemical & Biological Engineering, Iowa State University, Ames, IA, USA

Abstract An overview of the basic formulation and conceptual ideas needed for modeling polydisperse multiphase systems is provided. Special emphasis is given to systems exhibiting polydispersity in more than one internal coordinate. Such systems are described by a multivariate population balance equation, governing a number density function, which can be solved using sectional or moment methods. When the particle velocity is treated as a fluctuating quantity, the corresponding number density function is the one-point velocity density function used in kinetic theory. For this special case, a generalized population balance equation is employed to describe polydispersity in the velocity and other internal coordinates (such as the particle size.) Here, due to their flexibility in treating inhomogeneous flows, we focus on quadrature-based moment methods and show how moment transport equations can be derived from the generalized population balance equation for polydisperse multiphase flows. An example application to the one-dimensional spray equation is used to illustrate the modeling concepts.

1 Introduction

The governing equations needed to describe polydisperse multiphase flows are presented in this work. For clarity, our discussion of the governing equations will be limited to particulate systems. However, the reader familiar with multiphase flow modeling will recognize that our comments hold in a much more general context (Fox, 2006b). The starting point for our discussion will be the one-point number density function (NDF).

1.1 Number density functions

The dispersed phase is constituted by discrete elements. Each element of the dispersed phase is generally identified by a number of properties known as *coordinates*. Two elements are identical if they have identical values for their coordinates, otherwise elements are indistinguishable. Usually coordinates are classified as *internal* and *external*. External coordinates are spatial coordinates. Internal coordinates refer to properties of the elements such as their momenta (or velocities), their enthalpies (or temperatures), or their volumes, surface areas, or sizes. When using an Eulerian approach, the dispersed phase is generally described by a number density function (NDF). The NDF contains information about how the population of elements inside a control volume is distributed over the properties of interest.

Let us consider a population of dispersed entities such as solid particles inside a control volume located at the physical point $\mathbf{x} = (x_1, x_2, x_3)$ and of measure

$$\mathrm{d}\mathbf{x} = \mathrm{d}x_1 \mathrm{d}x_2 \mathrm{d}x_3.$$

Let $\boldsymbol{\xi} = (\xi_1, \xi_2, \ldots, \xi_N)$ be the internal-coordinate vector. The NDF $n_{\boldsymbol{\xi}}(\boldsymbol{\xi}; \mathbf{x}, t)$ is defined as the expected number of entities in the physical volume $\mathrm{d}\mathbf{x}$ and in the phase-space volume $\mathrm{d}\boldsymbol{\xi}$:

$$n_{\boldsymbol{\xi}} \, \mathrm{d}\mathbf{x} \, \mathrm{d}\boldsymbol{\xi}. \tag{1.1}$$

Note that $n_{\boldsymbol{\xi}}$ is a function of time t, space $\mathbf{x}$ and the internal-coordinate vector $\boldsymbol{\xi}$. Hereinafter we will include the explicit dependence on these variables only when required for clarity. Although stochastic algorithms can be used to simulate the evolution of a population of particles represented by the NDF, the NDF is not a random quantity. Rather it is the ensemble average of infinite number of realizations of the stochastic algorithm. The NDF is an average quantity of the dispersed phase. In fact, it has the mathematical characteristics of an averaged function: it is smooth and differentiable with respect to time, physical space and internal-coordinate space.

It is straightforward that the quantity

$$n_{\boldsymbol{\xi}}(\boldsymbol{\xi}) \, \mathrm{d}\boldsymbol{\xi} \tag{1.2}$$

represents the number density of dispersed entities contained in the phase-space volume $\mathrm{d}\boldsymbol{\xi}$ per unit of physical volume. If we integrate the NDF over all possible values of the internal-coordinate vector $\Omega_{\boldsymbol{\xi}}$ we obtain the total particle-number concentration $N(\mathbf{x}, t)$:

$$N = M_{\boldsymbol{\xi}}(0) \equiv \int_{\Omega_{\boldsymbol{\xi}}} n_{\boldsymbol{\xi}}(\boldsymbol{\xi}) \, \mathrm{d}\boldsymbol{\xi}, \tag{1.3}$$

which is the total number of entities per unit volume located at point $\mathbf{x}$ at time t and corresponds to zeroth moment of the density function, whereas

$$M_{\boldsymbol{\xi}}(\mathbf{k}) \equiv \int_{\Omega_{\boldsymbol{\xi}}} \xi_1^{k_1} \cdots \xi_N^{k_N} n_{\boldsymbol{\xi}}(\boldsymbol{\xi}) \, \mathrm{d}\boldsymbol{\xi}, \tag{1.4}$$

and $\mathbf{k} = (k_1, \ldots, k_N)$ is a vector containing the order of the moments with respect to each of the components of $\boldsymbol{\xi}$.

In order to introduce some particular concepts and properties, we consider a population of particles that can be described by using only one internal coordinate. Let us imagine that the particles can be characterized by their length L. Then $n_L(L; \mathbf{x}, t)$ is the length-based NDF, and $n_L \, \mathrm{d}L$ represents the expected number of particles with length between L and $L + \mathrm{d}L$. The total particle-number concentration can be calculated as

$$N = M_L(0) = \int_0^\infty n_L(L) \, \mathrm{d}L, \tag{1.5}$$

whereas a mean particle size can be defined as

$$L_{10} \equiv \frac{1}{N} \int_0^\infty L n_L(L) \, \mathrm{d}L = \frac{M_L(1)}{M_L(0)}. \tag{1.6}$$

This average property of the distribution is defined through the number density itself and it represents the mean particle size with respect to the number of particles in the system. Of course, other definitions are possible as it will become clearer below.

If we define the kth moment of the length-based NDF as

$$M_L(k) \equiv \int_0^\infty L^k n_L(L)\,\mathrm{d}L, \tag{1.7}$$

it is easy to see that a mean particle size can be defined as the ratio $M_L(k+1)/M_L(k)$ for any value of k. For example, the Sauter mean diameter is defined by setting $k = 2$:

$$L_{32} \equiv \frac{M_L(3)}{M_L(2)}, \tag{1.8}$$

and is useful for modeling processes such as mass transfer where the surface area is a key parameter.

If volume scales with the third power of length, we can relate particle volume V and particle length L through a volumetric shape factor k_v:

$$V = k_v L^3. \tag{1.9}$$

The resulting *volume density function* $V_L(L; \mathbf{x}, t)$ represents the volume of particles per unit spatial volume with lengths between L and $L + \mathrm{d}L$:

$$V_L \equiv k_v L^3 n_L. \tag{1.10}$$

For fractal-shaped particles, this volume cannot be used to compute the mass of the particle. Instead, the mass is related to the length through the *fractal dimension*. After normalizing V_L to unity, the *volume-fraction density function* $\alpha_V(L; \mathbf{x}, t)$ represents the volume fraction of particles with a specific length over the total particle volume:

$$\alpha_V \equiv \frac{L^3 n_L}{\int_0^\infty L^3 n_L(L)\,\mathrm{d}L}. \tag{1.11}$$

The mean particle length calculated from this volume-fraction density function is

$$L_{43} = \int_0^\infty \alpha_V(L) L\,\mathrm{d}L = \frac{M_L(4)}{M_L(3)}. \tag{1.12}$$

This definition of particle length is widely used for particulate systems.

Another very common internal coordinate for particulate systems is particle volume V. Using a definition very similar to Eq. (1.1), we can define the volume-based NDF as the expected number of particles with volume between V and $V + \mathrm{d}V$:

$$n_V\,\mathrm{d}\mathbf{x}\,\mathrm{d}V. \tag{1.13}$$

If volume and length scale with the third power as in Eq. (1.9), the relationship between the length-based and the volume-based NDF is straightforward. If fact, $n_L\,\mathrm{d}L$ quantifies the expected number of particles with length between L and $L + \mathrm{d}L$, which must be equal to the expected number of particles with volume between V and $V + \mathrm{d}V$ if Eq. (1.9) holds. Moreover it is easy to see that

$$\mathrm{d}V = 3k_v L^2 \,\mathrm{d}L. \tag{1.14}$$

Therefore, equality in the expected numbers requires

$$n_L(L)\,\mathrm{d}L = n_V(V)\mathrm{d}V \tag{1.15}$$

or, using Eqs. (1.9) and (1.14),

$$n_L(L)\,\mathrm{d}L = n_V(k_v L^3) 3k_v L^2 \,\mathrm{d}L, \tag{1.16}$$

from where we find that

$$n_L(L) = 3k_v L^2 n_V(k_v L^3), \tag{1.17}$$

which essentially tells us how to define n_L in terms of n_V. The same procedure can be used to find the NDF in terms of any well-defined function of L.

In the literature on particle aggregation and droplet coalescence, a mass-based NDF $n_m(m; \mathbf{x}, t)$ is often used. For these systems, it is possible to relate the mass of a particle to its characteristic length using

$$m = k_m L^{d_\mathrm{f}}, \tag{1.18}$$

where $1 < d_\mathrm{f} \leq 3$ is the *fractal dimension*. ($d_\mathrm{f} = 3$ for spherical particles and $d_\mathrm{f} \approx 1.8$ for diffusion-limited aggregates.) Using the procedure outline above, we can show that the length-based and mass-based NDF are related by

$$n_L(L) = d_\mathrm{f} k_m L^{d_\mathrm{f}-1} n_m(k_m L^{d_\mathrm{f}}). \tag{1.19}$$

Likewise, their moments are related by

$$M_m(k) = k_m^k M_L(d_\mathrm{f} k). \tag{1.20}$$

It is often the case that in closed systems the total particle mass is conserved, i.e., $M_m(1)$ is constant. In a length-based analysis, this corresponds to conservation of the fractional moment $M_L(d_\mathrm{f})$.

A special case of considerable interest occurs when the internal-coordinate vector is the particle velocity vector $\mathbf{U}_\mathrm{p}$. For multiphase flows, it will be necessary to distinguish between the velocities of the various phases. For example, for gas-solid flow we will denote the fluid velocity vector by $\mathbf{U}_\mathrm{f}$. However, for the present discussion, we can assume that the particles are traveling in a vacuum. In fact, particle velocity is a *special* internal coordinate since it is related to particle position (i.e., external coordinates). It is possible to define a velocity-based NDF $n_{\mathbf{U}_\mathrm{p}}(\mathbf{U}_\mathrm{p}; \mathbf{x}, t)$ that is parameterized by the velocity components $\mathbf{U}_\mathrm{p} = (U_{\mathrm{p}1}, U_{\mathrm{p}2}, U_{\mathrm{p}3})$.

In order to obtain the total particle-number concentration (i.e., number of particles per unit volume) it is sufficient to integrate over all possible values of particle velocity $\Omega_\mathbf{v}$:

$$N = M_{\mathbf{U}_\mathrm{p}}(0) \equiv \int_{\Omega_\mathbf{v}} n_{\mathbf{U}_\mathrm{p}}(\mathbf{U}_\mathrm{p})\,\mathrm{d}\mathbf{U}_\mathrm{p}, \tag{1.21}$$

which is just the zero-order moment of the velocity-based NDF. Because we have assumed that particles may have different velocities, we can calculate for each spatial location and at each time instant an average velocity $\mathbf{U}_{p10}$ using the first two moments of the velocity-based NDF:

$$\mathbf{U}_{p10} \equiv \frac{1}{N} \int_{\Omega_v} \mathbf{U}_p n_{\mathbf{U}_p}(\mathbf{U}_p)\, d\mathbf{U}_p = \frac{M_{\mathbf{U}_p}(1)}{M_{\mathbf{U}_p}(0)}, \tag{1.22}$$

which is analogous to the definition of the mean particle length reported in Eq. (1.6).

It is clear that the NDF n_{ξ} and the velocity-based NDF $n_{\mathbf{U}_p}$ have the same mathematical meaning, and can be treated in a similar framework. In what follows we will derive the transport equations for these NDFs, keeping separate the treatment for *"standard"* internal coordinates such as particle size, volume, area, or temperature, from *"special"* internal coordinates such as particle velocity. We will refer to the first transport equation as the population balance equation (PBE) and to the second one as a generalized population balance equation (GPBE). The reader should keep in mind that the key distinction between the two models is that for the PBE the velocity of the dispersed phase is assumed to be known (e.g., identical to the continuous phase), and thus the NDF can be treated as a scalar field.

1.2 Transport in phase space

We now consider the evolution of the NDF in time, internal-coordinate and physical space, and we show that the underlying mathematical structure is very similar for both the PBE and the GPBE.

Population balance equation The population balance equation (PBE) is a simple continuity statement written in terms of the NDF. It can be derived as a balance for particles in some fixed subregion of internal-coordinate and physical space (Ramkrishna, 2000). Let us consider a finite control volume in physical space $\Omega_\mathbf{x}$ and in phase space Ω_{ξ} with boundaries defined as $\partial\Omega_\mathbf{x}$ and $\partial\Omega_{\xi}$, respectively. The particle-number balance equation can be written as

$$\frac{\partial}{\partial t} \int_{\Omega_\mathbf{x}\Omega_{\xi}} n_{\xi}\, d\mathbf{x} d\xi + \int_{\Omega_{\xi}\partial\Omega_\mathbf{x}} n_{\xi} \mathbf{U}_p \cdot d\mathbf{A}_\mathbf{x} d\xi + \int_{\Omega_\mathbf{x}\partial\Omega_{\xi}} n_{\xi}\, \dot{\xi} \cdot d\mathbf{A}_{\xi} d\mathbf{x} = \int_{\Omega_\mathbf{x}\Omega_{\xi}} h_{\xi}\, d\mathbf{x} d\xi, \tag{1.23}$$

where $\mathbf{U}_p$ is the velocity vector for the particulate system, $\dot{\xi}$ is the continuous rate of change in the internal-coordinate space, and h_{ξ} is the discontinuous jump function representing discrete events.

The first term in Eq. (1.23) is an integral over the control volume and represents accumulation. The second term is a surface integral over the boundary of the control volume and represents the net flux of number density due to convection in physical space. Integrals over a control volume boundary are defined through the infinitesimal surface unit vector $d\mathbf{A}_\mathbf{x}$ with magnitude equal to the measure of the infinitesimal surface and direction defined by the vector normal to the surface itself. The third term represents instead convection in the internal-coordinate space. In fact, particles move in physical space because of the particle velocity $\mathbf{U}_p = d\mathbf{x}/dt$ and in the internal-coordinate space because of a internal-coordinate velocity $\dot{\xi} = d\xi/dt$. This vector contains in each component the rate of change of the ith internal coordinate for continuous processes. We generally refer to *continuous* processes when internal coordinates change continuously with a time scale much smaller than the one characterizing solutions to the PBE. Therefore, although all processes (at least at the molecular level) are discontinuous, these processes can be treated as continuous in Eq. (1.23).

If the Reynolds-Gauss theorem (Aris, 1962) is applied to Eq. (1.23) and the integrals on the boundary of the control volume are written in terms of integrals over the control volume itself, it is straightforward to obtain

$$\int_{\Omega_{\mathbf{x}}\Omega_{\boldsymbol{\xi}}} \left(\frac{\partial}{\partial t} n_{\boldsymbol{\xi}} + \nabla_{\mathbf{x}} \cdot \left(\mathbf{U}_{\mathrm{p}} n_{\boldsymbol{\xi}} \right) + \nabla_{\boldsymbol{\xi}} \cdot \left(\dot{\boldsymbol{\xi}} n_{\boldsymbol{\xi}} \right) - h_{\boldsymbol{\xi}} \right) \mathrm{d}\mathbf{x}\, \mathrm{d}\boldsymbol{\xi} = 0, \tag{1.24}$$

where $\nabla_{\mathbf{x}} = (\partial/\partial x_1, \partial/\partial x_2, \partial/\partial x_3)$ is the gradient vector in three-dimensional physical space and $\nabla_{\boldsymbol{\xi}} = (\partial/\partial \xi_1, \ldots, \partial/\partial \xi_N)$ is the gradient vector in N-dimensional internal-coordinate space. In order for Eq. (1.24) to be satisfied for any arbitrary control volumes $\Omega_{\mathbf{x}}$ and $\Omega_{\boldsymbol{\xi}}$, at every point $\mathbf{x}$ and $\boldsymbol{\xi}$ the integrand must satisfy the relation

$$\frac{\partial n_{\boldsymbol{\xi}}}{\partial t} + \nabla_{\mathbf{x}} \cdot \left(\mathbf{U}_{\mathrm{p}} n_{\boldsymbol{\xi}} \right) + \nabla_{\boldsymbol{\xi}} \cdot \left(\dot{\boldsymbol{\xi}} n_{\boldsymbol{\xi}} \right) = h_{\boldsymbol{\xi}}, \tag{1.25}$$

or, using Einstein notation (i.e., repeated Roman indices imply summation),

$$\frac{\partial n_{\boldsymbol{\xi}}}{\partial t} + \frac{\partial}{\partial x_i} \left(U_{\mathrm{p}i} n_{\boldsymbol{\xi}} \right) + \frac{\partial}{\partial \xi_i} \left(\dot{\xi}_i n_{\boldsymbol{\xi}} \right) = h_{\boldsymbol{\xi}}. \tag{1.26}$$

Note that although we treat $\mathbf{x}$ and $\boldsymbol{\xi}$ in the same manner, they are in fact different types of vectors. The vectors $\mathbf{x}$ and $\mathbf{U}_{\mathrm{p}}$ are the standard vectors for position and velocity used in continuum mechanics. The internal-coordinate vector $\boldsymbol{\xi}$, on the other hand, is a generalized vector of length N in the sense of linear algebra.

Generalized population balance equation By using a very similar approach to the one outlined above for the PBE, it is possible to derive a GPBE for a NDF that includes particle velocity as an internal variable. We will denote this general NDF as $n(\mathbf{U}_{\mathrm{p}}, \boldsymbol{\xi}; \mathbf{x}, t)$ (i.e., without subscripts on n). The simplest GPBE (i.e., velocity without other internal coordinates) is known as the Boltzmann equation and was first derived in the context of kinetic theory (Chapman and Cowling, 1961). The final form of the GPBE is

$$\frac{\partial n}{\partial t} + \nabla_{\mathbf{x}} \cdot \left(\mathbf{U}_{\mathrm{p}} n \right) + \nabla_{\mathbf{U}_{\mathrm{p}}} \cdot \left(\mathbf{F}_{\mathrm{p}} n \right) + \nabla_{\boldsymbol{\xi}} \cdot \left(\dot{\boldsymbol{\xi}} n \right) = h \tag{1.27}$$

where $\nabla_{\mathbf{U}_{\mathrm{p}}} = (\partial/\partial U_{\mathrm{p}1}, \partial/\partial U_{\mathrm{p}2}, \partial/\partial U_{\mathrm{p}3})$ is the gradient vector in three-dimensional particle velocity space and $\mathbf{F}_{\mathrm{p}} = \mathrm{d}\mathbf{U}_{\mathrm{p}}/\mathrm{d}t$ is the continuous rate of change of particle velocity, namely the force per unit mass acting on particles. The right-hand side h is again the discontinuous jump term, but now including discontinuous changes in particle momentum.

The first term on the left-hand side of Eq. (1.27) is the accumulation term, whereas the second one is a drift term that accounts for convection in physical space due to the particle velocity. The third term is the drift term representing convection in particle velocity space. In fact, a force acting on particles produces a uniform acceleration, or in other words a continuous change in particle velocity. The term on the right-hand side represents discontinuous jumps in the particle-velocity space and in internal-coordinate space due to discrete events, mainly collisions between particles, and for this reason it is also known as a collision integral. As is well known in statistical physics, the collision integral is not closed since it involves velocity correlations between two particles. Thus, a closure hypothesis must be introduced to define h in terms of n. When two particles collide they exchange momentum and therefore in a single instant they change their momenta (and therefore velocities) from one set of specific values to another one. In this respect the collision process is discrete and produces a finite and discontinuous change in particle-velocity space.

Now that we have introduced the GPBE transport equation, it is appropriate to discuss the problem of closure. In general, we say that Eq. (1.27) is *closed* if all of the terms in the equation can be computed from knowledge of $n(\mathbf{U}_\mathrm{p}, \boldsymbol{\xi}; \mathbf{x}, t)$ and its moments. Since the latter are computed from n, mention of the moments is redundant. Nevertheless, we mention it anyway to remind the reader that closure at the level of the GPBE includes as a subset closure at the level of moments. We will in fact assume that the GPBE of interest is closed and focus on methods for computing $n(\mathbf{U}_\mathrm{p}, \boldsymbol{\xi}; \mathbf{x}, t)$. Finally, we want to remind the reader that even if the GPBE is closed, this does not imply that the *moment equations* will be closed. In fact, in most problems of practical interest this will not be the case. Nevertheless, it is often useful to know the form of the (unclosed) moment transport equations. In the following, we give a short overview on how to find moment equations.

1.3 Moment equations

It was already mentioned that the PBE (Eq. 1.25) and the GPBE (Eq. 1.27) have a very similar structure. It is useful to derive from the transport equations some integral quantities of interest (namely moments), and to define their transport equations. For simplicity let us consider a PBE with only particle length as the internal coordinate. The PBE reported in Eq. (1.25) then becomes

$$\frac{\partial n_L}{\partial t} + \nabla_\mathbf{x} \cdot \left(\mathbf{U}_\mathrm{p} n_L\right) + \frac{\partial}{\partial L}\left(G_L n_L\right) = h_L, \tag{1.28}$$

where the continuous rate of change of particle length $\dot{L}$ is denoted by G_L and discrete events are denoted by h_L.

Applying the moment transform (Eq. 1.7) to Eq. (1.28) we obtain

$$\frac{\partial M_L(k)}{\partial t} + \nabla_\mathbf{x} \cdot \left(\mathbf{U}_\mathrm{p} M_L(k)\right) = k G_L(k-1) M_L(k-1) + h_L(k), \tag{1.29}$$

where the average particle growth rate is defined by

$$G_L(k) = \frac{\int_0^\infty G_L n_L(L) L^k \,\mathrm{d}L}{\int_0^\infty n_L(L) L^k \,\mathrm{d}L}, \tag{1.30}$$

and the moment transform of h_L by

$$h_L(k) = \int_0^\infty h_L L^k \,\mathrm{d}L. \tag{1.31}$$

If G_L is negative for $0 \leq L$, then an additional sink term may be needed in Eq. (1.29) to account for particle loss due to disappearance at the origin. This occurs, for example, in systems composed of evaporating droplets when the evaporation rate is proportional to surface area. In the equations below, for simplicity, we will assume that particle growth rate G_L is size independent so that $G_L(k) = G$. However, in general, the functional dependence of G_L on size must be provided by the modeler in order to apply Eq. (1.30).

The zero moment $M_L(0)$ is equivalent to the total particle-number concentration N and its transport equation follows from Eq. (1.29):

$$\frac{\partial N}{\partial t} + \nabla_{\mathbf{x}} \cdot \left(\mathbf{U}_{\mathrm{p}} N\right) = h_L(0). \tag{1.32}$$

The drift term has disappeared in Eq. (1.32) since the continuous growth of particle size does not change the total particle-number concentration. However, N is influenced by the rate of formation of particles (e.g., nucleation), and the rates of aggregation and breakage, which cause appearance and disappearance of particles. These processes are all contained in the source term $h_L(0)$.

The third moment $M_L(3)$ is related to the fraction of volume occupied by particles with respect to the suspending fluid and can be easily found from Eq. (1.29):

$$\alpha_{\mathrm{p}} = \int_0^\infty k_v L^3 n_L(L)\,\mathrm{d}L = k_v M_L(3), \tag{1.33}$$

where α_{p} indicates the volume fraction for the particle phase. The fluid phase is denoted by α_{f}, and $\alpha_{\mathrm{p}} + \alpha_{\mathrm{f}} = 1$. The resulting transport equation for the volume fraction of the disperse phase is

$$\frac{\partial \alpha_{\mathrm{p}}}{\partial t} + \nabla_{\mathbf{x}} \cdot \left(\mathbf{U}_{\mathrm{p}} \alpha_{\mathrm{p}}\right) = 3 k_v G M_L(2) + k_v h_L(3). \tag{1.34}$$

It is interesting to note that if the drift term is null (i.e., particles are not growing in size) and if there is no introduction of new particles into the system (i.e., the nucleation rate is null), the right-hand side of Eq. (1.34) is null, resulting is a continuity equation for the particle phase of the form

$$\frac{\partial \alpha_{\mathrm{p}}}{\partial t} + \nabla_{\mathbf{x}} \cdot \left(\mathbf{U}_{\mathrm{p}} \alpha_{\mathrm{p}}\right) = 0. \tag{1.35}$$

A very common case for which $h_L(3)$ is null but the nucleation rate is non-zero occurs when nucleation produces nuclei with zero size. For this case, in fact, only $h_L(0)$ is non-zero.

The process of finding moments starting from the PBE can be continued to arbitrary order. We should note that in most applications the resulting moment equations *will not be closed*. In other words, the moment equation of order k will involve higher-order moments.

The very same approach as described above for the PBE can be applied to the GPBE (Eq. 1.27). Applying the moment transform to Eq. (1.27) for the zero-order moment, the transport equation for the total particle-number concentration is obtained:

$$\frac{\partial N}{\partial t} + \nabla_{\mathbf{x}} \cdot \left(\mathbf{U}_{\mathrm{p}10} N\right) = h(0), \tag{1.36}$$

where the moment transform of the collision term is

$$h(0) \equiv \int_{\Omega_{\mathbf{U}_{\mathrm{p}}}} \int_{\Omega_{\boldsymbol{\xi}}} h(\mathbf{U}_{\mathrm{p}}, \boldsymbol{\xi})\,\mathrm{d}\mathbf{U}_{\mathrm{p}}\,\mathrm{d}\boldsymbol{\xi} \tag{1.37}$$

and (see Eq. 1.22) $\mathbf{U}_{p10}$ is the average particle velocity. This transport equation is very similar to Eq. (1.32). We can note that for the case where the particle size is constant ($h(0) = 0$), the relation $\alpha_p = k_v L^3 N$ will allow us to rewrite Eq. (1.36) as a balance for the particle volume fraction.

If the first-order particle-velocity moment transform is applied, we obtain

$$\frac{\partial \mathbf{U}_{p10} N}{\partial t} + \nabla_{\mathbf{x}} \cdot \left(\mathbf{U}_{p20} N \right) = \mathbf{F}_p(0) + \mathbf{h}(1), \tag{1.38}$$

where

$$\mathbf{U}_{p20} \equiv \frac{1}{N} \int_{\Omega_{U_p}} \mathbf{U}_p \mathbf{U}_p n(\mathbf{U}_p, \boldsymbol{\xi}) \, \mathrm{d}\mathbf{U}_p \, \mathrm{d}\boldsymbol{\xi}, \tag{1.39}$$

$$\mathbf{F}_p(0) \equiv \int_{\Omega_{U_p}} \mathbf{F}_p n(\mathbf{U}_p, \boldsymbol{\xi}) \, \mathrm{d}\mathbf{U}_p \, \mathrm{d}\boldsymbol{\xi}, \tag{1.40}$$

and

$$\mathbf{h}(1) \equiv \int_{\Omega_{U_p}} \int_{\Omega_{\xi}} \mathbf{U}_p h(\mathbf{U}_p, \boldsymbol{\xi}) \, \mathrm{d}\mathbf{U}_p \, \mathrm{d}\boldsymbol{\xi}. \tag{1.41}$$

Equation (1.38) is a transport equation for the mean particle velocity $\mathbf{U}_{p10}$ and has a structure very similar to the ones previously reported. Note that the term $\nabla_{\mathbf{x}} \cdot (\mathbf{U}_{p20} N)$ is a scalar product of the spatial gradient vector and the tensor produced by the diadic product of velocity vectors. The first term on the left-hand side represents accumulation whereas the second one represents convection of velocity in physical space. The first term on the right-hand side represents the acceleration produced by a force acting on the particles, whereas the second term represents the change of momentum due to discrete events such as particle collisions. We note in passing that nearly all of the terms in Eq. (1.38) are unclosed at the moment level. We can also note that for the case where the particle density and size are constant ($\mathbf{h}(1) = 0$), the relation $\rho_p \alpha_p = \rho_p k_v L^3 N$ will allow us to rewrite Eq. (1.38) as a balance for the particle-phase momentum ($\rho_p \alpha_p \mathbf{U}_{p10}$).

1.4 Physical-space transport

The PBE and the GPBE contain terms describing transport in physical space. These terms have to be treated with different approaches according to the type of flow. For example when dealing with laminar flow, particles move because of convection and molecular diffusion due to Brownian motion, whereas when dealing with turbulent flow fluctuations in n_{ξ} due to turbulent mixing must also be taken into account. We show how this works for two simple examples. Note that in the literature the distinction between laminar and turbulent flow in often not made when deriving models for multiphase systems, and hence it is unclear exactly what types of "fluctuations" are being modeled. We take the approach that the NDF and GPBE should first be well defined for laminar flows. Reynolds averaging of the resulting moment transport equation can then be used to model turbulent flows. In other words, in our definition of multiphase flow models turbulence is associated with random vorticity in velocity moments such as $\mathbf{U}_{p10}$ (Eq. 1.22).

Laminar flow In the case of laminar flow, particles move due to their velocity $\mathbf{U}_p$. It is interesting to note that also for laminar flow it could be necessary to refer to a generalized mean velocity. In fact, a population of particles with different sizes in laminar flow will have different velocities because of the different drag forces acting on each particle. Therefore, even in laminar flow, the population can show a distribution not only over particle size but also over particle velocity. Likewise, when particles are small enough, they will be subject to Brownian motion, which can be thought of as a random, chaotic component for particle velocity that has to be added to the mean velocity. This qualitative statement can become quantitative if particle size is compared to the mean free path for the molecules constituting the continuous phase. If particle size is much larger than the mean free path then Brownian motion is negligible and vice versa. When describing Brownian motion in a Lagrangian framework, the conditional particle velocity $\langle \mathbf{U}_\mathrm{p} | \boldsymbol{\xi} \rangle$ can be written as the sum of the mean component plus a fluctuating one, whereas in an Eulerian framework it can be represented by a flux of particles expressed as a size-dependent diffusion coefficient multiplied by the mean concentration gradient using the very same approach as used to define molecular properties (such as fluid viscosity) in kinetic theory. In this limit, Eq. (1.27) can be written in closed form as

$$\frac{\partial n_{\xi}}{\partial t} + \nabla_{\mathbf{x}} \cdot \left(\mathbf{U}_{\mathrm{p}0} n_{\xi} - \Gamma(\boldsymbol{\xi}) \nabla_{\mathbf{x}} n_{\xi} \right) + \nabla_{\xi} \cdot \left(\dot{\boldsymbol{\xi}} n_{\xi} \right) = h_{\xi}, \tag{1.42}$$

where $\Gamma(\boldsymbol{\xi})$ is the effective diffusion coefficient, which can be estimated, for example, by Nernst-Einstein theory (Bird et al., 2002). Note that when writing this equation, we implicitly assume that h_{ξ} can be expressed in terms of n_{ξ} so that the right-hand side is *closed*.

Turbulent flow In the case of turbulent flows all quantities fluctuate in a chaotic manner around their mean values. These fluctuations are due to the non-linear convection term in the mean-field transport equation for particle momentum. Obviously, if the mean-field particle momentum is turbulent, then so will be the higher-order moments. However, if the NDF n_{ξ} can be closed at the mean-field level as described by Eq. (1.42), then we can safely ignore the higher-order moments. In turbulent flows usually the Reynolds average is introduced (Pope, 2000). (A similar theory can be developed in the context of large-eddy simulations (LES) (Fox, 2003).) It consists of calculating ensemble-average quantities of interest (usually lower-order moments). Given a fluctuating property of a turbulent flow $\phi(\mathbf{x}, t)$, its Reynolds average at a fixed point in time and space is defined by

$$\langle \phi \rangle = \int_{-\infty}^{+\infty} \psi f_{\phi}(\psi) \, \mathrm{d}\psi, \tag{1.43}$$

where $f_{\phi}(\psi; \mathbf{x}, t)$ is the one-point probability density function (PDF) of $\phi(\mathbf{x}, t)$ defined by the following probability statement:

$$f_{\phi}(\psi) \, \mathrm{d}\psi = P[\psi \leq \phi(\mathbf{x}, t) < \psi + \mathrm{d}\psi]. \tag{1.44}$$

The probability on the right-hand side of this expression is computed from the ensemble of all random fields $\phi(\mathbf{x}, t)$ observable in a turbulent flow (Pope, 2000; Fox, 2003).

The random fields of primary interest in multiphase flows are the mean-field velocity of the dispersed phase $\mathbf{U}_{\mathrm{p}0}$ and the NDF n_{ξ} defined by the PBE (Eq. 1.42). We could also start with Eq. (1.25) wherein fluctuations around $\mathbf{U}_{\mathrm{p}0}$ are neglected. However, it is useful to start with Eq. (1.42) in order to illustrate the treatment of the diffusion term in a turbulent flow. We can define the one-point joint PDF of $\mathbf{U}_{\mathrm{p}0}$ and n_{ξ} as

$$f_{\mathbf{U}_{\mathrm{p}0}, n_{\xi}}(\mathbf{V}_{\mathrm{p}}, n_{\xi}^{*}) \, \mathrm{d}\mathbf{V}_{\mathrm{p}} \, \mathrm{d}n_{\xi}^{*} = P[(\mathbf{V}_{\mathrm{p}} \leq \mathbf{U}_{\mathrm{p}0}(\mathbf{x}, t) < \mathbf{V}_{\mathrm{p}} + \mathrm{d}\mathbf{V}_{\mathrm{p}}) \cap (n_{\xi}^{*} \leq n_{\xi}(\mathbf{x}, t) < n_{\xi}^{*} + \mathrm{d}n_{\xi}^{*})]. \tag{1.45}$$

Note that $\mathrm{d}n_{\xi}^{*}$ is a functional derivative and represents a positive perturbation of the entire function n_{ξ}^{*}. The function n_{ξ}^{*} represents an infinite number of scalars parameterized by $\boldsymbol{\xi}$. $\mathbf{V}_{\mathrm{p}}$ is the particle-velocity state space, which includes all the possible values of the particle velocity $\mathbf{U}_{\mathrm{p}0}$. Note that conceptually $f_{\mathbf{U}_{\mathrm{p}0},n_{\xi}}$ has the same meaning as the joint velocity, composition PDF used in turbulent reacting flows (Fox, 2003).

We can now define the Reynolds-averaged particle velocity as

$$\langle \mathbf{U}_{\mathrm{p}0} \rangle \equiv \iint_{-\infty}^{+\infty} \mathbf{V}_{\mathrm{p}} f_{\mathbf{U}_{\mathrm{p}0},n_{\xi}}(\mathbf{V}_{\mathrm{p}}, n_{\xi}^{*})\, \mathrm{d}\mathbf{V}_{\mathrm{p}}\, \mathrm{d}n_{\xi}^{*}. \tag{1.46}$$

Likewise, the Reynolds-average NDF is defined by

$$\langle n_{\xi} \rangle \equiv \iint_{-\infty}^{+\infty} n_{\xi}^{*} f_{\mathbf{U}_{\mathrm{p}0},n_{\xi}}(\mathbf{V}_{\mathrm{p}}, n_{\xi}^{*})\, \mathrm{d}\mathbf{V}_{\mathrm{p}}\, \mathrm{d}n_{\xi}^{*}. \tag{1.47}$$

It is also possible to define fluctuating components of each quantity as the difference between the instantaneous value and the Reynolds-average value:

$$\mathbf{u}_{\mathrm{p}0} = \mathbf{U}_{\mathrm{p}0} - \langle \mathbf{U}_{\mathrm{p}0} \rangle \tag{1.48}$$

and

$$n'_{\xi} = n_{\xi} - \langle n_{\xi} \rangle. \tag{1.49}$$

These quantities are used for Reynolds decomposition of higher-order terms:

$$\langle \mathbf{U}_{\mathrm{p}0} n_{\xi} \rangle = \langle \mathbf{U}_{\mathrm{p}0} \rangle \langle n_{\xi} \rangle + \langle \mathbf{u}_{\mathrm{p}0} n'_{\xi} \rangle, \tag{1.50}$$

where $\langle \mathbf{u}_{\mathrm{p}0} n'_{\xi} \rangle$ is called the *turbulent flux* of number-density fluctuations. Physically, this term denotes the advection of fluid elements with a given n'_{ξ} by velocity fluctuations, and thus enters the transport equation as a *spatial* flux.

The Reynolds-average NDF can be found from Eq. (1.42):

$$\frac{\partial \langle n_{\xi} \rangle}{\partial t} + \nabla_{\mathbf{x}} \cdot \left(\langle \mathbf{U}_{\mathrm{p}0} n_{\xi} \rangle - \Gamma(\boldsymbol{\xi}) \nabla_{\mathbf{x}} \langle n_{\xi} \rangle \right) + \nabla_{\xi} \cdot \left(\dot{\boldsymbol{\xi}} \langle n_{\xi} \rangle \right) = \langle h_{\xi} \rangle, \tag{1.51}$$

where we have assumed, for simplicity, that $\dot{\boldsymbol{\xi}}$ depends only on $\boldsymbol{\xi}$ and not, for example, on other scalar fields transported by the flow. When performing the Reynolds average, the internal-coordinate vector $\boldsymbol{\xi}$ is not affected. Thus, the Reynolds average commutes with time, space, and internal-coordinate gradients. See (Fox, 2003) for a discussion of this topic. If, for example, the growth rate depends on the concentration of a chemical species in the fluid phase, then this assumption would not hold. Using the Reynolds decomposition in Eq. (1.50) and a gradient-diffusion model for the turbulent flux (Fox, 2003), we can write

$$\langle \mathbf{U}_{\mathrm{p}0} n_{\xi} \rangle = \langle \mathbf{U}_{\mathrm{p}0} \rangle \langle n_{\xi} \rangle - \Gamma_{\mathrm{Tp}} \nabla_{\mathbf{x}} \langle n_{\xi} \rangle, \tag{1.52}$$

which closes the convection term in Eq. (1.51). The turbulent diffusivity Γ_{Tp} is found by solving a separate model for turbulent kinetic energy (k_{p}) and turbulent energy dissipation (ε_{p}) in the particle phase. Thus, using this approximation, Eq. (1.51) reduces to

$$\frac{\partial \langle n_{\xi} \rangle}{\partial t} + \nabla_{\mathbf{x}} \cdot \left(\langle \mathbf{U}_{\mathrm{p0}} \rangle \langle n_{\xi} \rangle \right) + \nabla_{\xi} \cdot \left(\dot{\xi} \langle n_{\xi} \rangle \right) = \nabla_{\mathbf{x}} \cdot \left(\Gamma_{\mathrm{Tp}} + \Gamma(\xi) \right) \nabla_{\mathbf{x}} \langle n_{\xi} \rangle + \langle h_{\xi} \rangle. \quad (1.53)$$

In high-Reynolds-number turbulent flows, the "molecular" diffusion term $\Gamma(\xi)$ will be negligible compared to Γ_{Tp} (Fox, 2003).

Except for the last term, Eq. (1.53) is closed. In many practical applications (e.g., particle aggregation), h_{ξ} will be quadratic in n_{ξ}. Thus, the closure problem associated with $\langle h_{\xi} \rangle$ is very similar to that faced when closing the chemical source term for turbulent reacting flows (Fox, 2003). For example, if the time scales of the phenomena accounted for in h_{ξ} are much slower than the turbulent time scales, then we can close $\langle h_{\xi} \rangle$ in terms of $\langle n_{\xi} \rangle$. In turbulent reacting flows, this is known as the slow-chemistry limit (Fox, 2003). However, in general, one must resort to the use of a *micromixing model* to capture interactions between the turbulent velocity field and the NDF.

2 Generalized Population Balance Equation

In this section, we introduce a more fundamental approach for deriving the GPBE starting from the dynamic equations for a system of discrete particles. We then illustrate how the moment transport equations are derived from the GPBE.

2.1 Granular systems

We will consider first a granular system of N_{p} particles in the absence of a surrounding fluid phase. The dynamics of the nth particle can be described by the position of its center of mass $\mathbf{X}^{(n)}$, its velocity $\mathbf{U}^{(n)}$, and its N internal coordinates $\xi^{(n)}$. Because particles have finite size, the volume occupied by a particle will be non-zero. Nevertheless, a far as the particle dynamics are concerned, we can label a particle by its center of mass, which is a point with no volume. We could use the particle momentum instead of the velocity. However, since the particle mass can be one of the internal coordinates, there is no lose of generality by using the velocity. The notion of internal coordinate is very general and can include particle mass, radius, geometry, angular momentum, etc. As we describe later, modeling how the internal coordinates change due to "growth" and collisions is usually non-trivial. For simplicity, we will assume for now that the particles have identical masses. The dynamics of the particles obey

$$\frac{\mathrm{d}\mathbf{X}^{(n)}}{\mathrm{d}t} = \mathbf{U}^{(n)}, \quad (2.1)$$

$$\frac{\mathrm{d}\mathbf{U}^{(n)}}{\mathrm{d}t} = \mathbf{A}^{(n)} + \mathbf{C}_{U}^{(n)} \quad (2.2)$$

and

$$\frac{\mathrm{d}\xi^{(n)}}{\mathrm{d}t} = \mathbf{G}^{(n)} + \mathbf{C}_{\xi}^{(n)}. \quad (2.3)$$

The terms on the right-hand sides represent, respectively, continuous acceleration $\mathbf{A}^{(n)}$ due to body forces, changes due to discontinuous collisions $\mathbf{C}_U^{(n)}$ and $\mathbf{C}_\xi^{(n)}$, and changes due to "growth" $\mathbf{G}^{(n)}$. In general, these terms depend on the complete set of variables for all particles, and are intrinsic functions of the volume of the system containing the particles. In other words, the system volume must be large enough to contain all particles with non-zero volumes. Equivalently, we can say that the terms are intrinsic functions of the particle number density. In order to simplify the notation, we will denote the set of all particle variables as follows:

$$\left\{\mathbf{X}^{(n)}\right\} \equiv \left\{\mathbf{X}^{(1)}, \ldots, \mathbf{X}^{(N_\mathrm{p})}\right\} \tag{2.4}$$

$$\left\{\mathbf{U}^{(n)}\right\} \equiv \left\{\mathbf{U}^{(1)}, \ldots, \mathbf{U}^{(N_\mathrm{p})}\right\} \tag{2.5}$$

and

$$\left\{\boldsymbol{\xi}^{(n)}\right\} \equiv \left\{\boldsymbol{\xi}^{(1)}, \ldots, \boldsymbol{\xi}^{(N_\mathrm{p})}\right\}. \tag{2.6}$$

Note that this system has a total of $N_\mathrm{f} = N_\mathrm{p} \times 3 \times 3 \times N$ degrees of freedom.

For each set of initial conditions, Eqs. (2.1)–(2.3) can be solved to find $\{\mathbf{X}^{(n)}\}$, $\{\mathbf{U}^{(n)}\}$ and $\{\boldsymbol{\xi}^{(n)}\}$. The initial conditions are randomly selected from known distribution functions, and we can assume that an infinite number of possible combinations exist. Each combination is called a *realization* of the granular flow and the set of all possible realizations forms an *ensemble*. Note that because the particles have finite size, they cannot be located at the same point: $\mathbf{X}^{(n)} \neq \mathbf{X}^{(m)}$ for $n \neq m$. Also, the collision operator $\mathbf{C}_U^{(n)}$ will generate chaotic trajectories and thus the particle positions will become uncorrelated after a relatively small number of collisions. In contrast, for particles suspended in a fluid the collisions are suppressed and correlations can be long lived and long range. We will make these concepts more precise when we introduce fluid-particle systems later. While the exact nature of the particle correlations is not a factor in the definition of the multi-particle joint PDF introduced below, it is important to keep in mind that they will have a strong influence on the particle dynamics.

Given that we have an infinite ensemble of realizations, we can define a *multi-particle joint PDF* by

$$\begin{aligned} f_{N_\mathrm{p}}\left(\{\mathbf{x}^{(n)}\}, \{\mathbf{V}^{(n)}\}, \{\boldsymbol{\eta}^{(n)}\}; t\right) \mathrm{d}\{\mathbf{x}^{(n)}\}\, \mathrm{d}\{\mathbf{V}^{(n)}\}\, \mathrm{d}\{\boldsymbol{\eta}^{(n)}\} \equiv \\ P\left[\cap_{n=1}^{N_\mathrm{p}} \left\{\left(\mathbf{x}^{(n)} < \mathbf{X}^{(n)}(t) \leq \mathbf{x}^{(n)} + \mathrm{d}\mathbf{x}^{(n)}\right) \cap \left(\mathbf{V}^{(n)} < \mathbf{U}^{(n)}(t) \leq \mathbf{V}^{(n)} + \mathrm{d}\mathbf{V}^{(n)}\right)\right.\right. \\ \left.\left. \cap \left(\boldsymbol{\eta}^{(n)} < \boldsymbol{\xi}^{(n)}(t) \leq \boldsymbol{\eta}^{(n)} + \mathrm{d}\boldsymbol{\eta}^{(n)}\right)\right\}\right], \end{aligned} \tag{2.7}$$

where the probability on the right-hand side is computed using all realizations in the ensemble, and the N_f state-space variables are $\{\mathbf{x}^{(n)}\}$, $\{\mathbf{V}^{(n)}\}$ and $\{\boldsymbol{\eta}^{(n)}\}$. The probability statement $P[A \cap B \cap C]$ is interpreted as the probability that events A, B and C occur together. The probability is computed as the fraction of all realizations in the ensemble for which the statement $A \cap B \cap C$ is true. Although it has too many degrees of freedom to be useful for modeling practical systems, in principle this joint PDF completely described the dynamics of the granular system.

The multi-particle joint PDF can be reduced to a *single-particle joint PDF* by integrating out all of the state variables for the other particles. Formally for the nth particle this yields

$$f_1^{(n)}\left(\mathbf{x}^{(n)}, \mathbf{V}^{(n)}, \eta^{(n)}; t\right) = \int_{m \neq n} f_{N_\mathrm{p}}\left(\{\mathbf{x}^{(m)}\}, \{\mathbf{V}^{(m)}\}, \{\eta^{(m)}\}; t\right) \mathrm{d}\{\mathbf{x}^{(m)}\}\, \mathrm{d}\{\mathbf{V}^{(m)}\}\, \mathrm{d}\{\eta^{(m)}\}, \tag{2.8}$$

where the integral is over all $m \neq n$. In the special case where all particles are *identically distributed*, the single-particle joint PDF will be the same for all n and we can denote it simply by f_1. One can argue that because the particle numbering system is arbitrary, the particles must be identically distributed. See Subramaniam (2000) for a discussion of examples where this does not hold. In any case, when deriving models for the NDF in practical systems, it will be convenient to assume that $f_1^{(n)} = f_1$. Note that unlike f_{N_p}, $f_1^{(n)}$ does not offer a complete description of the particle dynamics since two-particle (or more) correlations cannot in general be found from the single-particle PDF.

The general NDF is defined in terms of an expected value with respect to f_{N_p}:

$$\begin{aligned}
&n(\mathbf{U}_p, \boldsymbol{\xi}; \mathbf{x}, t) \\
&\equiv \sum_{n=1}^{N_\mathrm{p}} \left\langle \delta\left(\mathbf{X}^{(n)}(t) - \mathbf{x}\right) \delta\left(\mathbf{U}^{(n)}(t) - \mathbf{U}_\mathrm{p}\right) \delta\left(\boldsymbol{\xi}^{(n)}(t) - \boldsymbol{\xi}\right) \right\rangle \\
&= \sum_{n=1}^{N_\mathrm{p}} \int \delta\left(\mathbf{x}^{(n)} - \mathbf{x}\right) \delta\left(\mathbf{V}^{(n)} - \mathbf{U}_\mathrm{p}\right) \delta\left(\eta^{(n)} - \boldsymbol{\xi}\right) f_{N_\mathrm{p}}\, \mathrm{d}\{\mathbf{x}^{(m)}\}\, \mathrm{d}\{\mathbf{V}^{(m)}\}\, \mathrm{d}\{\eta^{(m)}\} \\
&= \sum_{n=1}^{N_\mathrm{p}} \int \delta\left(\mathbf{x}^{(n)} - \mathbf{x}\right) \delta\left(\mathbf{V}^{(n)} - \mathbf{U}_\mathrm{p}\right) \delta\left(\eta^{(n)} - \boldsymbol{\xi}\right) f_1^{(n)}\, \mathrm{d}\mathbf{x}^{(n)}\, \mathrm{d}\mathbf{V}^{(n)}\, \mathrm{d}\eta^{(n)} \\
&= \sum_{n=1}^{N_\mathrm{p}} f_1^{(n)}\left(\mathbf{x}, \mathbf{U}_\mathrm{p}, \boldsymbol{\xi}; t\right).
\end{aligned} \tag{2.9}$$

Recall that the random variables in this equation are the complete set of particle positions, velocities, and internal coordinates. The state-space variables $\mathbf{x}$, $\mathbf{U}_\mathrm{p}$ and $\boldsymbol{\xi}$ are fixed. Thus, if all particles are identically distributed, we find

$$n(\mathbf{U}_\mathrm{p}, \boldsymbol{\xi}; \mathbf{x}, t) = N_\mathrm{p} f_1\left(\mathbf{x}, \mathbf{U}_\mathrm{p}, \boldsymbol{\xi}; t\right). \tag{2.10}$$

The first line of Eq. (2.9) is the definition of n, and the second line is the definition of the expected value. In the third line, we integrate out all particle information expect for the nth particle. In the final line, we apply the definition of the multi-variate delta function. The other NDFs (i.e., $n_{\boldsymbol{\xi}}$ and $n_{\mathbf{U}_\mathrm{p}}$) can be found from Eq. (2.9) by integrating out the appropriate coordinates.

It is now evident from Eq. (2.9) that the general NDF is defined in terms of the following probability statement:

$$\begin{aligned}
&n(\mathbf{U}_\mathrm{p}, \boldsymbol{\xi}; \mathbf{x}, t)\, \mathrm{d}\mathbf{U}_\mathrm{p}\, \mathrm{d}\boldsymbol{\xi}\, \mathrm{d}\mathbf{x} \equiv \\
&\qquad P\left[\cup_{n=1}^{N_\mathrm{p}} \left\{\left(\mathbf{x} < \mathbf{X}^{(n)}(t) \leq \mathbf{x} + \mathrm{d}\mathbf{x}\right) \cap \left(\mathbf{U}_\mathrm{p} < \mathbf{U}^{(n)}(t) \leq \mathbf{U}_\mathrm{p} + \mathrm{d}\mathbf{U}_\mathrm{p}\right)\right.\right. \\
&\qquad\qquad \left.\left. \cap \left(\boldsymbol{\xi} < \boldsymbol{\xi}^{(n)}(t) \leq \boldsymbol{\xi} + \mathrm{d}\boldsymbol{\xi}\right)\right\}\right].
\end{aligned} \tag{2.11}$$

The probability statement $P[A \cup B \cup C]$ is interpreted as the probability that event A, or event B or event C occurs. The probability is computed as the fraction of all realizations in the ensemble for which the statement $A \cup B \cup C$ is true. The probability on the right-hand side is again defined with respect to the ensemble of all realizations of the granular flow. The dynamical behavior of the NDF will be determined by the right-hand sides of Eqs. (2.1)–(2.3).

2.2 Fluid-particle systems

The definition of the NDF derived above was carried out in the limit where the particles are surrounded by a vacuum. In the more general case where they are surrounded by a fluid, each particle will be accelerated by fluid stresses acting on its surface. The exact description of the resulting surface forces is very complicated and would involve the complete solution to the coupled fluid-particle system. For example, the fluid could obey the Navier-Stokes equation in the volume of the domain not occupied by particles. However, in order to define the NDF, we do not need to know the exact form of the surface forces. It suffices to know that they exist, and that in general they will depend not only on the surrounding fluid, but also on the complete set of random variables describing the particle system: $\{\mathbf{X}^{(n)}\}$, $\{\mathbf{U}^{(n)}\}$ and $\{\boldsymbol{\xi}^{(n)}\}$.

Formally, the presence of the fluid will generate a new continuous acceleration term in Eq. (2.2):

$$\frac{\mathrm{d}\mathbf{U}^{(n)}}{\mathrm{d}t} = \mathbf{A}_{\mathrm{p}}^{(n)} + \mathbf{A}^{(n)} + \mathbf{C}_{U}^{(n)}. \tag{2.12}$$

The fluid could also change the internal coordinate so that an additional "growth" term is needed. For the present discussion, we will ignore this possibility. In most descriptions of fluid-particle flows, $\mathbf{A}_{\mathrm{p}}^{(n)}$ is modeled as a drag term of the form

$$\mathbf{A}_{p}^{(n)} \propto C_{\mathrm{D}}^{(n)} \left|\mathbf{U}_{\mathrm{f}}^{(n)} - \mathbf{U}^{(n)}(t)\right| \left(\mathbf{U}_{\mathrm{f}}^{(n)} - \mathbf{U}^{(n)}(t)\right), \tag{2.13}$$

where $C_{\mathrm{D}}^{(n)}$ is a "drag coefficient", and $\mathbf{U}_{\mathrm{f}}^{(n)}$ is a representative fluid velocity in the "neighborhood" of $\mathbf{X}^{(n)}(t)$. In models for gas-solid flows, $\mathbf{U}_{\mathrm{f}}^{(n)}$ is called the "fluid velocity seen by the nth solid particle". In all cases, it is important to recognize that the statistical properties of $\mathbf{U}_{\mathrm{f}}^{(n)}$ will in general not be the same as for a Lagrangian fluid particle. Only in the limit of zero Stokes number where all solid particles exactly follow the fluid will the statistical properties be identical. This can occur, for example, for nanoparticles submerged in a viscous fluid. The correct choice of $\mathbf{U}_{\mathrm{f}}^{(n)}$ will be system dependent. However, a suitable choice might be the surface average of the fluid velocity at the edge of the momentum boundary layer surrounding the particle. It should be obvious that we cannot use the fluid velocity at the particle location $\mathbf{x} = \mathbf{X}^{(n)}(t)$ since, by definition, there is no fluid at that point. Using the momentum boundary layer to define $\mathbf{U}_{\mathrm{f}}^{(n)}$ has the advantage that the location will depend on the particle Reynolds number. It should be evident to the reader that the major challenge in developing a model for $\mathbf{A}_{\mathrm{p}}^{(n)}$ is to adequately account for its dependence on the other particles without having to solve for the full multi-particle joint PDF.

For the present discussion, we do not need to know the exact form of the model for the fluid velocity. Instead, we will simply assume that $\mathbf{U}_{\mathrm{f}}^{(n)}$ obeys

$$\frac{\mathrm{d}\mathbf{U}_{\mathrm{f}}^{(n)}}{\mathrm{d}t} = \mathbf{A}_{\mathrm{f}}^{(n)} \tag{2.14}$$

where the fluid acceleration $\mathbf{A}_\mathrm{f}^{(n)}$ depends on $\mathbf{A}_\mathrm{p}^{(n)}$ through the law of conservation of momentum. The set of random variables needed to describe a fluid-particle system with N_p particle will then be augmented to include the fluid velocity: $\{\mathbf{X}^{(n)}\}$, $\{\mathbf{U}^{(n)}\}$, $\{\mathbf{U}_\mathrm{f}^{(n)}\}$ and $\{\boldsymbol{\xi}^{(n)}\}$. However, the definition of the ensemble remains basically unchanged. A single realization in the ensemble is still defined by one possible set of initial conditions, but now extended to include $\{\mathbf{U}_\mathrm{f}^{(n)}(0)\}$. In practice, the initial fluid velocity field $\mathbf{U}(\mathbf{x}, 0)$ can be specified consistent with zero-slip boundary conditions on the particle surfaces. The surface terms $\{\mathbf{A}_\mathrm{p}^{(n)}\}$ can then be computed, and used with Eq. (2.13) to define $\{\mathbf{U}_\mathrm{f}^{(n)}(0)\}$.

Given the infinite ensemble of realizations, we can now define a *multi-particle-fluid joint PDF* by

$$\begin{aligned} f_{N_\mathrm{p}}&\left(\{\mathbf{x}^{(n)}\}, \{\mathbf{V}^{(n)}\}, \{\mathbf{V}_\mathrm{f}^{(n)}\}, \{\boldsymbol{\eta}^{(n)}\}; t\right) \mathrm{d}\{\mathbf{x}^{(n)}\}\mathrm{d}\{\mathbf{V}^{(n)}\}\mathrm{d}\{\mathbf{V}_\mathrm{f}^{(n)}\}\mathrm{d}\{\boldsymbol{\eta}^{(n)}\} = \\ &P\left[\cap_{n=1}^{N_\mathrm{p}}\left\{\left(\mathbf{x}^{(n)} < \mathbf{X}^{(n)}(t) \le \mathbf{x}^{(n)} + \mathrm{d}\mathbf{x}^{(n)}\right) \cap \left(\mathbf{V}^{(n)} < \mathbf{U}^{(n)}(t) \le \mathbf{V}^{(n)} + \mathrm{d}\mathbf{V}^{(n)}\right)\right.\right. \\ &\qquad \left.\left.\cap \left(\mathbf{V}_\mathrm{f}^{(n)} < \mathbf{U}_\mathrm{f}^{(n)}(t) \le \mathbf{V}_\mathrm{f}^{(n)} + \mathrm{d}\mathbf{V}_\mathrm{f}^{(n)}\right) \cap \left(\boldsymbol{\eta}^{(n)} < \boldsymbol{\xi}^{(n)}(t) \le \boldsymbol{\eta}^{(n)} + \mathrm{d}\boldsymbol{\eta}^{(n)}\right)\right\}\right], \end{aligned} \tag{2.15}$$

where the state-space variables for fluid velocity are $\{\mathbf{V}_\mathrm{f}^{(n)}\}$. Then, following the same steps used to obtain Eq. (2.9), we can define the particle-fluid NDF by

$$n(\mathbf{U}_\mathrm{p}, \mathbf{U}_\mathrm{f}, \boldsymbol{\xi}; \mathbf{x}, t) \equiv \sum_{n=1}^{N_\mathrm{p}} f_1^{(n)}\left(\mathbf{x}, \mathbf{U}_\mathrm{p}, \mathbf{U}_\mathrm{f}, \boldsymbol{\xi}; t\right), \tag{2.16}$$

which, in the limit of identically distributed particles, becomes

$$n(\mathbf{U}_\mathrm{p}, \mathbf{U}_\mathrm{f}, \boldsymbol{\xi}; \mathbf{x}, t) = N_p f_1\left(\mathbf{x}, \mathbf{U}_\mathrm{p}, \mathbf{U}_\mathrm{f}, \boldsymbol{\xi}; t\right). \tag{2.17}$$

Thus, we now have a mathematically consistent definition of the particle-fluid NDF in terms of a well-defined single-particle PDF. The extension of this result to cases where N_p changes with time due to, for example, aggregation and breakage is straightforward. One must simply condition on the value of N_p, and include a dynamical description of the processes that change the number of particles in the system. In practical applications, the major remaining challenge is to find an adequate model for the dynamics of $f_1^{(n)}$.

In order to understand the range of physics contained in the particle-fluid NDF, it is instructive to consider two limiting cases:

1. In the limit where the particles are completely uncoupled from the fluid, the particle-fluid NDF is separable:

$$n(\mathbf{U}_\mathrm{p}, \mathbf{U}_\mathrm{f}, \boldsymbol{\xi}; \mathbf{x}, t) = n(\mathbf{U}_\mathrm{p}, \boldsymbol{\xi}; \mathbf{x}, t) f_\mathrm{U}(\mathbf{U}_\mathrm{f}; \mathbf{x}, t) \tag{2.18}$$

where $f_\mathrm{U}(\mathbf{V}; \mathbf{x}, t)$ is the one-point PDF of the fluid defined by

$$f_\mathrm{U}(\mathbf{V}; \mathbf{x}, t) = \langle \delta\left(\mathbf{U}(\mathbf{x}, t) - \mathbf{V}\right)\rangle. \tag{2.19}$$

The ensemble used to compute the expected value on the right-hand side of Eq. (2.19) is the collection of all possible realizations of the fluid flow (in the absence of interaction with the particles). In laminar flow, f_{U} reduces to a delta function. In turbulent flow, f_{U} can be modeled using PDF methods. This limit would occur when the drag coefficient in Eq. (2.13) is null, and the presence of the particles can be ignored when solving for the fluid velocity.

2. In the limit where the particles all have the same velocity as the fluid and the fluid flow is not affected by the presence of the particles, the particle-fluid NDF becomes

$$\begin{aligned} n(\mathbf{U}_{\mathrm{p}}, \mathbf{U}_{\mathrm{f}}, \boldsymbol{\xi}; \mathbf{x}, t) &= \sum_{n=1}^{N_{\mathrm{p}}} f_1^{(n)}(\mathbf{x}, \mathbf{U}_{\mathrm{p}}, \boldsymbol{\xi}; t) f_{\mathrm{U}}(\mathbf{U}_{\mathrm{f}}; \mathbf{x}, t) \\ &= \sum_{n=1}^{N_{\mathrm{p}}} f_1^{(n)}(\mathbf{x}, \boldsymbol{\xi} | \mathbf{U}_{\mathrm{p}}; t) \delta(\mathbf{U}_{\mathrm{p}} - \mathbf{U}_{\mathrm{f}}) f_{\mathrm{U}}(\mathbf{U}_{\mathrm{f}}; \mathbf{x}, t), \end{aligned} \tag{2.20}$$

where $f_1^{(n)}(\mathbf{x}, \boldsymbol{\xi} | \mathbf{U}_{\mathrm{p}})$ is the one-particle conditional PDF given that $\mathbf{U}_{\mathrm{p}}^{(n)} = \mathbf{U}_{\mathrm{p}}$. Thus, the general NDF can be expressed as

$$n(\mathbf{U}_{\mathrm{p}}, \boldsymbol{\xi}; \mathbf{x}, t) = \sum_{n=1}^{N_{\mathrm{p}}} f_1^{(n)}(\mathbf{x}, \boldsymbol{\xi} | \mathbf{U}_{\mathrm{p}}; t) f_{\mathrm{U}}(\mathbf{U}_{\mathrm{p}}; \mathbf{x}, t) \tag{2.21}$$

or, for identically distributed particles, as

$$n(\mathbf{U}_{\mathrm{p}}, \boldsymbol{\xi}; \mathbf{x}, t) = N_{\mathrm{p}} f_1(\mathbf{x}, \boldsymbol{\xi} | \mathbf{U}_{\mathrm{p}}; t) f_{\mathrm{U}}(\mathbf{U}_{\mathrm{p}}; \mathbf{x}, t). \tag{2.22}$$

In words, the term $N_{\mathrm{p}} f_1(\mathbf{x}, \boldsymbol{\xi} | \mathbf{U}_{\mathrm{p}}; t)$ is the NDF for the internal coordinates at point $\mathbf{x}$ and time t *given* that all particles have the same velocity as the fluid. This limit would occur in a very dilute system when the drag coefficient in Eq. (2.13) is infinite.

The form of the particle-fluid NDF between these two limits will depend on the exact form of the surface-force models and the interstitial fluid dynamics. Thus, in theory, the only way to compute the NDF exactly is to simulate the complete particle-fluid system using direct simulations. Nevertheless, useful approximations for the NDF can be obtained by providing closures to the multi-particle statistics that arise in the transport equation for the NDF (i.e., in the GPBE).

In summary, we have demonstrated that by employing a well-defined multi-particle joint PDF it is straightforward to define the NDF in its various forms. Likewise, we have also shown how volume and time averages can be used to estimate the NDF from a single realization of the flow. Now we turn to the more difficult task of deriving the GPBE that describes the evolution of the NDF starting from the ordinary differential equations (ODEs) that define the particle positions, velocities, surface forces and internal coordinates.

2.3 Derivation of the GPBE

In this section we give a streamlined derivation of the GPBE for the particle-fluid NDF starting from the transport equation for multi-particle-fluid PDF. This will require three steps:

1. Starting from the ODEs for the particle properties (Eqs. 2.1, 2.3, 2.12 and 2.14), we first derive a generalized transport equation for $f_{N_{\mathrm{p}}}$.

2. Given the transport equation for f_{N_p}, we will find the transport equation for $f_1^{(n)}$ by integrating out the all degrees of freedom except those associated with the nth particle. Due to the loss of information, this step will generate several unclosed terms that must be modeled.
3. Finally, using the definition of the particle-fluid NDF given in Eq. (2.16), we will derive the GPBE.

Multi-particle joint PDF The general form of the transport equation for the multi-particle joint PDF is

$$\frac{\partial f_{N_\mathrm{p}}}{\partial t} + \sum_n \frac{\partial}{\partial \mathbf{x}^{(n)}} \cdot \left(\mathbf{V}^{(n)} f_{N_\mathrm{p}} \right) + \sum_n \frac{\partial}{\partial \mathbf{V}^{(n)}} \cdot \left[\left(\mathbf{A}_\mathrm{p}^{(n)} + \mathbf{A}^{(n)} \right) f_{N_\mathrm{p}} \right] + \sum_n \frac{\partial}{\partial \mathbf{V}_\mathrm{f}^{(n)}} \cdot \left(\mathbf{A}_\mathrm{f}^{(n)} f_{N_\mathrm{p}} \right) + \sum_n \frac{\partial}{\partial \boldsymbol{\eta}^{(n)}} \cdot \left(\mathbf{G}^{(n)} f_{N_\mathrm{p}} \right) = \mathcal{C}_{N_\mathrm{p}}. \tag{2.23}$$

The term on the right-hand side of this equation is the N_p-particle collision operator, which generates discontinuous changes in particle velocities $\mathbf{U}^{(n)}$ and internal coordinates $\boldsymbol{\xi}^{(n)}$. The first term on the left-hand side is accumulation of f_{N_p}. The remaining terms on the left-hand side represent continuous changes in particle-position space $\mathbf{x}^{(n)}$ and velocity space $\mathbf{V}^{(n)}$, fluid-velocity space $\mathbf{V}_\mathrm{f}^{(n)}$, and particle-internal-coordinate space $\boldsymbol{\eta}^{(n)}$. In order to simplify the notation, we have written the derivatives in a compact form that can be expanded. For example,

$$\sum_n \frac{\partial}{\partial \mathbf{x}^{(n)}} \cdot \left(\mathbf{V}^{(n)} f_{N_\mathrm{p}} \right) \equiv \sum_{n=1}^{N_\mathrm{p}} \sum_{i=1}^{3} \frac{\partial}{\partial x_i^{(n)}} \left(V_i^{(n)} f_{N_\mathrm{p}} \right). \tag{2.24}$$

The total number of independent variable appearing in Eq. (2.23) is thus quite large, and in fact too large for practical applications. However, as mentioned earlier, by coupling Eq. (2.23) with the Navier-Stokes equation to find the forces on the particles due to the fluid, the N_p-particle system is completely determined.

Although not written out explicitly, the reader should keep in mind that the phase-space "velocities" and collision term depend on the complete set of independent variables. For example, the surface terms depend on all of the state variables:

$$\mathbf{A}_\mathrm{p}^{(n)} \left(\{\mathbf{x}^{(n)}\}, \{\mathbf{V}^{(n)}\}, \{\mathbf{V}_\mathrm{f}^{(n)}\}, \{\boldsymbol{\eta}^{(n)}\} \right). \tag{2.25}$$

The only known way to determine these functions is to perform direct-numerical simulations of the fluid-particle system using all possible sets of initial conditions. Obviously, such an approach is intractable. We are thus led to reduce the number of independent variables and to introduce models that attempt to capture the "average" effect of multi-particle interactions.

Single-particle joint PDF The simplest, useful representation of the fluid-particle system is the single-particle joint PDF $f_1^{(n)}$ defined by integrating out all independent variables except those associated with the nth particle:

$$f_1^{(n)} \equiv \int_{m \neq n} f_{N_\mathrm{p}} \, \mathrm{d}\{\mathbf{x}^{(m)}\} \, \mathrm{d}\{\mathbf{V}^{(m)}\} \, \mathrm{d}\{\mathbf{V}_\mathrm{f}^{(m)}\} \, \mathrm{d}\{\boldsymbol{\eta}^{(m)}\}. \tag{2.26}$$

Starting from Eq. (2.23) and integrating out all variables except those associated with the nth particle yields

$$\frac{\partial f_1^{(n)}}{\partial t} + \frac{\partial}{\partial \mathbf{x}^{(n)}} \cdot \left(\mathbf{V}^{(n)} f_1^{(n)}\right) + \frac{\partial}{\partial \mathbf{V}^{(n)}} \cdot \left[\left(\langle \mathbf{A}_{\mathrm{p}}^{(n)} \rangle_1 + \langle \mathbf{A}^{(n)} \rangle_1\right) f_1^{(n)}\right] + \frac{\partial}{\partial \mathbf{V}_{\mathrm{f}}^{(n)}} \cdot \left(\langle \mathbf{A}_{\mathrm{f}}^{(n)} \rangle_1 f_1^{(n)}\right) + \frac{\partial}{\partial \boldsymbol{\eta}^{(n)}} \cdot \left(\langle \mathbf{G}^{(n)} \rangle_1 f_1^{(n)}\right) = \mathcal{C}_1^{(n)}, \quad (2.27)$$

where we have introduced the single-particle collision term defined by

$$\mathcal{C}_1^{(n)} \equiv \int_{m \neq n} \mathcal{C}_{N_{\mathrm{p}}} \, \mathrm{d}\{\mathbf{x}^{(m)}\} \, \mathrm{d}\{\mathbf{V}^{(m)}\} \, \mathrm{d}\{\mathbf{V}_{\mathrm{f}}^{(m)}\} \, \mathrm{d}\{\boldsymbol{\eta}^{(m)}\}, \quad (2.28)$$

and the single-particle conditional expected values defined, for example, by

$$\langle \mathbf{A}_{\mathrm{p}}^{(n)} \rangle_1 \equiv \frac{1}{f_1^{(n)}} \int_{m \neq n} \mathbf{A}_{\mathrm{p}}^{(n)} f_{N_{\mathrm{p}}} \, \mathrm{d}\{\mathbf{x}^{(m)}\} \, \mathrm{d}\{\mathbf{V}^{(m)}\} \, \mathrm{d}\{\mathbf{V}_{\mathrm{f}}^{(m)}\} \, \mathrm{d}\{\boldsymbol{\eta}^{(m)}\}. \quad (2.29)$$

Compared to Eq. (2.25), the left-hand side of this expression depends on a much smaller set of independent variables:

$$\langle \mathbf{A}_{\mathrm{p}}^{(n)} \rangle_1 \left(\mathbf{x}^{(n)}, \mathbf{V}^{(n)}, \mathbf{V}_{\mathrm{f}}^{(n)}, \boldsymbol{\eta}^{(n)}\right). \quad (2.30)$$

Thus, in principle, it should be simpler to deal with than its multi-particle counterpart. Nevertheless, due to the loss of direct information about particle-particle interactions, the single-particle conditional expected values must be represented by closures. The development of closures that yield accurate predictions of $f_1^{(n)}$ by solving Eq. (2.27) is a non-trivial task, and the forms of the closures will be highly dependent on the physical properties of the system.

GPBE The NDF is defined by Eq. (2.16), and evolves according to GPBE. Starting from Eq. (2.27), we first rewrite the independent variables as

$$\frac{\partial f_1^{(n)}}{\partial t} + \frac{\partial}{\partial \mathbf{x}} \cdot \left(\mathbf{U}_{\mathrm{p}} f_1^{(n)}\right) + \frac{\partial}{\partial \mathbf{U}_{\mathrm{p}}} \cdot \left[\left(\langle \mathbf{A}_{\mathrm{p}}^{(n)} \rangle_1 + \langle \mathbf{A}^{(n)} \rangle_1\right) f_1^{(n)}\right] + \frac{\partial}{\partial \mathbf{U}_{\mathrm{f}}} \cdot \left(\langle \mathbf{A}_{\mathrm{f}}^{(n)} \rangle_1 f_1^{(n)}\right) + \frac{\partial}{\partial \boldsymbol{\xi}} \cdot \left(\langle \mathbf{G}^{(n)} \rangle_1 f_1^{(n)}\right) = \mathcal{C}_1^{(n)}, \quad (2.31)$$

where terms are now evaluated at the phase-space location $(\mathbf{x}, \mathbf{U}_{\mathrm{p}}, \mathbf{U}_{\mathrm{f}}, \boldsymbol{\xi})$. Summing over all particles then yields the GPBE:

$$\frac{\partial n}{\partial t} + \frac{\partial}{\partial \mathbf{x}} \cdot (\mathbf{U}_{\mathrm{p}} n) + \frac{\partial}{\partial \mathbf{U}_{\mathrm{p}}} \cdot \left[\left(\langle \mathbf{A}_{\mathrm{p}} \rangle_1 + \langle \mathbf{A} \rangle_1\right) n\right] + \frac{\partial}{\partial \mathbf{U}_{\mathrm{f}}} \cdot (\langle \mathbf{A}_{\mathrm{f}} \rangle_1 n) + \frac{\partial}{\partial \boldsymbol{\xi}} \cdot (\langle \mathbf{G} \rangle_1 n) = \mathcal{S}_1. \quad (2.32)$$

In order to account for variable particle numbers, we generalize the collision term $\mathcal{S}_1$ to include changes in N_{p} due to nucleation, aggregation and breakage. These processes will also require models in order to close Eq. (2.32). This equation can be compared to Eq. (1.27), and it can be observed that they have the same general form. However, it is now clear that the GPBE cannot be solved until closures are provided for the conditional phase-space velocities ($\langle \mathbf{A}_{\rm p} \rangle_1$, $\langle \mathbf{A} \rangle_1$, $\langle \mathbf{A}_{\rm f} \rangle_1$ and $\langle \mathbf{G} \rangle_1$) and the discontinuous "source" term $\mathcal{S}_1$.

In general, we will consider computational models for cases where Eq. (2.32) appears in closed form. In particular, this implies that all of the "flux" terms ($\langle \mathbf{A}_{\rm p} \rangle_1$, $\langle \mathbf{A} \rangle_1$, $\langle \mathbf{A}_{\rm f} \rangle_1$, $\langle \mathbf{G} \rangle_1$ and $\mathcal{S}_1$) in the GPBE will depend only on the independent variables (t, $\mathbf{x}$, $\mathbf{U}_{\rm p}$, $\mathbf{U}_{\rm f}$ and $\boldsymbol{\xi}$); and on the NDF $n(\mathbf{U}_{\rm p}, \mathbf{U}_{\rm f}, \boldsymbol{\xi}; \mathbf{x}, t)$. We should note, however, that finding accurate closure models for these terms is a highly non-trivial task. In general, such models will be nonlinear and can involve convolution integrals (e.g., to describe aggregation and collision processes). Thus, the accurate numerical solution of Eq. (2.32) will also be non-trivial, and depend on the particular forms of the closure models.

Another formidable complication when solving Eq. (2.32) is the fact that the NDF often depends on a large number of independent variables. In order to reduce the dimensionality of the problem, it is sometimes possible to work with selected moments of the NDF by solving their transport equations. In general, these transport equations will not be closed and further modeling approximations are necessary. In the following section, we derive the most commonly used moment transport equations starting from Eq. (2.32).

2.4 Moment transport equations

In this section we will derive a few example transport equations for selected moments of the NDF starting from Eq. (2.32). Because the conditional phase-space velocities appearing in Eq. (2.32) have yet to be specified, we will not enter into the details on how these terms can be approximated at the level of moment closures. However, readers familiar with the kinetic theory of gases (Chapman and Cowling, 1961) will recognize that developing closures for the moments is a major challenge. The goal here is more modest. We are primarily interested in understanding the process of finding the moment transport equations starting from the GPBE from a mathematical perspective. However, as discussed later, our main goal is to illustrate that a *closed* GPBE is the proper starting point for developing models for polydisperse multiphase flows. In this context, it is important that the reader understand how the moment equations arise from the GPBE.

Phase-space integration Computing moments from Eq. (2.32) basically involves integration over some of the independent variables in the NDF (i.e., $\mathbf{U}_{\rm p}$, $\mathbf{U}_{\rm f}$ and $\boldsymbol{\xi}$) while keeping the other fixed (i.e., t and $\mathbf{x}$). In order to facilitate this task, it is useful to review some basic rules of integration.

1. **Treatment of time and space derivatives:** Let $g(\mathbf{U}_{\rm p}, \mathbf{U}_{\rm f}, \boldsymbol{\xi})$ be an arbitrary function of its variables. Then

$$\begin{aligned}\int g \left(\frac{\partial n}{\partial t} + \frac{\partial}{\partial \mathbf{x}} \cdot (\mathbf{U}_{\rm p} n) \right) \mathrm{d}\mathbf{U}_{\rm p}\, \mathrm{d}\mathbf{U}_{\rm f}\, \mathrm{d}\boldsymbol{\xi} &= \int \left(\frac{\partial g n}{\partial t} + \frac{\partial}{\partial \mathbf{x}} \cdot (\mathbf{U}_{\rm p} g n) \right) \mathrm{d}\mathbf{U}_{\rm p}\, \mathrm{d}\mathbf{U}_{\rm f}\, \mathrm{d}\boldsymbol{\xi} \\ &= \frac{\partial}{\partial t} \left(\int g n \, \mathrm{d}\mathbf{U}_{\rm p}\, \mathrm{d}\mathbf{U}_{\rm f}\, \mathrm{d}\boldsymbol{\xi} \right) + \frac{\partial}{\partial \mathbf{x}} \cdot \left(\int \mathbf{U}_{\rm p} g n \, \mathrm{d}\mathbf{U}_{\rm p}\, \mathrm{d}\mathbf{U}_{\rm f}\, \mathrm{d}\boldsymbol{\xi} \right). \end{aligned} \quad (2.33)$$

2. **Treatment of velocity derivatives:** With velocity derivatives, it is necessary to use integration by parts. For example,

$$\begin{aligned}\int g\frac{\partial}{\partial \mathbf{U}_{\mathrm{p}}}\cdot\left[\left(\langle \mathbf{A}_{\mathrm{p}}\rangle_1+\langle \mathbf{A}\rangle_1\right)n\right]\mathrm{d}\mathbf{U}_{\mathrm{p}}\,\mathrm{d}\mathbf{U}_{\mathrm{f}}\,\mathrm{d}\boldsymbol{\xi}\\ =\int \frac{\partial}{\partial \mathbf{U}_{\mathrm{p}}}\cdot\left[\left(\langle \mathbf{A}_{\mathrm{p}}\rangle_1+\langle \mathbf{A}\rangle_1\right)gn\right]\mathrm{d}\mathbf{U}_{\mathrm{p}}\,\mathrm{d}\mathbf{U}_{\mathrm{f}}\,\mathrm{d}\boldsymbol{\xi}\\ -\int \left(\langle \mathbf{A}_{\mathrm{p}}\rangle_1+\langle \mathbf{A}\rangle_1\right)\cdot\frac{\partial g}{\partial \mathbf{U}_{\mathrm{p}}}n\,\mathrm{d}\mathbf{U}_{\mathrm{p}}\,\mathrm{d}\mathbf{U}_{\mathrm{f}}\,\mathrm{d}\boldsymbol{\xi}.\end{aligned} \quad (2.34)$$

The first term on the right-hand side of this equation is a total derivative, so it can be integrated formally to find

$$\int \frac{\partial}{\partial \mathbf{U}_{\mathrm{p}}}\cdot\left[\left(\langle \mathbf{A}_{\mathrm{p}}\rangle_1+\langle \mathbf{A}\rangle_1\right)gn\right]\mathrm{d}\mathbf{U}_{\mathrm{p}}\,\mathrm{d}\mathbf{U}_{\mathrm{f}}\,\mathrm{d}\boldsymbol{\xi}=\left(\langle \mathbf{A}_{\mathrm{p}}\rangle_1+\langle \mathbf{A}\rangle_1\right)gn\big|_{\mathbf{U}_{\mathrm{p}}=\pm\infty}=0, \quad (2.35)$$

where the second term corresponds to flux normal to the surface bounding the velocity phase space. Because the velocity phase space extends to infinity, this flux must be null. Hence, the contributions arising due to the flux during integration by parts of the velocity variables (i.e., $\mathbf{U}_{\mathrm{p}}$ and $\mathbf{U}_{\mathrm{f}}$) are set to zero.

3. **Treatment of internal-coordinate derivatives:** With internal-coordinate derivatives, it is again necessary to use integration by parts:

$$\begin{aligned}\int g\frac{\partial}{\partial \boldsymbol{\xi}}\cdot\left(\langle \mathbf{G}\rangle_1 n\right)\mathrm{d}\mathbf{U}_{\mathrm{p}}\,\mathrm{d}\mathbf{U}_{\mathrm{f}}\,\mathrm{d}\boldsymbol{\xi}\\ =\int \frac{\partial}{\partial \boldsymbol{\xi}}\cdot\left(\langle \mathbf{G}\rangle_1 gn\right)\mathrm{d}\mathbf{U}_{\mathrm{p}}\,\mathrm{d}\mathbf{U}_{\mathrm{f}}\,\mathrm{d}\boldsymbol{\xi}-\int \langle \mathbf{G}\rangle_1\cdot\frac{\partial g}{\partial \boldsymbol{\xi}}n\,\mathrm{d}\mathbf{U}_{\mathrm{p}}\,\mathrm{d}\mathbf{U}_{\mathrm{f}}\,\mathrm{d}\boldsymbol{\xi}.\end{aligned} \quad (2.36)$$

The first term on the right-hand side of this equation is a total derivative, so it can be integrated formally. However, for internal coordinates the phase space does not usually extend to infinity. In order to see clearly what can happen, consider the case with one internal coordinate that is bounded by zero and infinity:

$$\int \frac{\partial}{\partial \xi}\cdot\left(\langle G\rangle_1 gn\right)\mathrm{d}\mathbf{U}_{\mathrm{p}}\,\mathrm{d}\mathbf{U}_{\mathrm{f}}\,\mathrm{d}\xi=\langle G\rangle_1 gn|_{\xi=\infty}-\langle G\rangle_1 gn|_{\xi=0}\,. \quad (2.37)$$

The final two terms on the right-hand side are the flux at infinity, which we can safely set to zero, and the flux at the origin. Depending on the forms of $\langle G\rangle_1$ and g, the flux at the origin may not be zero. For example, if g is non-zero when $\xi=0$, then the product $\langle G\rangle_1 n$ evaluated at $\xi=0$ would have to be zero in order for the flux term to cancel. Unfortunately, there are important applications where the flux term is non-zero, so one must pay attention to how the flux term is handled in the derivation of the moment equations. For example, if ξ represents the surface of evaporating droplets and $\langle G\rangle_1$ is constant (i.e., the evaporation rate is proportional to surface area), then n will be non-zero at $\xi=0$. Physically, the non-zero flux is due to the disappearance of droplets due to evaporation, and thus it cannot be neglected.

In summary, computing the moment transport equations starting from Eq. (2.32) involves integration over phase space using the rules described above for particular choices of g. In the following, we will assume that the flux term at the boundary of internal-coordinate phase space can be neglected. However, the reader should keep in mind that this assumption must be verified for particular cases.

Particle-number transport The total particle-number concentration $N(\mathbf{x}, t)$ corresponds to the zero-order moment of the NDF (i.e., $g = 1$), and is defined by

$$N \equiv \int n \, \mathrm{d}\mathbf{U}_\mathrm{p} \, \mathrm{d}\mathbf{U}_\mathrm{f} \, \mathrm{d}\boldsymbol{\xi}. \tag{2.38}$$

Its transport equation can be found from Eq. (2.32):

$$\frac{\partial N}{\partial t} + \frac{\partial}{\partial \mathbf{x}} \cdot (\mathbf{U}_N N) = \mathcal{S}_N, \tag{2.39}$$

where we have explicitly assumed that the flux at the boundary of internal-coordinate phase space is null. The number-average particle velocity is defined by

$$\mathbf{U}_N \equiv \frac{1}{N} \int \mathbf{U}_\mathrm{p} n \, \mathrm{d}\mathbf{U}_\mathrm{p} \, \mathrm{d}\mathbf{U}_\mathrm{f} \, \mathrm{d}\boldsymbol{\xi}, \tag{2.40}$$

and the particle-number source term by

$$\mathcal{S}_N \equiv \int \mathcal{S}_1 \, \mathrm{d}\mathbf{U}_\mathrm{p} \, \mathrm{d}\mathbf{U}_\mathrm{f} \, \mathrm{d}\boldsymbol{\xi}. \tag{2.41}$$

Note that the sign of the source term will depend on whether particles are created or destroyed in the system. Note also that the spatial transport term in Eq. (2.39) will generally not be closed unless, for example, all particles have identical velocities.

Particle-volume transport In order to derive a transport equation of the particle-volume fraction α_p, we will let ξ_1 equal to particle volume (V). The particle-volume fraction is then defined by

$$\alpha_\mathrm{p} \equiv \int \xi_1 n \, \mathrm{d}\mathbf{U}_\mathrm{p} \, \mathrm{d}\mathbf{U}_\mathrm{f} \, \mathrm{d}\boldsymbol{\xi}. \tag{2.42}$$

Starting from Eq. (2.32) with $g = \xi_1$, the transport equation for α_p is

$$\frac{\partial \alpha_\mathrm{p}}{\partial t} + \frac{\partial}{\partial \mathbf{x}} \cdot (\mathbf{U}_V \alpha_\mathrm{p}) = \langle G_V \rangle + \mathcal{S}_V \tag{2.43}$$

where the volume-average particle velocity is defined by

$$\mathbf{U}_V \equiv \frac{1}{\alpha_\mathrm{p}} \int \xi_1 \mathbf{U}_\mathrm{p} n \, \mathrm{d}\mathbf{U}_\mathrm{p} \, \mathrm{d}\mathbf{U}_\mathrm{f} \, \mathrm{d}\boldsymbol{\xi}. \tag{2.44}$$

For the case where all particles have the same volume, $\mathbf{U}_V = \mathbf{U}_N$. Note that like $\mathbf{U}_N$, the volume-average particle velocity will not usually be in closed form.

The particle-volume source terms are defined by

$$\langle G_V \rangle \equiv \int \langle G_1 \rangle_1 n \, \mathrm{d}\mathbf{U}_\mathrm{p} \, \mathrm{d}\mathbf{U}_\mathrm{f} \, \mathrm{d}\boldsymbol{\xi} \tag{2.45}$$

and

$$\mathcal{S}_V \equiv \int \xi_1 \mathcal{S}_1 \, \mathrm{d}\mathbf{U}_\mathrm{p} \, \mathrm{d}\mathbf{U}_\mathrm{f} \, \mathrm{d}\boldsymbol{\xi}. \tag{2.46}$$

The continuous contribution $\langle G_V \rangle$ could be due to material transfer from the fluid to the solid phase. On the other hand, the discontinuous term $\mathcal{S}_V$ might appear due to nucleation of particles with non-zero volume from the fluid phase. For systems where the particle volume is conserved, the right-hand side of Eq. (2.43) will be null. Finally, we should note that by definition, the particle-volume fraction cannot be greater than unity ($0 \leq \alpha_\mathrm{p} \leq 1$). In fact, because of the shapes of the particles usually do not allow the system to be completely occupied by the particle phase, the upper limit on α_p will often be less than unity. This constraint implies that the divergence of $\mathbf{U}_V$ must be non-negative when $\alpha_\mathrm{p} = 1$. As we shall see in later chapters, in two-fluid models this condition is often enforced by adding a pressure term to the transport equation for $\mathbf{U}_V$.

Particle-mass transport In order to derive a transport equation of the particle-mass density $\langle \rho_\mathrm{p} \rangle$, we will let ξ_2 equal to particle mass (m). The particle mass can be written as $m = \rho_\mathrm{p} V$ where ρ_p is the material density of the particle and V is its volume. Note that in addition to mass, either material density or volume could be included in the internal-coordinate vector. (In general, we will use mass and volume as the internal coordinates.) Thus, for example, fixing the particle masses to be equal does not imply that all particles have the same volume. The particle-mass density is then defined by

$$\langle \rho_\mathrm{p} \rangle \equiv \int \xi_2 n \, \mathrm{d}\mathbf{U}_\mathrm{p} \, \mathrm{d}\mathbf{U}_\mathrm{f} \, \mathrm{d}\boldsymbol{\xi}. \tag{2.47}$$

Starting from Eq. (2.32) with $g = \xi_2$, the transport equation for $\langle \rho_\mathrm{p} \rangle$ is

$$\frac{\partial \langle \rho_\mathrm{p} \rangle}{\partial t} + \frac{\partial}{\partial \mathbf{x}} \cdot \left(\mathbf{U}_m \langle \rho_\mathrm{p} \rangle \right) = \langle G_m \rangle + \mathcal{S}_m \tag{2.48}$$

where the mass-average particle velocity is defined by

$$\mathbf{U}_m \equiv \frac{1}{\langle \rho_\mathrm{p} \rangle} \int \xi_2 \mathbf{U}_\mathrm{p} n \, \mathrm{d}\mathbf{U}_\mathrm{p} \, \mathrm{d}\mathbf{U}_\mathrm{f} \, \mathrm{d}\boldsymbol{\xi}. \tag{2.49}$$

For the case where all particles have the same mass, $\mathbf{U}_m = \mathbf{U}_N$. Note that like $\mathbf{U}_N$ and $\mathbf{U}_V$, the mass-average particle velocity will not usually be in closed form.

The particle-mass source terms are defined by

$$\langle G_m \rangle \equiv \int \langle G_2 \rangle_1 n \, \mathrm{d}\mathbf{U}_\mathrm{p} \, \mathrm{d}\mathbf{U}_\mathrm{f} \, \mathrm{d}\boldsymbol{\xi} \tag{2.50}$$

and

$$\mathcal{S}_m \equiv \int \xi_2 \mathcal{S}_1 \, \mathrm{d}\mathbf{U}_\mathrm{p} \, \mathrm{d}\mathbf{U}_\mathrm{f} \, \mathrm{d}\boldsymbol{\xi}. \tag{2.51}$$

The continuous contribution $\langle G_m \rangle$ could be due to mass transfer from the fluid to the solid phase. On the other hand, the discontinuous term $\mathcal{S}_m$ might appear due to nucleation of particles with non-zero mass from the fluid phase. For systems where the particle mass is conserved, the right-hand side of Eq. (2.48) will be null.

More generally, since the total mass of the fluid-particle system is conserved, we can define the fluid-mass density $\langle \rho_\mathrm{f} \rangle$ by

$$\frac{\partial \langle \rho_\mathrm{f} \rangle}{\partial t} + \frac{\partial}{\partial \mathbf{x}} \cdot (\langle \mathbf{U}_\mathrm{f} \rangle \langle \rho_\mathrm{f} \rangle) = -\langle G_m \rangle - \mathcal{S}_m \tag{2.52}$$

where the mass-average fluid velocity $\langle \mathbf{U}_\mathrm{f} \rangle$ is defined by a conservation equation for the total momentum as discussed below. Note that $\langle \rho_\mathrm{p} \rangle$ and $\langle \rho_\mathrm{f} \rangle$ are not equal to the actual material densities ρ_p and ρ_f, respectively. Rather, they are equal to the mass of solid (fluid) per unit volume of the fluid-particle system. In order to distinguish between the two densities, when the material densities are constant we can relate them to the volume fractions α_p and α_f:

$$\rho_\mathrm{p} \alpha_\mathrm{p} = \langle \rho_\mathrm{p} \rangle \quad \text{and} \quad \rho_\mathrm{f} \alpha_\mathrm{f} = \langle \rho_\mathrm{f} \rangle . \tag{2.53}$$

The mixture-mass density can then be defined by

$$\rho_\mathrm{mix} \equiv \alpha_\mathrm{p} \rho_\mathrm{p} + \alpha_\mathrm{f} \rho_\mathrm{f}, \tag{2.54}$$

and satisfies the mixture continuity equation:

$$\frac{\partial \rho_\mathrm{mix}}{\partial t} + \frac{\partial}{\partial \mathbf{x}} \cdot (\mathbf{U}_\mathrm{mix} \rho_\mathrm{mix}) = 0 \tag{2.55}$$

where the mass-average mixture velocity is defined by

$$\mathbf{U}_\mathrm{mix} \equiv \frac{1}{\rho_\mathrm{mix}} \left(\alpha_\mathrm{p} \rho_\mathrm{p} \mathbf{U}_m + \alpha_\mathrm{f} \rho_\mathrm{f} \langle \mathbf{U}_\mathrm{f} \rangle \right) . \tag{2.56}$$

The transport equations for $\langle \rho_\mathrm{p} \rangle$ and $\langle \rho_\mathrm{f} \rangle$ are used in two-fluid models for multiphase flows.

Particle-momentum transport The particle-momentum density is defined by

$$\langle \rho_\mathrm{p} \mathbf{U}_\mathrm{p} \rangle \equiv \int \xi_2 \mathbf{U}_\mathrm{p} n \, \mathrm{d}\mathbf{U}_\mathrm{p} \, \mathrm{d}\mathbf{U}_\mathrm{f} \, \mathrm{d}\boldsymbol{\xi}, \tag{2.57}$$

where ξ_2 again corresponds to the particle mass. Starting from Eq. (2.32) with $g = \xi_2 \mathbf{U}_\mathrm{p}$, the transport equation for the particle-momentum density is

$$\frac{\partial \langle \rho_\mathrm{p} \mathbf{U}_\mathrm{p} \rangle}{\partial t} + \frac{\partial}{\partial \mathbf{x}} \cdot \langle \rho_\mathrm{p} \mathbf{U}_\mathrm{p} \mathbf{U}_\mathrm{p} \rangle = \langle \rho_\mathrm{p} \mathbf{A}_\mathrm{p} \rangle + \langle \rho_\mathrm{p} \mathbf{A} \rangle + \langle \mathbf{U}_\mathrm{p} G_m \rangle + \langle \rho_\mathrm{p} \mathbf{U}_\mathrm{p} \mathcal{S} \rangle \tag{2.58}$$

where the momentum-convection term is defined by

$$\langle \rho_\mathrm{p} \mathbf{U}_\mathrm{p} \mathbf{U}_\mathrm{p} \rangle \equiv \int \xi_2 \mathbf{U}_\mathrm{p} \mathbf{U}_\mathrm{p} n \, \mathrm{d}\mathbf{U}_\mathrm{p} \, \mathrm{d}\mathbf{U}_\mathrm{f} \, \mathrm{d}\boldsymbol{\xi}. \tag{2.59}$$

The vector dyadic products are defined component-wise: $(\mathbf{ab})_{ij} = a_i b_j$. Note that in general the momentum-average particle velocity will not usually be in closed form. From these definitions, we can identify three limiting cases:

1. If the particle mass (m) is constant, then $\langle \rho_\mathrm{p} \mathbf{U}_\mathrm{p} \rangle = mN\mathbf{U}_N$.
2. If the particle material density (ρ_p) is constant, then $\langle \rho_\mathrm{p} \mathbf{U}_\mathrm{p} \rangle = \rho_\mathrm{p} \alpha_\mathrm{p} \mathbf{U}_V$.
3. If m and ρ_p are constant, then $\langle \rho_\mathrm{p} \mathbf{U}_\mathrm{p} \rangle = \rho_\mathrm{p} \alpha_\mathrm{p} \langle \mathbf{U}_\mathrm{p} \rangle$.

These cases would suggest that the best averaged velocity to use depends on which internal coordinates are held constant.

The particle-momentum source terms appearing on the right-hand side of Eq. (2.58) are defined as follows. The first term:

$$\langle \rho_\mathrm{p} \mathbf{A}_\mathrm{p} \rangle \equiv \int \xi_2 \langle \mathbf{A}_\mathrm{p} \rangle_1 n \, \mathrm{d}\mathbf{U}_\mathrm{p} \, \mathrm{d}\mathbf{U}_\mathrm{f} \, \mathrm{d}\boldsymbol{\xi}, \tag{2.60}$$

describes surface forces on the particles due to the surrounding fluid. The second term:

$$\langle \rho_\mathrm{p} \mathbf{A} \rangle \equiv \int \xi_2 \langle \mathbf{A} \rangle_1 n \, \mathrm{d}\mathbf{U}_\mathrm{p} \, \mathrm{d}\mathbf{U}_\mathrm{f} \, \mathrm{d}\boldsymbol{\xi}, \tag{2.61}$$

is due to body forces (such as gravity) acting on individual particles. The third term:

$$\langle \mathbf{U}_\mathrm{p} G_m \rangle \equiv \int \mathbf{U}_\mathrm{p} \langle G_m \rangle n \, \mathrm{d}\mathbf{U}_\mathrm{p} \, \mathrm{d}\mathbf{U}_\mathrm{f} \, \mathrm{d}\boldsymbol{\xi}, \tag{2.62}$$

describes momentum added to the systems due to mass exchange from the liquid to the particle phase. The fourth term:

$$\langle \rho_\mathrm{p} \mathbf{U}_\mathrm{p} \mathcal{S} \rangle \equiv \int \xi_2 \mathbf{U}_\mathrm{p} \mathcal{S}_1 \, \mathrm{d}\mathbf{U}_\mathrm{p} \, \mathrm{d}\mathbf{U}_\mathrm{f} \, \mathrm{d}\boldsymbol{\xi}. \tag{2.63}$$

describes discontinuous changes in particle momentum due to collisions and particle nucleation. In general, none of these terms will appear in closed form (even if the GPBE is closed), and models must be provided to close Eq. (2.58).

Higher-order moment transport The moment equations that we have derived up this point are first order in any one variable. In order to describe fluctuations about the first-order moments, it is necessary to derive transport equations for second- and sometimes higher-order moments. Just as before, this is done starting from Eq. (2.32) with a particular choice for g. In order to illustrate how this is done, we will consider the function $g = \xi_2 U_{\mathrm{p}1}^2$ that results in the moment

$$\langle \rho_\mathrm{p} U_{\mathrm{p}1}^2 \rangle \equiv \int \xi_2 U_{\mathrm{p}1}^2 n \, \mathrm{d}\mathbf{U}_\mathrm{p} \, \mathrm{d}\mathbf{U}_\mathrm{f} \, \mathrm{d}\boldsymbol{\xi}, \tag{2.64}$$

where ξ_2 is again equal to particle mass. Note that this moment is one component of the particle kinetic energy, which is used to define the "solids temperature".

Starting from Eq. (2.32), the transport equation is

$$\frac{\partial \langle \rho_\mathrm{p} U_{\mathrm{p}1}^2 \rangle}{\partial t} + \frac{\partial}{\partial \mathbf{x}} \cdot \langle \rho_\mathrm{p} U_{\mathrm{p}1}^2 \mathbf{U}_\mathrm{p} \rangle = 2\langle \rho_\mathrm{p} U_{\mathrm{p}1} A_{\mathrm{p}1} \rangle + 2\langle \rho_\mathrm{p} U_{\mathrm{p}1} A_{\mathrm{p}1} \rangle + \langle U_{\mathrm{p}1}^2 G_m \rangle + \langle \rho_\mathrm{p} U_{\mathrm{p}1}^2 \mathcal{S} \rangle \tag{2.65}$$

where the convection term is defined by

$$\langle \rho_\mathrm{p} U_{\mathrm{p}1}^2 \mathbf{U}_\mathrm{p} \rangle \equiv \int \xi_2 U_{\mathrm{p}1}^2 \mathbf{U}_\mathrm{p} n \, \mathrm{d}\mathbf{U}_\mathrm{p} \, \mathrm{d}\mathbf{U}_\mathrm{f} \, \mathrm{d}\boldsymbol{\xi}. \tag{2.66}$$

Note that as is usually the case in moment methods, the convection term is unclosed and involves even higher-order moments. The remaining terms on the right-hand side of Eq. (2.65) are defined in a manner very similar to their counterparts in Eq. (2.58). Thus we will not discuss them in detail except to say that they will usually not appear in closed form.

3 An Example Application: The Spray Equation

There exists considerable interest in the development of numerical methods for simulating sprays of fine droplets (Hylkema and Villedieu, 1998; Hylkema, 1999; Miller and Bellan, 1999, 2000; Laurent et al., 2004; Réveillon et al., 2004). The principal physical processes that must be accounted for are (1) transport in real space, (2) droplet evaporation, (3) acceleration of droplets due to drag, and (4) coalescence of droplets leading to polydispersity. The major challenge in numerical simulations is to account for the strong coupling between these processes. Williams (Williams, 1958) proposed a relatively simple transport equation based on the kinetic theory that has proven to be a useful starting point for testing novel numerical methods for treating polydisperse, dense liquid sprays. In the literature, the Lagrangian Monte-Carlo approach (Dukowicz, 1980), also called Direct Simulation Monte-Carlo method (DSMC) (Bird, 1994), is generally considered to be the most accurate for solving Williams equation. However, its computational cost is high, especially in unsteady configurations, and the method is difficult to couple accurately with Eulerian descriptions of the gas phase. There is thus considerable impetus to develop moment-based methods for treating Williams equation. For simplicity, we consider only the droplet volume (v) and droplet velocity (U_p) in one spatial dimension (x) with a stagnant gas phase ($U_\mathrm{f} = 0$).

3.1 Williams spray equation

The Williams transport equation (Williams, 1958) for the joint volume, velocity number density function $n(v, U_\mathrm{p}; x, t)$ has the form of a GPBE:

$$\frac{\partial n}{\partial t} + \frac{\partial}{\partial x}(U_\mathrm{p} n) + \frac{\partial}{\partial v}(Rn) + \frac{\partial}{\partial U_\mathrm{p}}(Fn) = Q_\mathrm{coal}, \tag{3.1}$$

where R is the evaporation rate, F is the drag force acting on the droplet, and Q_coal is the coalescence term. A more general version of the spray equation would include the droplet temperature and molecular composition. For simplicity, we will use the Stokes drag model:

$$F(v, U_\mathrm{p}) = -\alpha v^{-2/3} U_\mathrm{p}. \tag{3.2}$$

For the evaporation rate, we will use

$$R(v) = -\beta v^{1/3}. \tag{3.3}$$

In these expressions, α and β are positive constants (Laurent et al., 2004). Note, however, that both F and R are nonlinear functions of v. At the level of the moment transport equations, we shall see below that this nonlinearity will lead to a closure problem. Also note that as the droplets evaporate and v goes to zero, the droplet velocity U_p must also approach zero in order that the drag (Eq. 3.2) remains bounded. Likewise, for the same velocity, the drag force will be larger on small droplets than on large droplets. In other words, small droplets will deaccelerate faster than large droplets. Thus it will be possible for a large droplet to catch up with and overtake a small droplet.

Using standard assumptions (Laurent et al., 2004), we can write $Q_{\mathrm{coal}} = Q^{-}_{\mathrm{coal}} + Q^{+}_{\mathrm{coal}}$ where

$$Q^{-}_{\mathrm{coal}} = -\int_{-\infty}^{+\infty}\int_{0}^{\infty} B(U_{\mathrm{p}}, U^{*}_{\mathrm{p}}, v, v^{*}) n(v, U_{\mathrm{p}}) n(v^{*}, U^{*}_{\mathrm{p}})\, \mathrm{d}v^{*}\, \mathrm{d}U^{*}_{\mathrm{p}}, \tag{3.4}$$

$$Q^{+}_{\mathrm{coal}} = \frac{1}{2}\int_{-\infty}^{+\infty}\int_{0}^{v} B(U^{\diamond}_{\mathrm{p}}, U^{*}_{\mathrm{p}}, v^{\diamond}, v^{*}) n(v^{\diamond}, U^{\diamond}_{\mathrm{p}}) n(v^{*}, U^{*}_{\mathrm{p}}) J\, \mathrm{d}v^{*}\, \mathrm{d}U^{*}_{\mathrm{p}}, \tag{3.5}$$

$v^{\diamond} = v - v^{*}$, $U^{\diamond}_{\mathrm{p}} = (vU_{\mathrm{p}} - v^{*}U^{*}_{\mathrm{p}})/(v - v^{*})$, and $J = (v/v^{\diamond})^{3}$ is the Jacobian of the transform $(v, U_{\mathrm{p}}) \rightarrow (v^{\diamond}, U^{\diamond}_{\mathrm{p}})$ with fixed $(v^{*}, U^{*}_{\mathrm{p}})$. The collision frequency function B is defined by

$$B(U_{\mathrm{p}}, U^{*}_{\mathrm{p}}, v, v^{*}) = E_{\mathrm{coal}}(|U_{\mathrm{p}} - U^{*}_{\mathrm{p}}|, v, v^{*}) \beta(v, v^{*}) |U_{\mathrm{p}} - U^{*}_{\mathrm{p}}|, \tag{3.6}$$

where E_{coal} is the coalescence efficiency probability, which, based upon the size of droplets and the relative velocity, discriminates between rebound and coalescence, and

$$\beta(v, v^{*}) = \pi \left[\left(\frac{3v}{4\pi}\right)^{1/3} + \left(\frac{3v^{*}}{4\pi}\right)^{1/3} \right]^{2}. \tag{3.7}$$

For simplicity, we will take $E_{\mathrm{coal}} = 0$ (no coalescence) or $E_{\mathrm{coal}} = 1$; however, any other functional form could be used. Note that droplets collide due to differences in their velocities ($|U_{\mathrm{p}} - U^{*}_{\mathrm{p}}|$). Such differences can arise due to initial conditions, or due to differences in the size-dependent drag force. Thus, we can expect that the volume v and the velocity U_{p} will become strongly correlated due both to the drag force and to coalescence.

Although the coalescence term is a relatively complex function of the NDF, at the level of the GPBE it is a closed term. Thus, in principle, the solution to Eq. (3.1) is a complete description of $n(v, U_{\mathrm{p}}; x, t)$. The direct method for solving Eq. (3.1) would be to discretize the three-dimensional space (v, U_{p}, x) using a finite-difference or a finite-volume approximation. This would result in a large system of coupled ordinary differential equations (ODE) in t for the NDF at each point in the discretized space. The coalescence term makes the direct approach particularly challenging because of the highly non-local interactions. Another approach is to treat x, v and U_{p} as random variables and to use a stochastic algorithm to *estimate* the NDF (Dukowicz, 1980). While this Lagrangian Monte-Carlo approach is generally considered to be more straightforward to implement than the direct approach, a large number of stochastic "particles" are required to control the estimation errors. In practice, stochastic methods provide adequate approximations for the lower-order moments of the NDF at reasonable computational cost. However, the statistical errors associated with locations in the tails of the NDF (i.e., points in (v, U_{p}, x) space where the NDF is small) make Lagrangian methods less attractive than direct methods for estimating the entire NDF. Fortunately, in many applications it suffices to have good estimates of the lower-order moments of the NDF. It is thus of interest to try to obtain such estimates by solving transport equations for the moments.

3.2 Moment transport equation

Moment transformation The moment transport equations corresponding to Eq. (3.1) are found using the procedure outlined earlier. For clarity, we will start by defining the integer moment transformation as

$$M(k_1, k_2; x, t) \equiv \int_{-\infty}^{+\infty} \int_0^{\infty} v^{k_1} U_{\mathrm{p}}^{k_2} n(v, U_{\mathrm{p}}; x, t)\, \mathrm{d}v\, \mathrm{d}U_{\mathrm{p}} \tag{3.8}$$

where k_1 and k_2 are non-negative integers. For the special cases where k_1 or k_2 are zero, we will define $M_v(k) \equiv M(k, 0)$ and $M_{\mathrm{p}}(k) \equiv M(0, k)$. Note that the moments depend on x and t, and thus are Eulerian quantities. In terms of this notation, some of the previously defined moments of interest are

$$N = M(0,0) = M_v(0) = M_{\mathrm{p}}(0) \quad \text{number density,}$$
$$\alpha_{\mathrm{p}} = M(1,0) = M_v(1) \quad \text{droplet volume fraction,}$$
$$U_N = M(0,1)/N \quad \text{number-average velocity,}$$

and

$$U_V = M(1,1)/\alpha_{\mathrm{p}} \quad \text{volume-average velocity.}$$

The reader should note that using the moments to represent the properties of the NDF represents a significant reduction in the number of unknowns that must be solved. For example, using the direct method, it would not be uncommon to use 100×100 finite-volume cells to discretize (v, U_{p}) space. In contrast, with moment methods it is unlikely that values for k_1 and k_2 larger than nine would be used. Thus, instead of 10^4 unknowns for a direct method, a moment method would require at most 100. However, as we shall see later, moment methods require closures for nonlinear terms that cannot be expressed as simple functions of the known moments.

Formally, the moment transformation of the spray equation yields

$$\frac{\partial M(k_1, k_2)}{\partial t} + \frac{\partial M(k_1, k_2+1)}{\partial x} = \beta \int_0^{\infty} v^{k_1} \frac{\partial}{\partial v} \left(v^{1/3} \int_{-\infty}^{+\infty} U_{\mathrm{p}}^{k_2} n\, \mathrm{d}U_{\mathrm{p}} \right) \mathrm{d}v$$
$$+ \alpha \int_{-\infty}^{+\infty} U_{\mathrm{p}}^{k_2} \frac{\partial}{\partial U_{\mathrm{p}}} \left(U_{\mathrm{p}} \int_0^{\infty} v^{k_1 - 2/3} n\, \mathrm{d}v \right) \mathrm{d}U_{\mathrm{p}} + \int_{-\infty}^{+\infty} \int_0^{\infty} v^{k_1} U_{\mathrm{p}}^{k_2} Q_{\mathrm{coal}}\, \mathrm{d}v\, \mathrm{d}U_{\mathrm{p}}. \tag{3.9}$$

From the second term on the left-hand side, we can already observe that the moment $M(k_1, k_2 + 1)$ will not be closed because it involves a higher-order moment $(k_2 + 1)$. We shall see that the closure of this term will determine whether or not the moment equations can describe the "crossing" of droplets with different sizes (Desjardins et al., 2006).

Evaporation and drag terms The terms on the right-hand side of Eq. (3.9) are also unclosed. For example, the evaporation term in Eq. (3.9) can be written as

$$\int_0^{\infty} v^{k_1} \frac{\partial}{\partial v} \left(v^{1/3} \int_{-\infty}^{+\infty} U_{\mathrm{p}}^{k_2} n\, \mathrm{d}U_{\mathrm{p}} \right) \mathrm{d}v = -\delta_{k_1,0} \delta_{k_2,0}\, v^{1/3} n_v \Big|_{v=0} - k_1 M(k_1 - 2/3, k_2) \tag{3.10}$$

where $\delta_{i,j}$ is the Kronecker delta. The first term on the right-hand side appears only in the transport equation for $M(0,0)$ and represents the loss of droplets due to evaporation. Note that because the evaporation rate is proportional to $v^{1/3}$, the droplet disappear in a *finite* amount of time. For this reason, the volume NDF n_v evaluated at $v = 0$ goes to infinity in such a manner as to make the product $v^{1/3} n_v$ nonzero. Note that because the volume NDF cannot be uniquely determined from the moments, the rate of loss of droplets due to evaporation is an unclosed term (Fox et al., 2006). The second term on the right-hand side of Eq. (3.10) is the rate of change of $M(k_1, k_2)$ due to evaporation. However, because k_1 and k_2 are integers, the moment $M(k_1 - 2/3, k_2)$ will be unclosed. For the same reason, the drag term in Eq. (3.9):

$$\int_{-\infty}^{+\infty} U_p^{k_2} \frac{\partial}{\partial U_p} \left(U_p \int_0^{\infty} v^{k_1 - 2/3} n \, dv \right) dU_p = -k_2 M(k_1 - 2/3, k_2), \tag{3.11}$$

will not be closed. Note that although Eqs. (3.10) and (3.11) both involve the moment $M(k_1 - 2/3, k_2)$, this need not be the case for other expressions for the drag (F) and evaporation rates (R).

Coalescence term Turning now to the coalescence term in Eq. (3.9), we will treat each of the two parts Q_{coal}^- and Q_{coal}^+ separately. The first part yields in a straightforward manner

$$\begin{aligned}
&\int_{-\infty}^{+\infty} \int_0^{\infty} v^{k_1} U_p^{k_2} Q_{\text{coal}}^- \, dv \, dU_p \\
&= - \int_{-\infty}^{+\infty} \int_0^{\infty} \int_{-\infty}^{+\infty} \int_0^{\infty} v^{k_1} U_p^{k_2} B(U_p, U_p^*, v, v^*) n(v, U_p) n(v^*, U_p^*) \, dv \, dU_p \, dv^* \, dU_p^* \\
&= -\frac{1}{2} \int_{-\infty}^{+\infty} \int_0^{\infty} \int_{-\infty}^{+\infty} \int_0^{\infty} \left(v^{k_1} U_p^{k_2} + v^{*k_1} U_p^{*k_2} \right) B(U_p, U_p^*, v, v^*) \\
&\qquad n(v, U_p) n(v^*, U_p^*) \, dv \, dU_p \, dv^* \, dU_p^*.
\end{aligned} \tag{3.12}$$

However, due to the complicated expression for B (Eq. 3.6), it is not at all obvious that Eq. (3.12) can be closed in terms of integer moments. The second part of the coalescence term requires a change in the order of integration, and a change of variables:

$$\begin{aligned}
&\int_{-\infty}^{+\infty} \int_{-\infty}^{+\infty} \int_0^{\infty} \int_0^{v} v^{k_1} U_p^{k_2} B(U_p^{\diamond}, U_p^*, v^{\diamond}, v^*) n(v^{\diamond}, U_p^{\diamond}) n(v^*, U_p^*) J \, dv^* \, dv \, dU_p^* \, dU_p \\
&= \int_{-\infty}^{+\infty} \int_{-\infty}^{+\infty} \int_0^{\infty} \int_{v^*}^{\infty} v^{k_1} U_p^{k_2} B(U_p^{\diamond}, U_p^*, v^{\diamond}, v^*) n(v^{\diamond}, U_p^{\diamond}) n(v^*, U_p^*) J \, dv \, dv^* \, dU_p^* \, dU_p \\
&= \int_{-\infty}^{+\infty} \int_0^{\infty} \int_{-\infty}^{+\infty} \int_0^{\infty} (v^* + v^{\diamond})^{k_1} \left(\frac{v^* U_p^* + v^{\diamond} U_p^{\diamond}}{v^* + v^{\diamond}} \right)^{k_2} B(U_p^{\diamond}, U_p^*, v^{\diamond}, v^*) n(v^{\diamond}, U_p^{\diamond}) \\
&\qquad n(v^*, U_p^*) \, dv^* \, dU_p^* \, dv^{\diamond} \, dU_p^{\diamond}.
\end{aligned} \tag{3.13}$$

Again, the final expression for the second part involves multiple integral and it is not obvious that it can be easily represented by integer moments.

Combining the two parts, the overall contribution to the moment transport equation due to coalescence becomes

$$\begin{aligned} Q_{\text{coal}}(k_1,k_2) &\equiv \int_{-\infty}^{+\infty}\int_0^{\infty} v^{k_1} U_{\text{p}}^{k_2} Q_{\text{coal}}\, \mathrm{d}v\, \mathrm{d}U_{\text{p}} \\ &= \frac{1}{2}\int_{-\infty}^{+\infty}\int_0^{\infty}\int_{-\infty}^{+\infty}\int_0^{\infty}\left[(v^*+v)^{k_1}\left(\frac{v^*U_{\text{p}}^*+vU_{\text{p}}}{v^*+v}\right)^{k_2} - v^{k_1}U_{\text{p}}^{k_2} - v^{*k_1}U_{\text{p}}^{*k_2}\right] \\ &\qquad B(U_{\text{p}},U_{\text{p}}^*,v,v^*)n(v,U_{\text{p}})n(v^*,U_{\text{p}}^*)\,\mathrm{d}v\,\mathrm{d}U_{\text{p}}\,\mathrm{d}v^*\,\mathrm{d}U_{\text{p}}^*. \end{aligned} \tag{3.14}$$

Aside from the fact that this expression is rather complicated, it does have important properties related to conservation of mass and momentum. To see this, consider the following three moments:

$$Q_{\text{coal}}(0,0) = -\frac{1}{2}\int_{-\infty}^{+\infty}\int_0^{\infty}\int_{-\infty}^{+\infty}\int_0^{\infty} B(U_{\text{p}},U_{\text{p}}^*,v,v^*)n(v,U_{\text{p}})n(v^*,U_{\text{p}}^*)\,\mathrm{d}v\,\mathrm{d}U_{\text{p}}\,\mathrm{d}v^*\mathrm{d}U_{\text{p}}^*, \tag{3.15}$$

$$Q_{\text{coal}}(1,0) = 0, \tag{3.16}$$

and

$$Q_{\text{coal}}(1,1) = 0. \tag{3.17}$$

The first expression (Eq. 3.15) gives the rate at which the number of droplets decreases with time due to coalescence. Note that this term is second order in the NDF, and thus will be small when $M(0,0)$ is small (i.e., for dilute sprays). The second expression (Eq. 3.16) states that the total volume of droplets is conserved during coalescence. Likewise, Eq. (3.15) states that the total momentum of the droplet phase is conserved during coalescence. For all other moments, the contribution due to coalescence will be nonzero. However, for dilute sprays, the coalescence term will be negligible when $M(0,0) \ll 1$ because the probability of collisions will be small relative to other processes.

Transport equation Collecting together all of the terms, the final expression for the moment transport equation for the spray equation is

$$\frac{\partial M(k_1,k_2)}{\partial t}+\frac{\partial M(k_1,k_2+1)}{\partial x} = -\delta_{k_1,0}\delta_{k_2,0}\psi_0-(\alpha k_2+\beta k_1)\,M(k_1-2/3,k_2)+Q_{\text{coal}}(k_1,k_2). \tag{3.18}$$

The terms on the left-hand side of this expression represent accumulation and transport, respectively. The terms of the right-hand side represent loss of droplets due to evaporation (ψ_0), drag (α), evaporation (β), and coalescence (Q_{coal}), respectively. Aside from the accumulation term, none of the other terms in Eq. (3.18) are closed. Thus, in order to proceed further, it will be necessary to provide closures for each of the unclosed terms.

3.3 Moment closures for the spray equation

Quadrature-based moment closures (McGraw, 1997; Marchisio and Fox, 2005) provide a systematic method for closing moment transport equations. The basic idea is to introduce N

quadrature weights (w_i) and $2N$ abscissas (v_i and u_i) defined such that the moments are given by

$$M(k_1, k_2) = \sum_{i=1}^{N} w_i v_i^{k_1} u_i^{k_2} \tag{3.19}$$

for $3N$ independent values of the pair (k_1, k_2), (in other words, $3N$ independent moments). The solution strategy then consists of solving the transport equations for the $3N$ moments, and inverting Eq. (3.19) to find the weights and abscissas whenever they are needed. Formally, we can denote the set of $3N$ moments as $\mathcal{M}$. The inversion of Eq. (3.19) can then be expressed as

$$\begin{aligned} w_i &= W_i(\mathcal{M}), \\ v_i &= V_i(\mathcal{M}), \\ u_i &= U_i(\mathcal{M}), \end{aligned} \tag{3.20}$$

where the right-hand sides are nonlinear functions of the moments $\mathcal{M}$. The question of how to determine these nonlinear functions is important (Wright et al., 2001; Fox, 2006a; Fox et al., 2006; Fox, 2007), and we shall return to it below. However the reader should note that for univariate NDFs the product-difference (PD) algorithm (McGraw, 1997) is the preferred method for finding the weights and abscissas for values of N up to approximately twenty.

The power of quadrature methods comes from their ability to approximate moments that are not included in the set $\mathcal{M}$. For example, in Eq. (3.18) the unclosed terms can be approximated by

$$M(k_1, k_2 + 1) \approx \sum_{i=1}^{N} w_i v_i^{k_1} u_i^{k_2+1}, \tag{3.21}$$

$$M(k_1 - 2/3, k_2) \approx \sum_{i=1}^{N} w_i v_i^{k_1-2/3} u_i^{k_2}, \tag{3.22}$$

and

$$Q_{\text{coal}}(k_1, k_2) \approx \frac{1}{2} \sum_{i=1}^{N} \sum_{j=1}^{N} w_i w_j \left[(v_i + v_j)^{k_1} \left(\frac{v_i u_i + v_j u_j}{v_i + v_j} \right)^{k_2} - v_i^{k_1} u_i^{k_2} - v_j^{k_1} u_j^{k_2} \right] B(u_i, u_j, v_i, v_j). \tag{3.23}$$

As discussed elsewhere (Fox et al., 2006), the evaporative flux ψ_0 can also be approximated using the weights and abscissas. As with standard quadrature methods, the accuracy of these approximations increases with increasing N (i.e., by using more quadrature nodes). In general, the evaporation and coalescence terms can be accurately approximated with $N = 2$–5 (Wright et al., 2001; Marchisio et al., 2003b,a; Fox, 2006a; Fox et al., 2006). However, in bivariate problems it may be necessary to use $N = 4$ or 9 in order to capture all lower-order moments correctly (Fox, 2007). In any case, the total number of moment transport equations that must be solved is $3N$, and is much smaller than the number of degrees of freedoms required for a direct method.

One-node closure Closure with $N = 1$ is usually not very accurate. However, it leads to simple-to-understand transport equations that will provide insight into why higher-order closures are needed. With one-node closure, there are three unknowns: w_1, v_1, and u_1; hence, we must choose three moments. The choice is not unique, but in order to control number, mass, and momentum we select $M(0,0)$, $M(1,0)$, and $M(1,1)$. The weights and abscissas are then related to the moments by

$$\begin{aligned} w_1 &= M(0,0), \\ v_1 &= M(1,0)/M(0,0), \\ u_1 &= M(1,1)/M(1,0). \end{aligned} \tag{3.24}$$

The corresponding transport equations are

$$\frac{\partial w_1}{\partial t} + \frac{\partial w_1 u_1}{\partial x} = -\psi_0 + \frac{1}{2} w_1^2 B(u_1, u_1, v_1, v_1), \tag{3.25}$$

$$\frac{\partial w_1 v_1}{\partial t} + \frac{\partial w_1 v_1 u_1}{\partial x} = -\beta w_1 v_1^{1/3}, \tag{3.26}$$

$$\frac{\partial w_1 v_1 u_1}{\partial t} + \frac{\partial w_1 v_1 u_1^2}{\partial x} = -(\alpha + \beta) w_1 v_1^{1/3} u_1. \tag{3.27}$$

Although Eq. (3.25) contains a coalescence term, from the definition of B (Eq. 3.6), we see that this term is null. Thus, the one-node closure completely neglects the effects of coalescence. Another problem with this closure can be understood by using Eqs. (3.26) and (3.27) to write a transport equation for u_1:

$$\frac{\partial u_1}{\partial t} + 2u_1 \frac{\partial u_1}{\partial x} = -\alpha v_1^{-2/3} u_1. \tag{3.28}$$

The left-hand side of this expression is the one-dimensional zero-pressure gas dynamics equation (Bouchut, 1994), which has the well-known property of generating "shocks" for particular initial conditions $u_1(x,0)$ (for any finite value of α including $\alpha = 0$). The classical example is $u_1(x,0) = \sin(x)$ for $0 \leq x \leq 2\pi$, which produces a shock at $x = \pi$ where w_1 approaches infinity. Such behavior is unphysical and would not be observed in Lagrangian Monte-Carlo simulations of a dilute spray. As we shall see below, such shocks are an artifact of assuming that the droplet velocity at any point x can take on only one value $u_1(x,t)$.

An example of shock formation for impinging droplets taken from Desjardins et al. (2006) is shown in Figure 1. In this example, two "packets" of droplets with opposite velocities are released at time zero. At $t = 0.225$ the packets start to overlap and, for the one-node closure, a singularity is formed in the number density ($M(0,0)$) at $x = 0.5$. For the one-node closure this singularity grows and all of the droplets remain stuck at $x = 0.5$. In reality, for a dilute spray (no coalescence) the packets should not interact with each other. Rather, they should simply continue moving with their original velocities – a phenomenon known as particle-trajectory crossing. This is not possible with the one-node closure because at $x = 0.5$ it is only possible to have one velocity ($u_1 = 0$, i.e., the average velocity). As noted above, this is an artifact of the closure and does not represent reality.

The one-node closure is essentially the description used in two-fluid models for gas-solid flows (Agrawal et al., 2001; Kaufmann et al., 2004). The principal difference is that the two-fluid model is derived based on the kinetic theory of gases near equilibrium (i.e., close to a Boltzmann

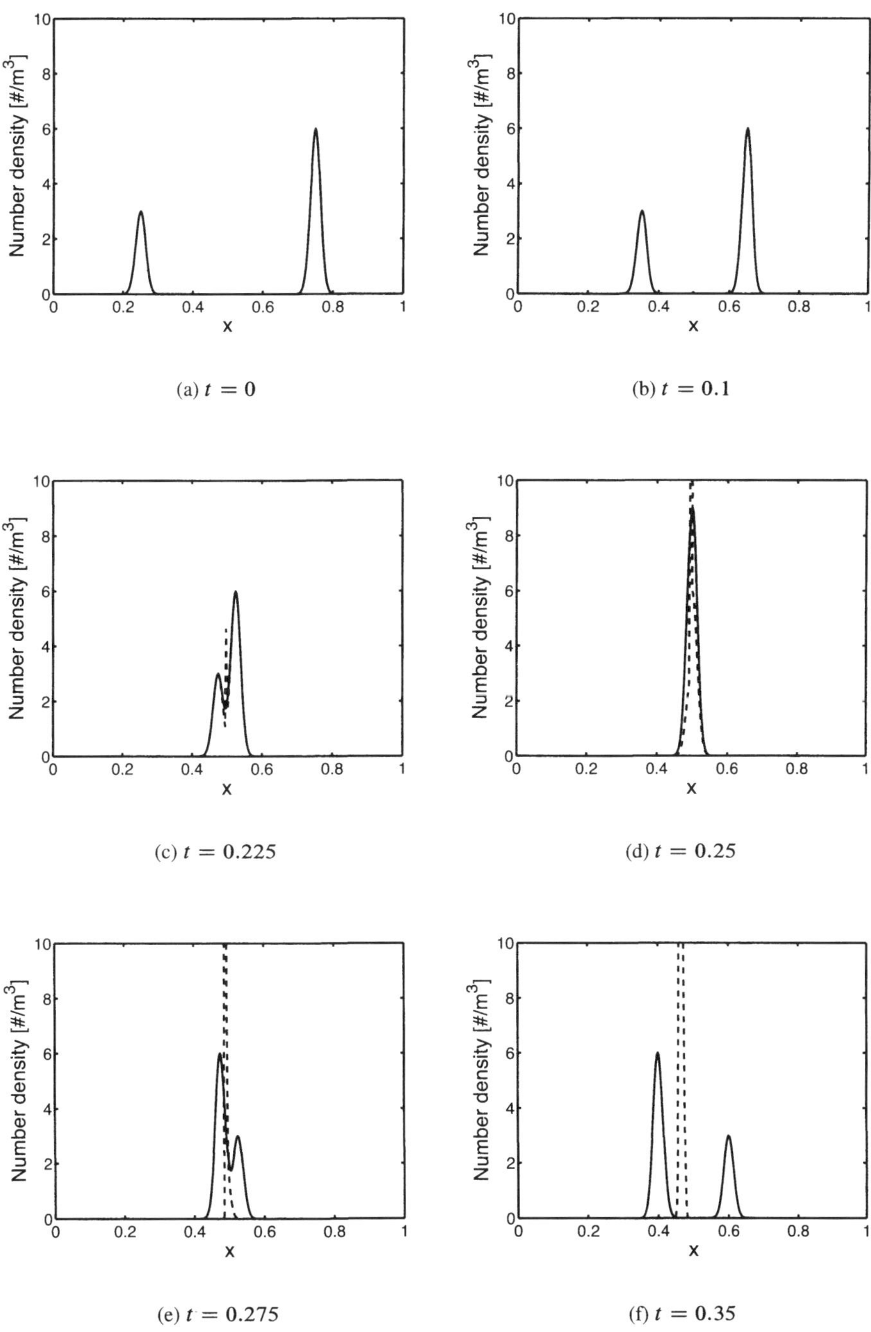

(a) $t = 0$

(b) $t = 0.1$

(c) $t = 0.225$

(d) $t = 0.25$

(e) $t = 0.275$

(f) $t = 0.35$

Figure 1. Time evolution of number density for impinging droplets with $\alpha = 0$. – –, one-node closure; —–, two-node closure.

distribution for velocity). Since the Boltzmann distribution is determined from the collision term (which is negligible in dilute flows), it cannot represent the highly non-equilibrium behavior seen in the dilute limit. In fact, in the dilute limit the velocity NDF is best represented by a collection of delta functions (as is done with quadrature methods (Marchisio and Fox, 2005)). Indeed, because the velocity is single-valued at every spatial location in a two-fluid simulation, such models cannot predict particle-trajectory crossings. For this reason, the mesoscale structures seen in two-fluid simulations of dilute gas-solid flows (Agrawal et al., 2001; Kaufmann et al., 2004) are most likely modeling artifacts. In summary, if an Eulerian model cannot predict particle-trajectory crossings as seen in Figure 1, then it cannot be trusted to predict any of the velocity statistics correctly in more complicated flows.

Two-node closure In order to predict particle-trajectory crossing, we must allow the local velocity to be multi-valued. The simplest model with this characteristic is the two-node closure. With $N = 2$ we must choose six moments for closure. A simple choice that works whenever $u_1 \neq u_2$ is

$$\begin{aligned} & & M(1,0) &= w_1 v_1 + w_2 v_2, \\ M(0,0) &= w_1 + w_2, & M(1,1) &= w_1 v_1 u_1 + w_2 v_2 u_2, \\ M(0,1) &= w_1 u_1 + w_2 u_2, & M(1,2) &= w_1 v_1 u_1^2 + w_2 v_2 u_2^2, \\ & & M(1,3) &= w_1 v_1 u_1^3 + w_2 v_2 u_2^3. \end{aligned} \tag{3.29}$$

The attraction for using this set of moments is that the PD algorithm (McGraw, 1997) can be used to find $(w_i v_i, u_i)$ from the last four moments (right column). The first two moments then form a linear system for w_i that is nonsingular as long as $u_1 \neq u_2$. Once w_i is known, it is straightforward to compute v_i. The six moment transport equations needed to find the weights and abscissas are given by

$$\frac{\partial M(0,0)}{\partial t} + \frac{\partial M(0,1)}{\partial x} = -\psi_0 + Q_{\text{coal}}(0,0), \tag{3.30}$$

$$\frac{\partial M(0,1)}{\partial t} + \frac{\partial M(0,2)}{\partial x} = -\alpha M(-2/3,1) + Q_{\text{coal}}(0,1), \tag{3.31}$$

$$\frac{\partial M(1,0)}{\partial t} + \frac{\partial M(1,1)}{\partial x} = -\beta M(1/3,0), \tag{3.32}$$

$$\frac{\partial M(1,1)}{\partial t} + \frac{\partial M(1,2)}{\partial x} = -(\alpha + \beta) M(1/3,1), \tag{3.33}$$

$$\frac{\partial M(1,2)}{\partial t} + \frac{\partial M(1,3)}{\partial x} = -(2\alpha + \beta) M(1/3,2) + Q_{\text{coal}}(1,2), \tag{3.34}$$

$$\frac{\partial M(1,3)}{\partial t} + \frac{\partial M(1,4)}{\partial x} = -(3\alpha + \beta) M(1/3,3) + Q_{\text{coal}}(1,3). \tag{3.35}$$

On the left-hand side, the two moments $M(0,2)$ and $M(1,4)$ are closed by writing them in terms of the weights and abscissas:

$$M(0,2) = w_1 u_1^2 + w_2 u_2^2, \tag{3.36}$$

$$M(1,4) = w_1 v_1 u_1^4 + w_2 v_2 u_2^4. \tag{3.37}$$

The terms of the right-hand side are closed in an analogous manner.

The numerical scheme used to solve Eqs. (3.30)–(3.35) employs a kinetic transport scheme to evaluate the spatial fluxes (Perthame, 1990; Desjardins et al., 2006). A first-order, explicit, finite-volume scheme for these equations can be written for the set of moments

$$\mathcal{M} = \begin{bmatrix} M(0,0) & M(0,1) & M(1,0) & M(1,1) & M(1,2) & M(1,3) \end{bmatrix}^{\mathrm{T}}$$

as

$$\mathcal{M}_i^{n+1} = \mathcal{M}_i^n - \frac{\Delta t}{\Delta x}\left[G(\mathcal{M}_i^n, \mathcal{M}_{i+1}^n) - G(\mathcal{M}_{i-1}^n, \mathcal{M}_i^n)\right] + \Delta t S_i^n \tag{3.38}$$

where n is the time step, i is the grid node, S is the right-hand side of Eqs. (3.30)–(3.35), and G is the flux function. Using the velocity abscissas, we can determine whether a quadrature node is moving left to right, or right to left. The flux function can then be expressed as

$$G(\mathcal{M}_i, \mathcal{M}_{i+1}) = H^+(\mathcal{M}_i) + H^-(\mathcal{M}_{i+1}) \tag{3.39}$$

where

$$\begin{aligned} H^+(\mathcal{M}) &= w_1 \max(u_1, 0)\begin{bmatrix} 1 \\ u_1 \\ v_1 \\ v_1 u_1 \\ v_1 u_1^2 \\ v_1 u_1^3 \end{bmatrix} + w_2 \max(u_2, 0)\begin{bmatrix} 1 \\ u_2 \\ v_2 \\ v_2 u_2 \\ v_2 u_2^2 \\ v_2 u_2^3 \end{bmatrix}, \\ H^-(\mathcal{M}) &= w_1 \min(u_1, 0)\begin{bmatrix} 1 \\ u_1 \\ v_1 \\ v_1 u_1 \\ v_1 u_1^2 \\ v_1 u_1^3 \end{bmatrix} + w_2 \min(u_2, 0)\begin{bmatrix} 1 \\ u_2 \\ v_2 \\ v_2 u_2 \\ v_2 u_2^2 \\ v_2 u_2^3 \end{bmatrix}. \end{aligned} \tag{3.40}$$

Higher-order flux schemes can also be developed to limit numerical diffusion (Desjardins et al., 2007). However, the key characteristic of the flux function is that the quadrature method provides a realizable set of weights and abscissas at every grid node that can be used to determine the node velocities.

Results for the impinging-droplets problem computed with the two-node closure are shown in Figure 1. It can be observed that the two packets pass through each other unchanged, exactly reproducing the Lagrangian predictions. As shown in Desjardins et al. (2006, 2007), the two-node closure does a remarkable job of reproducing the Lagrangian statistics for a number of different non-equilibrium flows (e.g., particle-jet crossing in two dimensions, particles rebounding off of walls). It is also noteworthy that the quadrature-based closures can handle particle-particle collisions (i.e., the term $Q_{\mathrm{coal}}(k_1, k_2)$ in the moment transport equation). Thus they can be used to investigate flows away from the dilute limit. In fact, if a valid transport equation for the NDF can be formulated in the dense limit, then quadrature-based closures can also be used there.

Despite its remarkable performance for the impinging-droplets problem, the two-node closure is not without its shortcomings. For example, in Figure 2 we modify the impinging-droplets problem slightly by placing a third packet of droplets in the center of the domain with zero velocity. As can be seen from the figure, the two-node closure works well at points where at most two

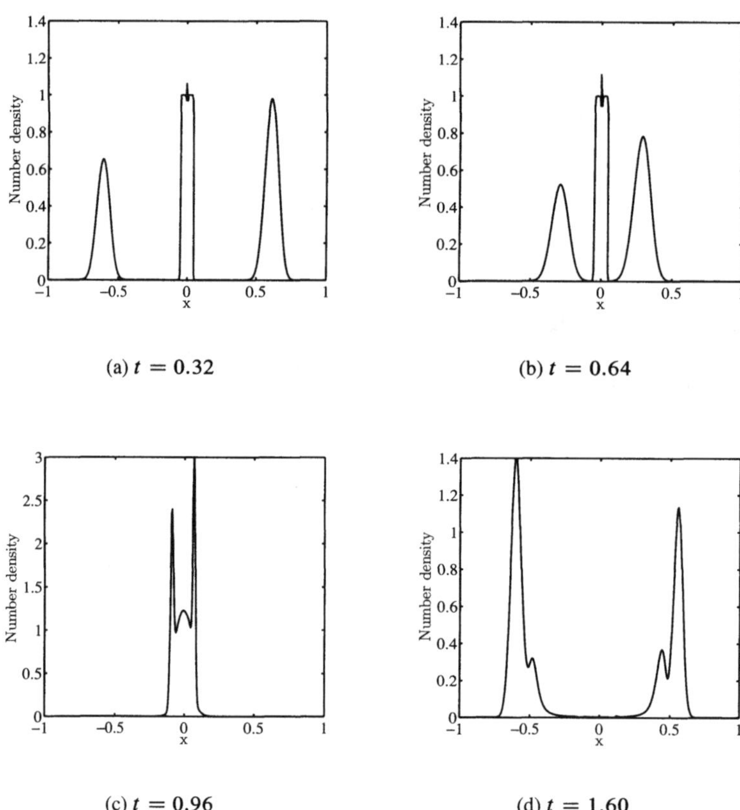

Figure 2. Time evolution of number density for droplets impinging on a stationary packet using the two-node closure.

packets are overlapping. At points where all three packets overlap, the two-node closure cannot represent the three distinct velocities. Instead, it replaces the three velocities (and weights) with two velocities that have the same moments up to third order, and the higher-order moments are dependent on these lower-order moments. (With three velocities and three weights the moments up to fifth order are independent.) The result is that the two packets pass through each other, but their shapes and velocities are deformed, and the third packet is carried away with the two moving packets. In the end, none of the moments of the velocity NDF are predicted correctly in comparison to Lagrangian simulations of the same flow.

Three-node closure In order to correctly predict the NDF with three overlapping packets, we must allow the local velocity to take on three values. The simplest model with this characteristic is the three-node closure. With $N = 3$ we must choose nine moments for closure. In addition to

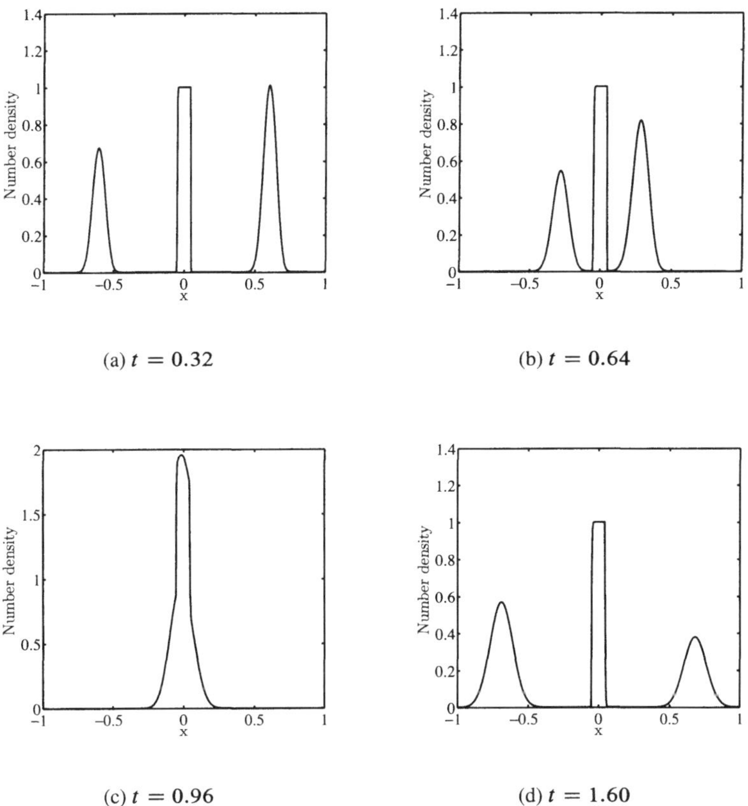

(a) $t = 0.32$ (b) $t = 0.64$

(c) $t = 0.96$ (d) $t = 1.60$

Figure 3. Time evolution of number density for droplets impinging on a stationary packet using the three-node closure.

the six moments in Eq. (3.29), we can add $M(0,2)$, $M(1,4)$, and $M(1,5)$:

$$\begin{aligned} M(0,0) &= w_1 + w_2 + w_3, \\ M(0,1) &= w_1 u_1 + w_2 u_2 + w_3 u_3, \\ M(0,2) &= w_1 u_1^2 + w_2 u_2^2 + w_3 u_3^2, \end{aligned} \qquad \begin{aligned} M(1,0) &= w_1 v_1 + w_2 v_2 + w_3 v_3, \\ M(1,1) &= w_1 v_1 u_1 + w_2 v_2 u_2 + w_3 v_3 u_3, \\ M(1,2) &= w_1 v_1 u_1^2 + w_2 v_2 u_2^2 + w_3 v_3 u_3^2, \\ M(1,3) &= w_1 v_1 u_1^3 + w_2 v_2 u_2^3 + w_3 v_3 u_3^3, \\ M(1,4) &= w_1 v_1 u_1^4 + w_2 v_2 u_2^4 + w_3 v_3 u_3^4, \\ M(1,5) &= w_1 v_1 u_1^5 + w_2 v_2 u_2^5 + w_3 v_3 u_3^5. \end{aligned} \tag{3.41}$$

As before, the PD algorithm can be used to find $(w_i v_i, u_i)$ from the last six moments (right column). The first three moments then form a linear system for w_i that is nonsingular as long as $u_1 \neq u_2 \neq u_3$. Once w_i is known, it is again straightforward to compute v_i. We should note, however, that although the moments in Eq. (3.41) allow for easy inversion, they may not be the best choice for accurately describing the correlation between v and U_p since many of the lower-order cross moments are not included in the moment set (Fox, 2007).

The moment transport equations and the numerical algorithm for the three-node closure are direct extensions of those used for the two-node closure. The application of the three-node closure to the three-packet problem is shown in Figure 3. We can observe that with three nodes the impinging packets traverse the domain successfully and the stationary packet remains unchanged in its original position. It can also be observed that the first-order numerical scheme generates numerical diffusion only for the moving packets. As noted earlier, the numerical diffusion can be greatly reduced by using a higher-order kinetic scheme (Desjardins et al., 2007).

The examples given above emphasize the importance of correctly treating the kinetic transport term in the GPBE, and the corresponding unclosed spatial flux term in the moment transport equation. In comparison, treatment of the drag, evaporation and coalescence terms is relatively straightforward because they involve only local interactions. From previous work (Fox et al., 2006), it is known that a four-node closure yields good results for the terms involving local interactions. For the kinetic term in one spatial dimension, the extension to a four-node closure is straightforward and follows the steps used in the three-node closure given above. Likewise, the extension to two and three spatial dimensions follows the same steps (Desjardins et al., 2006, 2007). However, the question of how best to choose the velocity moments when the velocity vector has more than one component is still open.

For example, in order to treat collisions in a manner that allows for isotropy of velocity components, it appears to be necessary to choose moment sets that are symmetric with respect to the velocity components. This constraint places a condition on which values of N can be used (Fox, 2007). Another technical difficulty in cases with multiple spatial dimensions is the inversion from moments to weights and abscissas. Unlike in the one-dimensional problems described above where the PD algorithm can be used, in two or three dimensions the moment sets will not lend themselves to the PD algorithm. It will thus be necessary to devise efficient inversion algorithms for multi-dimensional cases. Nevertheless, despite these open questions, the ability of quadrature methods to describe polydisperse systems ranging from dilute to dense interacting particles makes them a promising approach for developing practical models for multiphase systems based on the moment transport equation.

4 Summary and Conclusions

In this work we have reviewed the basic definitions and formulation of the governing equations needed to model polydisperse multiphase flows. In general, multiphase flows have polydispersity in both the velocity of the dispersed phase and in the internal coordinates that describe the thermochemical state of the different phases. In other words, it is possible to observe particles with a range of velocities, and a range of sizes, temperatures, etc., which can be described by a number density function (NDF). In cases where the velocity can be treated as locally single-valued (e.g., low Stokes number particulate flows), it is often possible to treat the polydispersity of the internal coordinates as scalar fields that are advected with a single velocity (e.g., the continuous-phase fluid velocity). In any case, the starting point for describing polydisperse multiphase flows is the transport equation for the NDF. When the velocity is single valued, this transport equation is commonly known as a population balance equation (PBE); otherwise, it is known by several different names, depending on the application. Here, we refer to it as the generalized population balance equation (GPBE).

From the modeling perspective, an important property of the GPBE is whether or not it appears in closed form. In other words, can it in principle be solved directly to find the NDF? If not, the first modeling step will be to close the GPBE. The closure of the collision integral in the Boltzmann equation is a well-known example of this procedure. Once this important step has been completed, the next modeling step is to decide whether or not the GPBE can be solved using a direct approach, such as a sectional method or Lagrangian Monte-Carlo simulations. In many multiphase flow applications, the direct approach will not be tractable and a moment closure must be attempted. In general, the moment transport equation will not appear in closed form, and its closure represents a significant challenge. Ideally, the statistics predicted by a moment closure should agree as closely as possible to those that would be found from a direct approach. For systems described by a PBE, quadrature-based closures have proven to be a powerful and accurate approach. Systems described by a GPBE are more challenging due to the kinetic term, but it has recently been shown that quadrature-based closures also work for such systems. In this work, the application of quadrature-based moment closures to the GPBE was illustrated using the one-dimensional spray equation. The resulting predictions for the velocity statistics are quite promising, but future work is needed to determine the range of applicability of such closures to other polydisperse multiphase flows.

Bibliography

K. Agrawal, P. N. Loezos, M. Syamlal, and S. Sundaresan (2001). The role of mesoscale structures in rapid gas-solid flows. *Journal of Fluid Mechanics*, 445:151–185.

R. Aris (1962). *Vectors, Tensors, and the Basic Equations of Fluid Mechanics*. Prentice-Hall, Englewood Cliffs, USA.

G. A. Bird (1994). Molecular gas dynamics and the direct simulation of gas flows. *Oxford Science Publications*, 42.

R. B. Bird, W. E. Stewart, and E. N. Lightfoot (2002). *Transport Phenomena*. John Wiley & Sons, New York, USA, second edition.

F. Bouchut (1994). On zero pressure gas dynamics. In *Advances in Kinetic Theory and Computing*, pages 171–190. World Scientific Publishing, River Edge, USA.

S. Chapman and T. G. Cowling (1961). *The Mathematical Theory of Non-Uniform Gases*. Cambridge University Press, Cambridge, United Kingdom.

O. Desjardins, R. O. Fox, and P. Villedieu (2006). A quadrature-based moment closure for the Williams spray equation. In *Proceedings of the 2006 Summer Program*, pages 223–234. Center for Turbulence Research (ctr.stanford.edu), Stanford, USA.

O. Desjardins, R. O. Fox, and P. Villedieu (2007). A quadrature-based moment method for dilute fluid-particle flows. *Journal of Computational Physics*, (submitted).

J. K. Dukowicz (1980). A particle-fluid numerical model for liquid sprays. *Journal of Computational Physics*, 35(2):229–253.

R. O. Fox (2003). *Computational Models for Turbulent Reacting Flows*. Cambridge University Press, Cambridge, United Kingdom.

R. O. Fox (2006a). Bivariate direct quadrature method of moments for coagulation and sintering of particle populations. *Journal of Aerosol Science*, 37:1562–1580.

R. O. Fox (2006b). CFD models for analysis and design of chemical reactors. In *Advances in Chemical Engineering*, volume 31, pages 231–305, Elsevier, Amsterdam, The Netherlands.

R. O. Fox (2007). Optimal moment sets for multivariate direct quadrature method of moments. *Journal of Aerosol Science*, (submitted).

R. O. Fox, F. Laurent, and M. Massot (2006). Numerical simulation of polydisperse, dense liquid sprays in an Eulerian framework: direct quadrature method of moments and multi-fluid method. *Journal of Computational Physics*, (submitted).

J. Hylkema (1999). *Modélisation cinétique et simulation numérique d'un brouillard dense de gouttelettes. Application aux propulseurs à poudre*. PhD thesis, ENSAE.

J. Hylkema and P. Villedieu (1998). A random particle method to simulate coalescence phenomena in dense liquid sprays. In *Lecture Notes in Physics*, volume 515, pages 488–493, Arcachon, France.

A. Kaufmann, O. Simonin, and T. Poinsot (2004). Direct numerical simulation of particle-laden homogeneous isotropic turbulent flows using a two-fluid model formulation. In *Proceedings of 5th International Conference on Multiphase Flow (ICMF'04)*.

F. Laurent, M. Massot, and P. Villedieu (2004). Eulerian multi-fluid modeling for the numerical simulation of coalescence in polydisperse dense liquid sprays. *Journal of Computational Physics*, 194:505–543.

D. L. Marchisio and R. O. Fox (2005). Solution of population balance equations using the direct quadrature method of moments. *Journal of Aerosol Science*, 36:43–73.

D. L. Marchisio, J. T. Pikturna, R. O. Fox, R. D. Vigil, and A. A. Barresi (2003a). Quadrature method of moments for population-balance equations. *AIChE Journal*, 49:1266–1276.

D. L. Marchisio, R. D. Vigil, and R. O. Fox (2003b). Quadrature method of moments for aggregation-breakage processes. *Journal of Colloid and Interface Science*, 258:322–334.

R. McGraw (1997). Description of aerosol dynamics by the quadrature method of moments. *Aerosol Science and Technology*, 27:255–265.

R. S. Miller and J. Bellan (1999). Direct numerical simulation of a confined three-dimensional gas mixing layer with one evaporating hydrocarbon-droplet-laden stream. *Journal of Fluids Mechanics*, 384:293–338.

R. S. Miller and J. Bellan (2000). Direct numerical simulation and subgrid analysis of a transitional droplet laden mixing layer. *Physics of Fluids*, 12(3):650–671.

B. Perthame (1990). Boltzmann type schemes for compressible Euler equations in one and two space dimensions. *SIAM Journal of Numerical Analysis*, 29:1–19.

S. B. Pope (2000). *Turbulent Flows*. Cambridge University Press, Cambridge, United Kingdom.

D. Ramkrishna (2000). *Population Balances*. Academic Press, San Diego, USA.

J. Réveillon, C. Péra, M. Massot, and R. Knikker (2004). Eulerian analysis of the dispersion of evaporating polydispersed sprays in a statistically stationary turbulent flow. *Journal of Turbulence*, 5(1):1–27.

S. Subramaniam (2000). Statistical representation of a spray as a point process. *Physics of Fluids*, 12:2413–2431.

F. A. Williams (1958). Spray combustion and atomization. *Physics of Fluids*, 1:541–545.

D. L. Wright, R. McGraw, and D. E. Rosner (2001). Bivariate extension of the quadrature method of moments for modeling simultaneous coagulation and sintering particle populations. *Journal of Colloid and Interface Science*, 236:242–251.

Quadrature Method of Moments for Poly-Disperse Flows

Daniele L. Marchisio

Department of Material Science and Chemical Engineering, Politecnico di Torino, Torino, Italy

Abstract. An overview of different moment methods for the solution of the generalized population balance equation is presented and in particular the use of methods based on quadrature approximations is discussed. Firstly the quadrature method of moments is derived for a simple spatially homogeneous, single-phase system, and some mathematical issues related to the algorithm to derive the quadrature approximation and its degree of accuracy are discussed. Then the method is extended and implemented for the simulation of spatially heterogeneous multi-phase flows. Eventually the direct quadrature method of moments is derived and its use for the solution of mono-, bi- and multi-variate population balance equations is explained. Particular attention is also devoted to how to couple these methods with multi-phase models, such as the mixture model and the multi-fluid model.

1 Generalized Population Balance Equation

The multi-phase systems discussed in this Chapter are constituted by a continuous primary phase and a dispersed secondary phase. The continuous primary phase is characterized by its velocity $\mathbf{U}_\mathrm{f}$ and other important properties such as temperature and composition, usually grouped into a single vector $\boldsymbol{\phi}$, and generally these quantities are function of space $\mathbf{x} = (x_1, x_2, x_3)$ and time t. As already reported the secondary phase is dispersed into the continuous phase and therefore it is constituted by particles, droplets or bubbles. The elements of the secondary disperse phase (that in what follows will be referred to as particles) are also characterized by their velocity $\mathbf{U}_\mathrm{p}$ and by some important properties such as composition, temperature, particle volume, particle surface area, etc. All these properties are usually grouped into the so-called internal property vector $\boldsymbol{\xi}$ to distinguish it from the external properties of the particles (i.e., position and velocity).

Since each particle of the secondary disperse phase can assume different values of velocity and of internal coordinate vector a Number Density Function (NDF) is usually defined, in such a way that the following quantity:

$$n\left(\mathbf{U}_\mathrm{p},\boldsymbol{\xi};\mathbf{x},t\right)\mathrm{d}\mathbf{U}_\mathrm{p}\mathrm{d}\boldsymbol{\xi} \tag{1.1}$$

represents the expected number of particles per unit volume (i,e., within an infinitesimal volume d$\mathbf{x}$ around the point $\mathbf{x}$) at time instant t, with velocity values between $\mathbf{U}_\mathrm{p}$ and $\mathbf{U}_\mathrm{p} + \mathrm{d}\mathbf{U}_\mathrm{p}$ and internal coordinate vector values between $\boldsymbol{\xi}$ and $\boldsymbol{\xi} + \mathrm{d}\boldsymbol{\xi}$.

The NDF is a very important quantity since contains all the information needed to characterize the population of particles. From the NDF it is moreover possible to derive some integral property of the population. For example, if the first internal coordinate is particle size $\xi_1 = L$, and if k_v is the volume shape factor, it is easy to show that the following quantity:

$$\varepsilon_{\mathrm{p}}(\mathbf{x},t) = k_{\mathrm{v}} \int_{\Omega_{\mathbf{U}_\mathrm{p}}} \int_{\Omega_{\boldsymbol{\xi}}} n(\mathbf{U}_{\mathrm{p}},\boldsymbol{\xi};\mathbf{x},t) L^3 \mathrm{d}\mathbf{U}_{\mathrm{p}} \mathrm{d}\boldsymbol{\xi} \tag{1.2}$$

is the volume fraction occupied by the disperse phase, where Ω_{Up} and Ω_{ξ} define all the possible values of particle velocity and internal coordinate vector and where the multi-dimensional integral on the right-hand side of Eq. (1.2) represents the mixed moment of order three with respect to particle size and of order zero with respect to the other variables. Of course the volume fractions of the continuous and disperse phases are defined so that they sum to unity (i.e., $\varepsilon_{\mathrm{f}} + \varepsilon_{\mathrm{p}} = 1$).

The evolution of the NDF is governed by the Generalized Population Balance Equation (GPBE), that using the Einstein notation (repeated indices imply summation) results in what follows:

$$\begin{aligned}\frac{\partial n(\mathbf{U}_{\mathrm{p}},\boldsymbol{\xi};\mathbf{x},t)}{\partial t} + \frac{\partial}{\partial x_i}\left[U_{\mathrm{p},i} n(\mathbf{U}_{\mathrm{p}},\boldsymbol{\xi};\mathbf{x},t)\right] - \frac{\partial^2}{\partial x_i^2}\left[\Gamma_{\mathrm{x}} n(\mathbf{U}_{\mathrm{p}},\boldsymbol{\xi};\mathbf{x},t)\right] = \\ -\frac{\partial}{\partial U_{\mathrm{p},i}}\left[\dot{U}_{\mathrm{p},i} n(\mathbf{U}_{\mathrm{p}},\boldsymbol{\xi};\mathbf{x},t)\right] - \frac{\partial}{\partial \xi_i}\left[G_i n(\mathbf{U}_{\mathrm{p}},\boldsymbol{\xi};\mathbf{x},t)\right] \\ -\frac{\partial^2}{\partial \xi_i \partial \xi_j}\left[\Gamma_{ij} n(\mathbf{U}_{\mathrm{p}},\boldsymbol{\xi};\mathbf{x},t)\right] + H(\mathbf{U}_{\mathrm{p}},\boldsymbol{\xi};\mathbf{x},t)\end{aligned} \tag{1.3}$$

where $U_{\mathrm{p},i}$ is the ith component of the particle velocity vector $\mathbf{U}_{\mathrm{p}}$, Γ_{x} is the spatial diffusion coefficient, $\dot{U}_{\mathrm{p},i}$ is the rate of continuous change of the ith component of particle velocity $\mathbf{U}_{\mathrm{p}}$ resulting from the balance of forces per unit mass (i.e., acceleration) acting on a particle with velocity $\mathbf{U}_{\mathrm{p}}$ and internal coordinate vector $\boldsymbol{\xi}$, G_i is the rate of continuous change of the ith component of the internal coordinate vector ξ_i, Γ_{ij} is the dispersion coefficient relative to the rates of continuous change for the ith and j^{th} internal coordinates.

The first term on the left-hand side of Eq. (1.3) is the accumulation term, whereas the second one represents convection in physical space, in fact, particles move from point to point because of their velocity $\mathbf{U}_{\mathrm{p}}$. The third term on the left-hand side represents instead particle diffusion due to Brownian motions or other phenomena (e.g., turbulent fluctuations).

The first term on the right-hand side accounts for the continuous change in particle velocity (i.e., acceleration or deceleration) due to the resulting forces acting on each single particle and it is usually related to fluid-particle forces such as, drag force, virtual mass force, Basset force, Saffman or lift force and rotational force (Crowe, Sommerfeld and Tsuji, 1998).

The second term represents instead the rate of continuous change of the internal coordinate vector. Usually this term refers to molecular processes, since only processes that occur on the molecular level, appear as continuous processes when observed at the particle length- and time-scales. For example, during precipitation and crystallization, the internal coordinate particle size $\xi_1 = L$, continuously increases with a rate $G_1 = \mathrm{d}L / \mathrm{d}t$ caused by continuous diffusion of solute molecules from the bulk of the super-saturated solution to the solid particles. This continuous growth rate is usually referred to as the molecular growth rate (Dirksen and Ring, 1991). On the opposite, during particle dissolution, diffusion of solute molecules from the particle surface to the bulk of the solution causes a continuous reduction of the size of the particles that is in this case is quantified by a continuous (negative) dissolution rate. If for example one of the internal coordinates is particle surface area, it can continuously change due to permanence of the particle at high temperature, and in this case the rate of change of particle surface area can be written in terms of a sintering/spheroidization rate (Rosner and Yu, 2001). Analogously if particle enthalpy is included in the internal coordinate vector, heat exchange of the particle with the surrounding fluid will result in a continuous change of particle enthalpy, and therefore it will be possible to define a continuous rate of change of particle enthalpy.

The third term on the right-hand side of Eq. (1.3) accounts for the dispersion growth rate. In fact, in many practical applications, the continuous rate of change of particle internal coordinate (i.e., especially particle size) can be affected by random fluctuations around a mean value. For example, the molecular growth rate of a particle in a super-saturated solution depends upon the solute concentration (that is generally included in the continuous composition vector $\boldsymbol{\phi}$) and by particle size L. However, inevitably some defect in the crystalline structure of the growing crystal causes fluctuations of the growth rate around the mean value G_1. These fluctuations are mathematically represented by a dispersion coefficient Γ_{11}, that acts as a diffusion coefficient of particle size in the particle size space (Jones and Larson, 1999).

The last term on the right-hand side is the discontinuous event term, that quantifies the rate of change in particle velocity $\mathbf{U}_p$ and in internal coordinate vector ξ, due to discrete events. The definition of discrete event depends on the length- and time-scale of observation, since at the molecular level all events are discrete! With discrete events (in opposition to continuous events) we refer here to processes that in an infinitesimally small interval of time (in comparison with the time- and length-scale of the particle) produce a finite change in particle velocity $\mathbf{U}_p$ and internal coordinate vector ξ. For example, particle collision (and subsequent adhesion) can cause, in an infinitesimally small interval of time (the collision time), a finite change in particle velocity and particle size. Usually the discrete event term takes into account for nucleation (namely introduction of new particles into the system), aggregation (also know as agglomeration or coalescence) and break up.

Further discussion concerning this term requires that some simplifications are introduced. For example, let us imagine that the NDF is expressed only as a function of time, space and particle size L. In this case the discontinuous event term can be written as follows:

$$\begin{aligned} H(L;\mathbf{x},t) = {} & J_o(\mathbf{x},t)\delta(L-L_o) + \frac{L^2}{2}\int_0^L \frac{\beta\left(\left(L^3-\lambda^3\right)^{1/3},\lambda\right)}{\left(L^3-\lambda^3\right)^{2/3}} n\left(\left(L^3-\lambda^3\right)^{1/3};\mathbf{x},t\right) n(\lambda;\mathbf{x},t)\,\mathrm{d}\lambda \\ & - n(L;\mathbf{x},t)\int_0^\infty \beta(L,\lambda)n(\lambda;\mathbf{x},t)\,\mathrm{d}\lambda + \int_L^\infty a(\lambda)b(L|\lambda)n(\lambda;\mathbf{x},t)\,d\lambda - a(L)n(L;\mathbf{x},t) \end{aligned} \tag{1.4}$$

where the first term quantifies the nucleation of new particles with rate of formation (per unit volume and unit time) equal to J_o and under the assumption of mono-disperse nuclei with size equal to L_0. The second and third terms represent instead the formation and disappearance of particles of size L because of aggregation of smaller and bigger particles. Here the underlying hypotheses are that aggregation is a secondary process (i.e., resulting from the collision of only two particles) and that the frequency of aggregation can be quantified in terms of the aggregation kernel β that is generally assumed to be a function of particle size, in addition to other properties of the particulate system as well as some properties of the continuous phase (e.g., composition, temperature, etc.)

The fourth and fifth terms in Eq. (1.4) define instead the rate of appearance and disappearance of particles of size L, due to break up of bigger and smaller particles. In this case the breakage process is assumed to be of first order, namely proportional to the concentration of particles of size L, and it is kinetically quantified by the breakage kernel a and the daughter distribution function b. The breakage kernel represents the frequency of disruption of particles whereas the daughter distribution function identifies the internal coordinate (e.g., size, volume, surface area) distribution of particles during a fragmentation event.

The literature in this field is very vast and a comprehensive description is not possible in this context. However, readers interested in the subject can consult the book of Ramkrishna (2000) for a general discussion on the derivation of the discontinuous event term and on the functional form of the aggregation and breakage kernels. In the case of precipitation and crystallization the most common kernels can be found in several works (e.g., Marchisio, Barresi and Garbero, 2002; Marchisio, Rivautella, and Barresi, 2006) as well as in the case of aerosol and solid nano-particles formation and evolution in gaseous media (Zucca et al., 2006). Other interesting solid-liquid systems are colloidal suspensions, where a number of different aggregation and breakage kernels have been applied (Wang et al., 2005; Marchisio, Vigil and Fox, 2003b; Marchisio et al., 2006a) as well as in the case of gas-liquid and liquid-liquid systems (Sanyal et al., 2005; Andersson and Andersson, 2006; Jaworski et al., 2007). Moreover, particular attention must be paid to choice of the daughter distribution function as shown in the work of Diemer and Olson (2002).

Since the terms on the left-hand side of the GPBE identify time and spatial transport, whereas the terms of the right-hand side include continuous and discrete changes in particle velocity and internal coordinates, generally Eq. (1.3) is written as follows:

$$\frac{\partial n(\mathbf{U}_p,\xi;\mathbf{x},t)}{\partial t}+\frac{\partial}{\partial x_i}\left[U_{p,i}n(\mathbf{U}_p,\xi;\mathbf{x},t)\right]-\frac{\partial^2}{\partial x_i^2}\left[\Gamma_x n(\mathbf{U}_p,\xi;\mathbf{x},t)\right]=S(\mathbf{U}_p,\xi;\mathbf{x},t) \qquad (1.5)$$

where all the terms on the right-hand side are grouped into a single source term S. This notation will be particularly useful when the solution of this equation in Computational Fluid Dynamics (CFD) codes will be discussed. In these codes in fact, time accumulation and physical transport due to convection and diffusion are usually already taken care of by the code itself and treated in a finite volume framework. The solution of the GPBE is then reduced to the proper treatment of the source term S.

2 The Quadrature Method of Moments (QMOM)

Let us now consider a numerical scheme for the solution of the GPBE based on the derivation of transport equations for the moments of the NDF and on the employment of a quadrature approximation for the solution of the so-called closure problem, as explained in the pioneering work of Hulburt and Katz (1964). In what follows theoretical and practical aspects are presented, starting with the solution of spatially homogeneous and mono-variate problems, extending then the discussion to the solution of spatially heterogeneous and multi-phase systems.

2.1 QMOM for mono-variate spatially homogeneous and single-phase problems

A mono-variate NDF is characterized by a single internal coordinate, namely the internal coordinate vector is reduced to a single scalar $\boldsymbol{\xi} \equiv \xi$. Moreover the simplification hypothesis of spatial homogeneity and the reduction to the study of single-phase problems, reduce the GPBE to the following partial differential equation:

$$\frac{\partial n(\xi;t)}{\partial t} = S(\xi;t) \tag{2.1}$$

where the NDF is now function only of the single internal coordinate ξ and of time t. The quadrature method of moments (QMOM) is based on the idea of solving the GPBE in terms of the moments of the NDF and the moment of order k of the NDF is defined as follows:

$$m_k(t) = \int_0^{+\infty} n(\xi;t)\xi^k \mathrm{d}\xi\,. \tag{2.2}$$

Some of the moments of the NDF have particular physical meanings, for example the moment of order zero represents the total particle number density, namely the total number of particles per unit volume. Other moments can also have different physical meanings and for example, if the internal coordinate is particle size $\xi = L$ then the second order moment and the third order moment are related to the total particle surface area and total particle volume, through the following relationships:

$$A_{\mathrm{T}} = \int_0^{+\infty} n(L;t)k_{\mathrm{A}}L^2\mathrm{d}L = k_{\mathrm{A}}m_2\,, \tag{2.3}$$

$$V_{\mathrm{T}} = \int_0^{+\infty} n(L;t)k_{\mathrm{V}}L^3\mathrm{d}L = k_{\mathrm{V}}m_3\,, \tag{2.4}$$

where k_{A} and k_{V} are the surface and volume shape factors. Moreover the moments of the distribution can be used to define different average particle sizes, for example the number-averaged particle size is defined as the ratio between the first and zero moment:

$$d_{10} = \frac{\int_0^{+\infty} Ln(L)\mathrm{d}L}{\int_0^{+\infty} n(L)\mathrm{d}L} = \frac{m_1}{m_0}\,, \tag{2.5}$$

the area-averaged particle size, or Sauter diameter, is defined as the ratio between the third and second moment:

$$d_{32} = \frac{\int_0^{+\infty} L^3 n(L)\,\mathrm{d}L}{\int_0^{+\infty} L^2 n(L)\,\mathrm{d}L} = \frac{m_3}{m_2}, \tag{2.6}$$

whereas the volume-averaged particle size is defined as the ratio between the fourth and third moment:

$$d_{43} = \frac{\int_0^{+\infty} L^4 n(L)\,\mathrm{d}L}{\int_0^{+\infty} L^3 n(L)\,\mathrm{d}L} = \frac{m_4}{m_3}. \tag{2.7}$$

If the moment transform is now applied to the GPBE for a general internal coordinate ξ in the form reported in Eq. (2.1), the following equation is obtained:

$$\frac{\mathrm{d}m_k(t)}{\mathrm{d}t} = \overline{S}_k = \int_0^{+\infty} S(\xi;t)\xi^k \mathrm{d}\xi, \tag{2.8}$$

where $\overline{S}_k$ is the moment transform of order k of the source term. The problem is said to be closed if this quantity can be written in terms of the moments themselves, otherwise a closure is needed. For example, if a population of particles characterized by an internal coordinate ξ is undergoing a continuous process, characterized by a continuous rate of change $G = \mathrm{d}\xi/\mathrm{d}t$ (also known as growth rate), the characteristic population balance equation is:

$$\frac{\partial n(\xi;t)}{\partial t} + \frac{\partial}{\partial \xi}\left[G(\xi)n(\xi;t)\right] = 0, \tag{2.9}$$

and if the growth rate is independent on the internal coordinate[1], the evolution equation of the kth moment is:

$$\frac{\mathrm{d}m_k(t)}{\mathrm{d}t} = kG\int_0^{+\infty} n(\xi;t)\xi^{k-1}\mathrm{d}\xi = kGm_{k-1}, \tag{2.10}$$

and in other words the problem is closed since, any moment of the NDF is defined by an evolution equation whose source term can be written as a function of lower-order moments. However, very often the growth rate is a function of the internal coordinate. For example, during crystallization and precipitation processes, when dealing with diffusion-controlled or

[1] There is another underlying hypothesis behind this simple derivation: the flux of the NDF at the integration limits must be null. This is almost always the case, but in some important applications this condition in not verified. For example during particle dissolution or droplet evaporation, the flux of the NDF at $\xi = 0$ (with ξ being particle size or volume) is not null.

surface nucleation-controlled molecular growth processes, the resulting molecular growth rate is definitely size-dependent. Typical functional forms are based on a power law as follows:

$$G(\xi)=G_o\xi^{\alpha}, \tag{2.11}$$

and the evolution of the kth moment becomes now:

$$\frac{dm_k(t)}{dt}=kG_o m_{k+\alpha-1} \tag{2.12}$$

that for $\alpha=-1$, results in some unclosed terms (i.e., some evolution equation is written in terms of moments that are not known).

Another interesting example is the case of pure particle aggregation. If the internal coordinate ξ is particle mass, the evolution equation for the moment of order k^{th} is:

$$\frac{\mathrm{d}m_k(t)}{\mathrm{d}t}=\frac{1}{2}\int_0^{\infty}\xi^k\int_0^{\infty}\beta(\xi-\xi',\xi')n(\xi-\xi';t)n(\xi';t)\mathrm{d}\xi'\mathrm{d}\xi-\int_0^{\infty}\xi^k n(\xi;t)\int_0^{\infty}\beta(\xi,\xi')n(\xi';t)\mathrm{d}\xi'\mathrm{d}\xi \tag{2.13}$$

that can be written in a closed form only if the aggregation kernel is independent from the internal coordinate; in fact, under this simplification hypothesis the equation becomes:

$$\frac{\mathrm{d}m_k(t)}{\mathrm{d}t}=\frac{a_o}{2}\int_0^{\infty}\xi^k\int_0^{\infty}n(\xi-\xi';t)n(\xi';t)\mathrm{d}\xi'\mathrm{d}\xi-a_o\int_0^{\infty}\xi^k n(\xi;t)\int_0^{\infty}n(\xi';t)\mathrm{d}\xi'\mathrm{d}\xi\,. \tag{2.14}$$

If a change of variable (i.e., ξ - ξ' = u) is applied the set of equation for k = 0, 1, and 2, becomes as follows:

$$\frac{\mathrm{d}m_0}{\mathrm{d}t}=-\frac{1}{2}a_o m_0^2;\quad \frac{\mathrm{d}m_1}{\mathrm{d}t}=0;\quad \frac{\mathrm{d}m_2}{\mathrm{d}t}=a_o m_1^2\,, \tag{2.15}$$

that, as it can be clearly seen, is written in closed form (i.e., the evolution equation of each moment has a source term written in terms of known moments). In general however, when the aggregation kernel depends on the value of the internal coordinate ξ the problem is not closed, since it is not possible to write the source term for the kth moment in terms of lower-order moments.

For all these cases where the problem is unclosed a closure is needed. Here with closure we refer to a numerical scheme able to calculate with high accuracy the source term of the moment of order k^{th} simply by using the set of available moments (i.e., m_0, m_1, $m_2,\ldots$, m_{k-1}, m_k). A very simple way to overcome this problem is to calculate it through an interpolation formula:

$$\overline{S}_k=\int_0^{+\infty}S(\xi)\xi^k\mathrm{d}\xi\approx\sum_{\alpha=1}^{n}w_{\alpha}S(\xi_{\alpha})\xi_{\alpha}^k\,, \tag{2.16}$$

where w_α and ξ_α are respectively the weights and the nodes of the interpolation formula and n is the number of nodes used for the interpolation. The accuracy of an interpolation equation is quantified by its degree of accuracy. The degree of accuracy is equal to d if the interpolation formula is exact when the integrand function is a polynomial of order smaller or equal to d and moreover there exists at least one polynomial of order $d+1$ that makes the interpolation equation not exact.

At this point the only requirement for the interpolation equation is to have distinct nodes, but nothing else is said concerning the position of these nodes. Among interpolation equations we can have formulae based on equally-spaced nodes (Newton-Cotes) or nodes coincident to the zeros of orthogonal polynomials (Gaussian quadrature). Standard interpolation equations have a low degree of accuracy (of about n-1) whereas quadrature approximations are particularly interesting because have a degree of accuracy equal to $2n$-1.

As it has been already reported the source term for the moment of order k^{th} is the integral of the source term of the population balance equation, and an important characteristic in the functional form of these integrals is that it is always possible to extract from the integral the NDF, as follows:

$$\overline{S}_k = \int_0^{+\infty} S(\xi)\xi^k \mathrm{d}\xi = \int_0^{+\infty} f(\xi)n(\xi)\mathrm{d}\xi, \tag{2.17}$$

where $f(\xi)$ groups all the terms left out after highlighting the presence of the NDF. Since all the source terms we will be dealing with when solving population balance equations can be written in this way, then the NDF can be used as a weight function in the integral, and therefore the following theorem that defines a Gaussian quadrature can be used (Press et al., 2002). Given the equation:

$$\int_0^{+\infty} f(\xi)n(\xi)\mathrm{d}\xi = \sum_{\alpha=1}^{n} w_\alpha f(\xi_\alpha) + R_n(f), \tag{2.18}$$

if $n(\xi) \geq 0$ in the integration interval $(0,+\infty)$ and all the moments m_k of the NDF $n(\xi)$ exist, it is a Gaussian quadrature if the nodes ξ_α coincide with the zeros of the polynomial $P_n(\xi)$ of order n orthogonal in the interval with respect to the NDF $n(\xi)$.

The main implication that the approximation is a Gaussian quadrature is that its degree of accuracy is equal to $2n$-1, or in other words the approximation is exact ($R_n = 0$) anytime the integrand function $f(\xi)$ is a polynomial of order equal or smaller than $2n$-1. Let us now define the family of polynomials orthogonal to the NDF, and let us derive an algorithm to find these polynomials and to calculate their roots (or in order words to calculate the quadrature approximation). A polynomial is said to be orthogonal in the interval $(0,+\infty)$ with respect to the weight function $n(\xi)$ (i.e., the NDF), if it belongs to the family of polynomials satisfying the following relationship (Dette and Studden, 1997):

$$\int_0^{+\infty} n(\xi)P_n(\xi)P_m(\xi)\mathrm{d}\xi \begin{cases} = 0 & \text{if} \quad n \neq m \\ \neq 0 & \text{if} \quad n = m \end{cases}, \tag{2.19}$$

whereas it is said to be ortho-normal when:

$$\int_0^{+\infty} n(\xi)P_n(\xi)P_m(\xi)\mathrm{d}\xi \begin{cases} = 0 & \text{if} \quad n \neq m \\ = 1 & \text{if} \quad n = m \end{cases}. \tag{2.20}$$

Moreover, orthogonal polynomials can be written in terms of a recursive relationship as follows:

$$P_{-1}(\xi)=0;\ P_0(\xi)=1;\ P_{i+1}(\xi)=(\xi-\alpha_i)P_i(\xi)-\beta_i P_{i-1}(\xi), \tag{2.21}$$

where the coefficients α_i and β_i can be calculated imposing ortho-normality between polynomials. For example, for $i = 1$:

$$P_0(\xi)=1;\ P_1(\xi)=\xi-\alpha_0;\ P_2(\xi)=(\xi-\alpha_1)(\xi-\alpha_0)-\beta_1=\xi^2-(\alpha_0+\alpha_1)\xi+\alpha_0\alpha_1-\beta_1 \tag{2.22}$$

and imposing ortho-normality between these three polynomials (assuming $m_0 = 1$ as it will be clearer below) we obtain:

$$\alpha_0=m_1;\ \alpha_1=\frac{m_1^3-2m_1m_2+m_3}{m_2-m_1^2};\ \beta_1=m_2-m_1^2. \tag{2.23}$$

Once the coefficients of the polynomial are known its roots can be easily calculated by using any of the available numerical methods. However, since the algorithms for finding roots of high order polynomials are sometimes difficult to converge, a numerical procedure based on the recursive equation can be used. In particular, if the recursive formula reported in Eq. (2.21) is written as:

$$\xi P_i(\xi)=\beta_i P_{i-1}(\xi)+\alpha_i P_i(\xi)+P_{i+1}(\xi), \tag{2.24}$$

then the following system written in matricial form is obtained:

$$\xi\begin{pmatrix}P_0\\P_1\\\vdots\\P_{n-2}\\P_{n-1}\end{pmatrix}=\begin{pmatrix}\alpha_0 & 1 & 0 & 0 & 0\\ \beta_1 & \alpha_1 & 1 & 0 & 0\\ 0 & \ddots & \ddots & \ddots & 0\\ 0 & 0 & \beta_{n-2} & \alpha_{n-2} & 1\\ 0 & 0 & 0 & \beta_{n-1} & \alpha_{n-1}\end{pmatrix}\begin{pmatrix}P_0\\P_1\\\vdots\\P_{n-2}\\P_{n-1}\end{pmatrix}+P_n, \tag{2.25}$$

where the roots of the polynomial of order n are calculated by imposing $P_n(\xi) = 0$, resulting in a simple eigen-value/eigen-vector problem. If fact, from this equation it is immediate to see that the roots of the polynomial are the eigen-values of the matrix reported above. Moreover, since the eigen-values do not change after a symmetric transformation, they are identical to the ones of the matrix reported below:

$$\begin{pmatrix}\alpha_0 & \sqrt{\beta_1} & 0 & 0 & 0\\ \sqrt{\beta_1} & \alpha_1 & \sqrt{\beta_2} & 0 & 0\\ 0 & \ddots & \ddots & \ddots & 0\\ 0 & 0 & \sqrt{\beta_{n-2}} & \alpha_{n-2} & \sqrt{\beta_{n-1}}\\ 0 & 0 & 0 & \sqrt{\beta_{n-1}} & \alpha_{n-1}\end{pmatrix}, \tag{2.26}$$

that presents the advantage of being tri-diagonal and symmetric. The sequence of calculations just explained constitutes the algorithm used to calculate the quadrature approximation from the moments of the NDF. This algorithm is known as Product-Difference algorithm (Gordon, 1968) and can be summarized as follows. Firstly, from the first $2n$ moments of the NDF a matrix **A** is constructed. This matrix is constructed by forcing the first column to be equal to one, for the first element, and then zero for the rest of the rows. The second column is constructed by forcing the element in the ith row to be equal to $(-1)^i m_i$. The other elements are then calculated through a product and difference scheme,with the element in the ith row and in the jth column equalto $a_{1,j-1}a_{i+1,j-2}-a_{1,j-2}a_{i+1,j-1}$. Moreover, since we are interested in the construction of the polynomial ortho-normal to the NDF, m_0 can be assumed to be equal to one in this calculation. The final **A** matrix is then:

$$\mathbf{A}=\begin{pmatrix} 1 & m_0=1 & m_1 & m_2-m_1^2 & \vdots \\ 0 & -m_1 & -m_2 & m_2m_1-m_3 & \vdots \\ 0 & m_2 & m_3 & \vdots & \vdots \\ \vdots & \vdots & \vdots & \vdots & a_{1,j-1}a_{i+1,j-2}-a_{1,j-2}a_{i+1,j-1} \\ 0 & (-1)^i m_i & (-1)^i m_{i+1} & \vdots & \vdots \\ \vdots & \vdots & \vdots & \vdots & \vdots \\ \vdots & \vdots & (-1)^{2n-2} m_{2n-1} & 0 & 0 \\ 0 & (-1)^{2n-1} m_{2n-1} & 0 & 0 & 0 \\ 0 & 0 & 0 & 0 & 0 \end{pmatrix}. \quad (2.27)$$

The elements of this matrix can then be used to calculate the coefficients of the polynomials as follows

$$\alpha_i=\zeta_{2i}-\zeta_{2i+1};\quad \beta_i=\zeta_{2i-1}\zeta_{2i}, \quad (2.28)$$

with $i = 0,\ldots, n$-1 and where

$$\zeta_0=0;\quad \zeta_{i-1}=\frac{a_{1,i+1}}{a_{1,i}a_{1,i-1}}\quad i=2,\ldots,2n. \quad (2.29)$$

Once the coefficients of the polynomials are calculated, the matrix of rank n reported in Eq. (2.26) can be determined. As already reported, the n eigen-values of the matrix are the roots of the polynomial of order n ortho-normal to the NDF or in other words the nodes of the quadrature approximation ξ_i. The weights w_i of the quadrature approximation can be determined from the first component of the eigen-vector (p_1) as follows:

$$w_i=m_0\left(p_1\right)_i^2. \quad (2.30)$$

Summarizing this method is based on the simple idea of solving the GPBE in terms of $2n$ moments of the NDF ($m_0,\ldots,m_{2n-1}$). The closure problem is overcome by resorting to a quadrature approximation of order n, resulting in the following equation:

$$\frac{\mathrm{d}m_k(t)}{\mathrm{d}t} = \overline{S}_k = \int_0^{+\infty} S(\xi)\xi^k \mathrm{d}\xi = \int_0^{+\infty} f(\xi)n(\xi)\mathrm{d}\xi \approx \sum_{\alpha=1}^{n} w_\alpha f(\xi_\alpha). \tag{2.31}$$

This method, known as Quadrature Method of Moments (QMOM), has been first introduced for the simulation of aerosol evolution under size-dependent growth rates (McGraw, 1997) and since then it has been tested and validated under different operating conditions, for modeling particle aggregation (Barret and Webb, 1998; Marchisio et al., 2003), particle aggregation and simultaneous breakage (Marchisio, Vigil and Fox, 2003a) and in many other practical cases including particle nucleation, growth, and dissolution. During this validation two important factors emerged. The first one is that in general the higher is the number of nodes of the quadrature approximation the better is the approximation of the source term (see Eq. 2.31). The second is that it is recommended not to go beyond a certain number of nodes (e.g., n = 5 or 6) and consequently beyond a certain number of tracked moments (i.e., ten or twelve). However, it should be highlighted that for the most important applications in chemical engineering three nodes (n = 3) resulted in very accurate predictions, when compared with more sophisticated methods.

2.2 Extension to spatially heterogeneous problems

Let us now extend the results obtained in the previous paragraph, to a spatially heterogeneous system. In this case the GPBE contains the velocity of the particles $\mathbf{U}_\mathrm{p}$ (see Eq. 1.3) whose rate of change is expressed by the force per unit mass (i.e., acceleration) acting on the particle. This force is the result of different contributions, namely the drag force, the virtual mass forces, the Basset force, and the Saffman force. The drag force is usually considered the most important and can be calculated as follows:

$$\dot{\mathbf{U}}_\mathrm{p} = \frac{\mathrm{d}\mathbf{U}_\mathrm{p}}{\mathrm{d}t} = C_\mathrm{D}\frac{\pi L^2}{4}\frac{\rho_\mathrm{p}}{2}\left|\mathbf{U}_\mathrm{f} - \mathbf{U}_\mathrm{p}\right|\left(\mathbf{U}_\mathrm{f} - \mathbf{U}_\mathrm{p}\right), \tag{2.32}$$

where L is the particle size, ρ_p is the particle density, and C_D is the drag coefficient that is usually calculated through experimentally based relationships as a function of the particle Reynolds number [see for example Crowe, Sommerfeld and Tsuji (1998)]. From Eq. (2.32) it is possible to define a characteristic particle relaxation time τ_p, that defines the time required for the particle to adapt to the local fluid velocity:

$$\frac{\mathrm{d}\mathbf{U}_\mathrm{p}}{\mathrm{d}t} = \frac{1}{\tau_\mathrm{p}}\left(\mathbf{U}_\mathrm{f} - \mathbf{U}_\mathrm{p}\right), \tag{2.33}$$

and based on this characteristic time it is possible to define the Stokes number as the ratio between the characteristic particle time and the characteristic fluid time:

$$\mathrm{St} = \frac{\tau_\mathrm{p}}{\tau_\mathrm{f}}, \tag{2.34}$$

when St is much smaller than one, for example in the case of very small particles (i.e., $L \ll 0$), particles immediately adapt to the local flow field and therefore $\mathbf{U}_\mathrm{p} = \mathbf{U}_\mathrm{f}$. This is the infinite drag limit ($C_D \to +\infty$) where the system is defined as dilute. Under these conditions all the particles will have the same velocity and therefore the NDF will not be a function of particle velocity. In the opposite situation (i.e., dense system) the Stokes number is greater than one and the particle velocity is not equal to the fluid velocity.

Let us first consider the small Stokes number limit. As already reported under these conditions the velocity of the particles is the same for all the particles and is equal to the fluid velocity. The NDF is for the mono-variate and spatially heterogeneous problem under study only function of the unique internal coordinate ξ, and of space $\mathbf{x}$ and time t. The GPBE written in terms of the moments of the NDF, becomes:

$$\frac{\partial m_k(\mathbf{x},t)}{\partial t} + \nabla_{\mathbf{x}} \cdot \left[\mathbf{U}_{\mathrm{f}} m_k(\mathbf{x},t)\right] - \frac{\partial^2}{\partial x_i^2}\left[\Gamma_{\mathrm{x}} m_k(\mathbf{x},t)\right] = \overline{S}_k(\mathbf{x},t), \tag{2.35}$$

where the moments are convected with the fluid velocity $\mathbf{U}_{\mathrm{f}}$, and undergo spatial diffusion according to a diffusion coefficient Γ_{x}. As already stated all the other relevant phenomena related to the evolution of the population of particles (i.e., particle nucleation, molecular growth, aggregation, breakage) are included in the source term on the right-hand side of Eq. (2.35).

If the flow is turbulent than the moments of the NDF must be treated as random variables, as it happens for all the other turbulent variables, such as species concentrations and enthalpy. In the case of turbulent flows it is convenient to define average and fluctuating variables. For example, if the Reynolds average is used, it is possible to define a Reynolds-averaged moment of order k of the NDF and a fluctuating one[2]:

$$m_k = \langle m_k \rangle + m_k' . \tag{2.36}$$

If the internal coordinate is particle size $\xi = L$, in the case of particle nucleation, molecular growth, aggregation and breakage, the Reynolds-averaged population balance equation becomes:

$$\begin{aligned} &\frac{\partial \langle m_k \rangle}{\partial t} + \frac{\partial}{\partial x_i}\left[\langle U_{\mathrm{f},i}\rangle \langle m_k \rangle\right] - \frac{\partial^2}{\partial x_i^2}\left[(\Gamma + \Gamma_{\mathrm{t}})\langle m_k \rangle\right] = \langle J_{\mathrm{o}}\rangle (L_{\mathrm{o}})^k + k\int_0^{+\infty} G(L)\langle n(L)\rangle L^{k-1}\mathrm{d}L \\ &+\frac{1}{2}\int_0^{+\infty}\int_0^{+\infty}\left(L^3+\lambda^3\right)^{k/3}\beta(L,\lambda)\langle n(L)n(\lambda)\rangle \mathrm{d}\lambda \mathrm{d}L - \int_0^{+\infty}\int_0^{+\infty} L^k \beta(L,\lambda)\langle n(\lambda)n(L)\rangle \mathrm{d}\lambda\mathrm{d}L \\ &+\int_0^{+\infty}\int_0^{+\infty} L^k a(\lambda) b(L|\lambda)\langle n(\lambda)\rangle \mathrm{d}\lambda\mathrm{d}L - \int_0^{+\infty} L^k a(L)\langle n(L)\rangle \mathrm{d}L \end{aligned} \quad , \tag{2.37}$$

where $\langle U_{\mathrm{f},i}\rangle$ is the ith component of the Reynolds-averaged fluid velocity $\langle \mathbf{U}_{\mathrm{f}}\rangle$, Γ is the particle diffusion coefficient due to Brownian motions, Γ_{t} is the particle turbulent diffusivity and $\langle J_{\mathrm{o}}\rangle$ is the Reynolds-averaged nucleation rate. It is important to highlight here that since aggregation is a second order process a new closure problem is generated. In fact, the following assumption:

[2] This is the so-called RANS approach. In a very similar way it would be possible to define a filtered moment and a fluctuating one following the Large Eddy Simulation (LES) approach.

$$\beta(L,\lambda)\langle n(L)n(\lambda)\rangle \approx \beta(L,\lambda)\langle n(L)\rangle\langle n(\lambda)\rangle, \tag{2.38}$$

is valid only if the characteristic time for aggregation is much longer than the characteristic time for particle turbulent fluctuations (Marchisio et al., 2006a). If the quadrature approximation is applied and if w_α and L_α are the weights and nodes of the quadrature approximation corresponding to the Reynolds-averaged moments of the NDF, and symmetric breakage is considered, the following equation is obtained:

$$\begin{aligned}\frac{\partial\langle m_k\rangle}{\partial t}+\frac{\partial}{\partial x_i}\left[\langle U_{\mathrm{f},i}\rangle\langle m_k\rangle\right]-\frac{\partial^2}{\partial x_i^2}\left[(\Gamma+\Gamma_{\mathrm{t}})\langle m_k\rangle\right]=\langle J_{\mathrm{o}}\rangle(L_{\mathrm{o}})^k+k\sum_{\alpha=1}^{n}G(L_\alpha)w_\alpha L_\alpha^{k-1}\\+\frac{1}{2}\sum_{\alpha=1}^{n}\sum_{\gamma=1}^{n}\left[\left(L_\alpha^3+L_\gamma^3\right)^{k/3}-L_\alpha^k-L_\gamma^k\right]\beta_{\alpha\gamma}w_\alpha w_\gamma+\sum_{\alpha=1}^{n}L_\alpha^k a_\alpha w_\alpha\left(2^{3-k/3}-1\right)\end{aligned} \tag{2.39}$$

where $\beta_{\alpha\gamma}=\beta(L_\alpha,L_\gamma)$ and where $a_\alpha=a(L_\alpha)$. It is interesting to highlight that with particle size as internal coordinate the Reynolds-averaged volume fraction occupied by the solid particles is calculated through the third moment of the NDF:

$$\langle\varepsilon_{\mathrm{p}}\rangle(\mathbf{x},t)=k_{\mathrm{v}}\int_0^{+\infty}\langle n\rangle(L;\mathbf{x},t)L^3\mathrm{d}L=k_{\mathrm{v}}\langle m_3\rangle, \tag{2.40}$$

and in the case of pure aggregation and breakage with no nucleation and growth the evolution of the volume fraction is governed by the following equation:

$$\frac{\partial\langle\varepsilon_{\mathrm{p}}\rangle}{\partial t}+\frac{\partial}{\partial x_i}\left[\langle U_{\mathrm{f},i}\rangle\langle\varepsilon_{\mathrm{p}}\rangle\right]-\frac{\partial^2}{\partial x_i^2}\left[(\Gamma+\Gamma_{\mathrm{t}})\langle\varepsilon_{\mathrm{p}}\rangle\right]=0, \tag{2.41}$$

since the source terms for the third moment disappear because of the constraint of conservation of total particle volume (as can be easily verified from Eq. 2.39 by replacing k = 3). Usually these equations are solved together with the continuity and the Reynolds-averaged Navier-Stokes (RANS) equations for the continuous phase:

$$\begin{aligned}&\frac{\partial}{\partial t}\left[\langle\varepsilon_{\mathrm{f}}\rangle\right]+\frac{\partial}{\partial x_i}\left[\langle\varepsilon_{\mathrm{f}}\rangle\langle U_{\mathrm{f},i}\rangle\right]=0\\&\frac{\partial}{\partial t}\left[\langle\varepsilon_{\mathrm{f}}\rangle\langle U_{\mathrm{f},i}\rangle\right]+\frac{\partial}{\partial x_j}\left[\langle\varepsilon_{\mathrm{f}}\rangle\langle U_{\mathrm{f},i}\rangle\langle U_{\mathrm{f},j}\rangle\right]=\frac{\mu_{\mathrm{f}}}{\rho_{\mathrm{f}}}\frac{\partial^2}{\partial x_j\partial x_j}\left[\langle\varepsilon_{\mathrm{f}}\rangle\langle U_{\mathrm{f},i}\rangle\right]-\frac{\langle\varepsilon_{\mathrm{f}}\rangle}{\rho_{\mathrm{f}}}\frac{\partial\langle p\rangle}{\partial x_i}-\frac{\partial}{\partial x_j}\left[\langle\varepsilon_{\mathrm{f}}\rangle\langle u'_{\mathrm{f},i}u'_{\mathrm{f},j}\rangle\right]\end{aligned}, \tag{2.42}$$

where as already reported $\langle\varepsilon_{\mathrm{f}}\rangle = 1 - \langle\varepsilon_{\mathrm{p}}\rangle$ is the Reynolds-averaged volume fraction occupied by the fluid (i.e., the continuous phase), $u'_{\mathrm{f},i}$ is the ith component of fluctuating fluid velocity, μ_{f} is the fluid viscosity, ρ_{f} is the fluid density, $\langle p\rangle$ is the Reynolds-averaged fluid pressure. Since usually the small Stokes number limit is verified for small and dilute particulate systems (i.e., $\langle\varepsilon_{\mathrm{p}}\rangle \ll \langle\varepsilon_{\mathrm{f}}\rangle$), under these operating conditions Eq. (2.42) is solved assuming $\langle\varepsilon_{\mathrm{f}}\rangle$ = 1. As it is well known the Reynolds stress tensor $\langle u'_{\mathrm{f},i}\, u'_{\mathrm{f},j}\rangle$ that appears as last term on the right-hand side of Eq. (2.45), can be modelled by resorting to the Bousinnesq hypothesis (Valerio et al., 1998):

$$\left\langle u'_{\mathrm{f},i} u'_{\mathrm{f},j} \right\rangle = -\frac{\mu_{\mathrm{t}}}{\rho_{\mathrm{f}}} \frac{\partial \left\langle U_{\mathrm{f},i} \right\rangle}{\partial x_j}, \tag{2.43}$$

where the turbulent viscosity μ_t is calculated through the turbulent kinetic energy k and the turbulent dissipation rate ε, by using the following relationship:

$$\mu_{\mathrm{t}} = \rho_{\mathrm{f}} C_{\mu} \frac{k^2}{\varepsilon}. \tag{2.44}$$

where C_{μ} is a constant usually taken equal to 0.09. The Reynolds stress tensor can also be calculated by solving appropriate algebraic or transport equations, generating what is known as Reynolds Stress Models (RMS) as explained by Valerio et al. (1998) and in the book of Pope (2000)[3]. The turbulent viscosity of the continuous phase also defines the particle turbulent diffusion coefficient, that in the small Stokes number limit is usually defined through a turbulent Schimdt number (Baldyga and Orciuch, 2001):

$$\Gamma_{\mathrm{t}} = \frac{C_{\mu}}{Sc^{\mathrm{t}}} \frac{k^2}{\varepsilon}. \tag{2.45}$$

QMOM has been used for the simulation of many spatially heterogeneous solid-fluid systems for small Stokes numbers, in many reactor configurations, such as confined impinging jet reactors (Gavi et al., 2007), stirrer tanks (Marchisio et al., 2006b), and Taylor-Couette reactors (Marchisio, Barresi and Fox, 2001) always resulting in predictions in good agreement with experimental data.

2.3 Extension to multi-phase problems

In the opposite situation when the Stokes number is not small, solid particles do not have enough time to relax to the local flow field of the continuous phases. In this case solid particles move with a different velocity from the fluid and moreover the velocity of the particles is different from particle to particle; as a consequence the NDF is a function of space $\mathbf{x}$, time t, of the unique internal coordinate ξ, and of particle velocity $\mathbf{U}_{\mathrm{p}}$. The GPBE becomes:

$$\begin{aligned} &\frac{\partial n(\mathbf{U}_{\mathrm{p}},\xi;\mathbf{x},t)}{\partial t} + \frac{\partial}{\partial x_i}\left[U_{\mathrm{p},i} n(\mathbf{U}_{\mathrm{p}},\xi;\mathbf{x},t)\right] - \frac{\partial^2}{\partial x_i^2}\left[\Gamma_{\mathrm{x}} n(\mathbf{U}_{\mathrm{p}},\xi;\mathbf{x},t)\right] = \\ &-\frac{\partial}{\partial U_{\mathrm{p},i}}\left[\dot{U}_{\mathrm{p},i} n(\mathbf{U}_{\mathrm{p}},\xi;\mathbf{x},t)\right] - \frac{\partial}{\partial \xi}\left[G n(\mathbf{U}_{\mathrm{p}},\xi;\mathbf{x},t)\right] - \frac{\partial^2}{\partial \xi^2}\left[\Gamma_{\xi} n(\mathbf{U}_{\mathrm{p}},\xi;\mathbf{x},t)\right] + H(\mathbf{U}_{\mathrm{p}},\xi;\mathbf{x},t) \end{aligned} \tag{2.46}$$

where G is the continuous rate of change of the internal coordinate ξ, and Γ_{ξ} is the dispersion coefficient for the internal coordinate, whereas the other terms and notation have the usual

[3] Generally the treatment of turbulence for these multi-phase systems characterized by small Stokes numbers follows the standard single-phase approach. In fact, as already reported these conditions are verified for very dilute multi-phase systems, where the effect of the disperse phase on the flow and turbulence fields of the continuous phase is neglected.

meaning already explained in Eq. (1.3). Let us now define the moment of order k with respect to the internal coordinate ξ (and of order zero with respect to particle velocity) as follows:

$$m_k(\mathbf{x},t) = \int_0^{+\infty}\int_0^{+\infty} n(\mathbf{U}_\mathrm{p},\xi;\mathbf{x},t)\xi^k \mathrm{d}\xi \mathrm{d}\mathbf{U}_\mathrm{p} . \tag{2.47}$$

If the moment transform is now applied to the GPBE (see Eq. 2.46) the resulting transport equation for the moment of order k with respect to the internal coordinate ξ becomes:

$$\frac{\partial m_k(\mathbf{x},t)}{\partial t} + \frac{\partial}{\partial x_i}\left[U^k_{\mathrm{p},i} m_k(\mathbf{x},t)\right] - \frac{\partial^2}{\partial x_i^2}\left[\Gamma_\mathrm{x} m_k(\mathbf{x},t)\right] = \overline{S}_k(\mathbf{x},t), \tag{2.48}$$

where the first three terms in the left-hand side of Eq. (2.46) representing transport in time and space are mainly unchanged, the first term on the right-hand side (i.e., particle acceleration) is equal to zero, whereas the last three terms are now grouped together into the moment transform of order k of the source term $\overline{S}_k$. It is very important to highlight that because the NDF is now a function of both, internal coordinate and particle velocity, each moment of the NDF must be convected with a different average velocity. In fact, when deriving Eq. (2.48) it is possible to show that the velocity with which the moment of order k is convected is:

$$\mathbf{U}^k_\mathrm{p} = \frac{\int_0^{+\infty}\int_0^{+\infty} n(\mathbf{U}_\mathrm{p},\xi;\mathbf{x},t)\mathbf{U}_\mathrm{p}\xi^k \mathrm{d}\xi \mathrm{d}\mathbf{U}_\mathrm{p}}{\int_0^{+\infty}\int_0^{+\infty} n(\mathbf{U}_\mathrm{p},\xi;\mathbf{x},t)\xi^k \mathrm{d}\xi \mathrm{d}\mathbf{U}_\mathrm{p}}, \tag{2.49}$$

where it is easy to recognize in the denominator of Eq. (2.49) the moment of order k defined in Eq. (2.47). This is a very important factor in the simulation of poly-disperse flows, in fact, only adopting a different velocity for each moment of the NDF, the fact that different particles are characterized by different velocities, can be correctly described.

An efficient way to solve the problem posed in Eq. (2.49) is to assume a conditional relationship between particle internal coordinate and particle velocity. This is a very common approach when the internal coordinate is particle size $\xi = L$. In fact, in many cases particle velocity is controlled by fluid-particle interactions, that in turns are determined by particle size. In these cases it is generally possible to assume that particles within a certain size range move with a certain velocity, and only particles with different size move with different velocity. Under this hypothesis the average velocity for the moment of order k can be written as follows:

$$\mathbf{U}^k_\mathrm{p} = \frac{\int_0^{+\infty}\int_0^{+\infty} n(L;\mathbf{x},t)\delta\left[\mathbf{U}_\mathrm{p} - \mathbf{U}_\mathrm{p}(L)\right]\mathbf{U}_\mathrm{p} L^k \mathrm{d}L \mathrm{d}\mathbf{U}_\mathrm{p}}{\int_0^{+\infty}\int_0^{+\infty} n(L;\mathbf{x},t)\delta\left[\mathbf{U}_\mathrm{p} - \mathbf{U}_\mathrm{p}(L)\right] L^k \mathrm{d}L \mathrm{d}\mathbf{U}_\mathrm{p}} = \frac{\int_0^{+\infty} n(L;\mathbf{x},t)\mathbf{U}_\mathrm{p}(L) L^k \mathrm{d}L}{m_k(\mathbf{x},t)} \approx \frac{\sum_{\alpha=1}^{n} w_\alpha \mathbf{U}_\mathrm{p}(L_\alpha) L^k_\alpha}{m_k(\mathbf{x},t)}, \tag{2.50}$$

where $\mathbf{U}_\mathrm{p}(L)$ is the relationship between particle velocity and particle size, where again the integral can be solved by resorting to the quadrature approximation, and where a new NDF is obtained simply by integrating out particle velocity:

$$n(L;\mathbf{x},t)=\int_{0}^{+\infty} n(\mathbf{U}_{\mathrm{p}},L;\mathbf{x},t)\mathrm{d}\mathbf{U}_{\mathrm{p}} \,. \tag{2.51}$$

The conditional relationship between particle velocity and size can be derived making use of an algebraic slip formulation. The basic assumption of the algebraic slip formulation is that a local equilibrium between the continuous and the disperse phases should be reached over a short spatial length scale. Following the approach proposed by Manninen, Taivassalo, and Kallio (1996) the slip velocity, namely the velocity difference between the particle and the fluid, can be calculated as follows:

$$\mathbf{U}_{\mathrm{p}}(L)=\mathbf{U}_{\mathrm{f}}+\frac{(\rho_{\mathrm{f}}-\rho_{\mathrm{m}})L^2}{18\mu C_D}\mathbf{g}, \tag{2.52}$$

where C_{D} is the drag coefficient already introduced in Eq. (2.32). This is the so-called drift flux model that is the simplest algebraic slip formulation, in which the acceleration of the particle is given by gravity (i.e., reported as **g** in Eq. 2.52). This approach is usually coupled with continuity and momentum balance equations for the mixture constituted by the continuous (primary) and the disperse (secondary) phases and the overall approach is known as the mixture model (Fluent, 2004). This approach has been successfully applied to the simulation of turbulent liquid-liquid dispersions undergoing breakage in kinetics static mixers and we remand readers interested in the subject to the original work (Jaworsky et al., 2007).

3 The Direct Quadrature Method of Moments (DQMOM)

Since the quadrature approximation transforms any integral of the form:

$$\int_{0}^{\infty} n(\xi)f(\xi)\mathrm{d}\xi \approx \sum_{\alpha=1}^{n} w_\alpha f(\xi_\alpha), \tag{3.1}$$

into a simple summation of n terms, this is equivalent to an assumption on the functional form of the NDF with a summation of n delta Dirac functions:

$$n(\xi;t)=\sum_{\alpha=1}^{n} w_\alpha(t)\delta\left[\xi-\xi_\alpha(t)\right], \tag{3.2}$$

where as already reported w_α are the weights and ξ_α are the nodes of the quadrature approximation. The use of the quadrature approximation is equivalent to the assumption that the real population of particles is constituted by an equivalent population of particles with number densities equal to the weights w_α and with characteristics values of the internal coordinate equal to the nodes ξ_α. With QMOM the number densities and the characteristic values of the internal coordinates are calculated by resorting to the product-difference algorithm, and the population balance equation is solved through the moments of the NDF. However, as it will be explained in the following paragraphs, it is also possible to directly track the evolution of the quadrature approximation, or in other words it is possible to directly track the evolution of this equivalent population of particles, represented by the quadrature approximation, that shares with the real population a number of moments. This method requires the solution of evolution/transport equations for weights and nodes of the quadrature approximation rather than moments of the NDF and is called Direct Quadrature Method of Moments (DQMOM). In the following paragraphs the method will be firstly presented for a simple mono-variate spatially homogeneous system, and then results will be extended to spatially heterogeneous systems, to bi-variate population balance equations, and eventually to real poly-disperse multi-phase systems.

3.1 DQMOM for mono-variate spatially homogeneous and single-phase problems

Let us consider again a simple problem where spatial gradients for the properties of interest are ignored and the multiphase system is considered as a pseudo single-phase system. If the functional form reported in Eq. (3.2) is inserted into the GPBE (see Eq. 2.1) the following expression is found:

$$\sum_{\alpha=1}^{n} \frac{\partial}{\partial t}\left[w_\alpha \delta(\xi-\xi_\alpha)\right] = S(\xi;t), \tag{3.3}$$

and after rearranging and some simple calculations the following expression is found:

$$\sum_{\alpha=1}^{n} \delta(\xi-\xi_\alpha)\left[\frac{dw_\alpha}{dt}\right] - \sum_{\alpha=1}^{n} \delta'(\xi-\xi_\alpha)\left[\frac{\mathrm{d}(w_\alpha\xi_\alpha)}{\mathrm{d}t} - \xi_\alpha \frac{\mathrm{d}w_\alpha}{\mathrm{d}t}\right] = S(\xi;t), \tag{3.4}$$

and if the following notation is used:

$$\frac{\mathrm{d}w_\alpha}{\mathrm{d}t} = a_\alpha; \qquad \frac{\mathrm{d}(w_\alpha\xi_\alpha)}{\mathrm{d}t} = b_\alpha, \tag{3.5}$$

the following equation is derived:

$$\sum_{\alpha=1}^{n} \delta(\xi-\xi_\alpha) a_\alpha - \sum_{\alpha=1}^{n} \delta'(\xi-\xi_\alpha)\left[b_\alpha - \xi_\alpha a_\alpha\right] = S(\xi;t), \tag{3.6}$$

and after rearranging we find:

$$\sum_{\alpha=1}^{n}\left[\delta(\xi-\xi_\alpha) + \delta'(\xi-\xi_\alpha)\xi_\alpha\right] a_\alpha - \sum_{\alpha=1}^{n} \delta'(\xi-\xi_\alpha) b_\alpha = S(\xi). \tag{3.7}$$

The application of the moment transform and the knowledge of the following simple rules valid for Dirac functions:

$$\begin{aligned} &\int_0^{+\infty} \xi^k \delta(\xi-\xi_\alpha)\, d\xi = \xi_\alpha^k \\ &\int_0^{+\infty} \xi^k \delta'(\xi-\xi_\alpha)\, d\xi = -k\xi_\alpha^{k-1} \end{aligned}, \tag{3.8}$$

results in the following linear system:

$$(1-k)\sum_{\alpha=1}^{n}\xi_\alpha^k a_\alpha + k\sum_{\alpha=1}^{n}\xi_\alpha^{k-1}b_\alpha = \int_0^{+\infty} S(\xi)\xi^k \mathrm{d}\xi \approx \overline{S}_k^{(n)}, \tag{3.9}$$

where the last term (i.e., $\overline{S}_k^{(n)}$) indicates that the moment transform of the source term is evaluated with a quadrature approximation of order n. Equation (3.9) allows the calculation of the source terms for the evolution equations for weights and abscissas (a_α and b_α), with the knowledge of the moment transform of the source terms of some of the moments. For example, if the evolution of the first four moments of the NDF (m_0, m_1, m_2, m_3) is simulated by tracking a quadrature approximation with two nodes (n = 2), by using DQMOM the following four transport equations must be solved:

$$\frac{\mathrm{d}w_1}{\mathrm{d}t} = a_1; \frac{\mathrm{d}w_2}{\mathrm{d}t} = a_2; \frac{\mathrm{d}(w_1\xi_1)}{\mathrm{d}t} = b_1; \frac{\mathrm{d}(w_2\xi_2)}{\mathrm{d}t} = b_2, \tag{3.10}$$

where the four source terms a_1, a_2, b_1 and b_2 are calculated by solving the linear system reported in Eq. (3.9) solved for k = 0, 1, 2, and 3, resulting in:

$$\begin{array}{lllll} a_1 & + a_2 & & & = \overline{S}_0 \\ & & b_1 & + b_2 & = \overline{S}_1 \\ -\xi_1^2 a_1 & -\xi_2^2 a_2 & +2\xi_1 b_1 & +2\xi_2 b_2 & = \overline{S}_2 \\ -2\xi_1^3 a_1 & -2\xi_2^3 a_2 & +3\xi_1^2 b_1 & +3\xi_2^2 b_2 & = \overline{S}_3 \end{array}, \tag{3.11}$$

that in matricial form becomes:

$$\begin{pmatrix} 1 & 1 & 0 & 0 \\ 0 & 0 & 1 & 1 \\ -\xi_1^2 & -\xi_2^2 & 2\xi_1 & 2\xi_2 \\ -2\xi_1^3 & -2\xi_2^3 & 3\xi_1^2 & 3\xi_2^2 \end{pmatrix} \begin{pmatrix} a_1 \\ a_2 \\ b_1 \\ b_2 \end{pmatrix} = \begin{pmatrix} \overline{S}_0 \\ \overline{S}_1 \\ \overline{S}_2 \\ \overline{S}_3 \end{pmatrix}. \tag{3.12}$$

In general the evolution of $2n$ moments (m_0, m_1, m_2, …, m_{2n-2}, m_{2n-1}) is simulated by resorting to a quadrature approximation with n nodes and directly solving the following transport equations:

$$\frac{\mathrm{d}w_1}{\mathrm{d}t} = a_1, \ldots, \frac{\mathrm{d}w_n}{\mathrm{d}t} = a_n; \frac{\mathrm{d}(w_1\xi_1)}{\mathrm{d}t} = b_1, \ldots, \frac{\mathrm{d}(w_n\xi_n)}{\mathrm{d}t} = b_n. \tag{3.13}$$

As already stated the source terms are calculated by solving the following linear system, obtained by imposing in Eq. (3.13) k = 0, 1, 2, …, $2n$-2, $2n$-1 and resulting in:

$$\begin{pmatrix} 1 & \cdots & 1 & 0 & \cdots & 0 \\ 0 & \cdots & 0 & 1 & \cdots & 1 \\ -\xi_1^2 & \cdots & -\xi_n^2 & 2\xi_1 & \cdots & 2\xi_n \\ \vdots & \vdots & \vdots & \vdots & \vdots & \vdots \\ -2(1-n)\xi_1^{2n-1} & \cdots & -2(1-n)\xi_n^{2n-1} & (2n-1)\xi_1^{2n-2} & \cdots & (2n-1)\xi_n^{2n-2} \end{pmatrix} \begin{pmatrix} a_1 \\ \vdots \\ a_n \\ b_1 \\ \vdots \\ b_n \end{pmatrix} = \begin{pmatrix} \overline{S}_0 \\ \overline{S}_1 \\ \vdots \\ \vdots \\ \overline{S}_{2n-2} \\ \overline{S}_{2n-1} \end{pmatrix}. \quad (3.14)$$

Equation (3.14) can be solved if the source term vector is known and if the linear system is non-singular. A closer inspection of the matrix shows that the determinant of the matrix is null and therefore the linear system is singular, if two nodes are identical. This is the main source of instability in DQMOM simulations. When this happens usually perturbation of the nodes and other numerical strategies can be employed to overcome the singularity, as explained in Marchisio and Fox (2005). As far as the source term vector is concerned, in the case of nucleation, growth, aggregation and breakage, and if the internal coordinate is the particle size $\xi = L$, the moment transform of order k of the source term becomes:

$$\begin{aligned} \overline{S}_k = {} & J_0 (L_0)^k + k\int_0^{+\infty} G(L)n(L)L^{k-1}\mathrm{d}L + \frac{1}{2}\int_0^{+\infty}\int_0^{+\infty}\left(L^3+\lambda^3\right)^{k/3}\beta(L,\lambda)n(L)n(\lambda)\mathrm{d}\lambda\mathrm{d}L \\ & -\int_0^{+\infty}\int_0^{+\infty} L^k\beta(L,\lambda)n(\lambda)n(L)\mathrm{d}\lambda\mathrm{d}L + \int_0^{+\infty}\int_0^{+\infty} L^k a(\lambda)b(L|\lambda)n(\lambda)\mathrm{d}\lambda\mathrm{d}L - \int_0^{+\infty} L^k a(L)n(L)\mathrm{d}L \end{aligned}, \quad (3.15)$$

that again after applying the quadrature approximation results in the source term reported in Eq. (2.39).

3.2 Extension to spatially heterogeneous problems

The extension of the approach to spatially heterogeneous problems is straightforward. The only difference stands in the fact that since spatial gradients are not null, their derivative terms do not disappear in the derivation. However they can be grouped together into some correction terms transforming the linear system for the calculation of the source terms into the following one (Marchisio and Fox, 2005):

$$\begin{pmatrix} 1 & \cdots & 1 & 0 & \cdots & 0 \\ 0 & \cdots & 0 & 1 & \cdots & 1 \\ -\xi_1^2 & \cdots & -\xi_n^2 & 2\xi_1 & \cdots & 2\xi_n \\ \vdots & \vdots & \vdots & \vdots & \vdots & \vdots \\ -2(1-n)\xi_1^{2n-1} & \cdots & -2(1-n)\xi_n^{2n-1} & (2n-1)\xi_1^{2n-2} & \cdots & (2n-1)\xi_n^{2n-2} \end{pmatrix} \begin{pmatrix} a_1 \\ \vdots \\ a_n \\ b_1 \\ \vdots \\ b_n \end{pmatrix} = \begin{pmatrix} \overline{S}_0 \\ \overline{S}_1 \\ \vdots \\ \vdots \\ \overline{S}_{2n-2}+\overline{C}_{2n-2} \\ \overline{S}_{2n-1}+\overline{C}_{2n-1} \end{pmatrix} \quad (3.16)$$

where the correction term can be easily calculated through the following expression:

$$\overline{C}_k = k(k-1)\sum_{\alpha=1}^{N}\left[w_\alpha \xi_\alpha^{k-2}\Gamma_x \frac{\partial \xi_\alpha}{\partial x_i}\frac{\partial \xi_\alpha}{\partial x_i}\right]. \quad (3.17)$$

From this equation it is possible to see that the correction term is not null only when spatial gradients of node values are present and when there is a diffusion mechanism involved. In fact, if the population of particles undergoes, convection, molecular growth, aggregation and breakage, but there is no spatial diffusion involved (for example in the case of large particles when both Brownian motions and turbulent fluctuations are not effective) the correction term is null since Γ_x is zero. Moreover the correction term is null for $k = 0$ and $k = 1$, meaning that it is needed to correctly predict the evolution of moments of the NDF of order equal to or greater than two.

The solution of this linear system provides the source term values for the nodes of the quadrature approximation that are now transported according to the following transport equations:

$$\begin{aligned} &\frac{\partial w_\alpha}{\partial t} + \frac{\partial}{\partial x_i}\left[U_{\mathrm{p},i} w_\alpha\right] - \frac{\partial^2}{\partial x_i^2}\left[\Gamma_{\mathrm{x}} w_\alpha\right] = a_\alpha \\ &\frac{\partial (w_\alpha \xi_\alpha)}{\partial t} + \frac{\partial}{\partial x_i}\left[U_{\mathrm{p},i}(w_\alpha \xi_\alpha)\right] - \frac{\partial^2}{\partial x_i^2}\left[\Gamma_{\mathrm{x}}(w_\alpha \xi_\alpha)\right] = b_\alpha \end{aligned} . \tag{3.18}$$

As previously explained, if the Stokes number is very low and the particles follow the fluid, then the particle velocity $\mathbf{U}_\mathrm{p}$ appearing in Eq. (3.18) can be replaced by the fluid velocity $\mathbf{U}_\mathrm{f}$. Moreover, in the case of turbulent flows, Reynolds-averaged values for moments of the NDF can be defined as well as for weights and nodes of the quadrature approximation, and the final transport equations become:

$$\begin{aligned} &\frac{\partial \langle w_\alpha \rangle}{\partial t} + \frac{\partial}{\partial x_i}\left[\langle U_{\mathrm{f},i}\rangle \langle w_\alpha \rangle\right] - \frac{\partial^2}{\partial x_i^2}\left[(\Gamma + \Gamma_\mathrm{t})\langle w_\alpha \rangle\right] = a_\alpha \\ &\frac{\partial \langle w_\alpha \xi_\alpha \rangle}{\partial t} + \frac{\partial}{\partial x_i}\left[\langle U_{\mathrm{f},i}\rangle \langle w_\alpha \xi_\alpha \rangle\right] - \frac{\partial^2}{\partial x_i^2}\left[(\Gamma + \Gamma_\mathrm{t})\langle w_\alpha \xi_\alpha \rangle\right] = b_\alpha \end{aligned} , \tag{3.19}$$

where the particle velocity has been replaced by the Reynolds-averaged fluid velocity (calculated through the solution of the Navier-Stokes equation for the fluid as reported in Eq. 2.42) and the spatial diffusion coefficient has been replaced by the summation of the diffusion coefficient due to Brownian motions and of the particle turbulent diffusion coefficient.

3.3 DQMOM for bi-variate homogeneous problems

The problems treated so far were all written in terms of an NDF function of time, space, particle velocity and of only one internal coordinate. However, as stated in the very beginning real problems require the use of NDFs that depend on more than one internal coordinate. Let us consider now a simple spatially homogeneous system where the NDF is defined in terms of two internal coordinates ξ_1 and ξ_2, representing two relevant properties of the population of particles (i.e., mass, surface area, size, fractal dimension,…)

The resulting population balance equation is:

$$\frac{\partial n(\xi;t)}{\partial t}=-\frac{\partial}{\partial \xi_1}\left[G_1 n(\xi;t)\right]-\frac{\partial}{\partial \xi_2}\left[G_2 n(\xi;t)\right]-\frac{\partial^2}{\partial \xi_1^2}\left[\Gamma_{11} n(\xi;t)\right]-\frac{\partial^2}{\partial \xi_2^2}\left[\Gamma_{22} n(\xi;t)\right]$$
$$-\frac{\partial^2}{\partial \xi_1 \partial \xi_2}\left[\Gamma_{12} n(\xi;t)\right]-\frac{\partial^2}{\partial \xi_2 \partial \xi_1}\left[\Gamma_{21} n(\xi;t)\right]+H(\xi;t) , \quad (3.20)$$

where the first two terms on the right-hand side represents continuous changes on the internal coordinates, whereas the following three terms are internal coordinate dispersion terms, and eventually the last term represents discrete events such as, nucleation, aggregation and breakage. As previously explained the population balance equation can be written in a more compact form:

$$\frac{\partial n(\xi;t)}{\partial t}=S(\xi;t), \quad (3.21)$$

where the source term S now groups together all the phenomena affecting the evolution of the particulate system. Also for this bi-variate system we can define a quadrature approximation that is equivalent to the following assumption on the functional form of the NDF:

$$n(\xi;t)=\sum_{\alpha=1}^{n} w_\alpha(t)\delta\left[\xi_1-\xi_{1\alpha}(t)\right]\delta\left[\xi_2-\xi_{2\alpha}(t)\right], \quad (3.22)$$

where now w_α are the weights of the nodes characterized by the abscissa values $\xi_{1\alpha}$ for the first internal coordinate and by the abscissa values $\xi_{2\alpha}$ for the second one. Again if this expression is replaced into the population balance equation:

$$\sum_{\alpha=1}^{n}\delta(\xi_1-\xi_{1\alpha})\delta(\xi_2-\xi_{2\alpha})\left[\frac{\mathrm{d}w_\alpha}{\mathrm{d}t}\right]-\sum_{\alpha=1}^{n}\delta'(\xi_1-\xi_{1\alpha})\delta(\xi_2-\xi_{2\alpha})\left[\frac{\mathrm{d}w_\alpha\xi_{1\alpha}}{\mathrm{d}t}-\xi_{1\alpha}\frac{\mathrm{d}w_\alpha}{\mathrm{d}t}\right]$$
$$-\sum_{\alpha=1}^{n}\delta(\xi_1-\xi_{1\alpha})\delta'(\xi_2-\xi_{2\alpha})\left[\frac{\mathrm{d}w_\alpha\xi_{2\alpha}}{\mathrm{d}t}-\xi_{2\alpha}\frac{\mathrm{d}w_\alpha}{\mathrm{d}t}\right]=S(\xi_1,\xi_2;t) , \quad (3.23)$$

and then if the following notation is used:

$$\frac{\mathrm{d}w_\alpha}{\mathrm{d}t}=a_\alpha;\frac{\mathrm{d}(w_\alpha\xi_{1\alpha})}{\mathrm{d}t}=b_{1\alpha};\frac{\mathrm{d}(w_\alpha\xi_{2\alpha})}{\mathrm{d}t}=b_{2\alpha}, \quad (3.24)$$

the following equation is obtained:

$$\sum_{\alpha=1}^{n}\delta(\xi_1-\xi_{1\alpha})\delta(\xi_2-\xi_{2\alpha})a_\alpha-w_\alpha\sum_{\alpha=1}^{n}\delta'(\xi_1-\xi_{1\alpha})\delta(\xi_2-\xi_{2\alpha})b_{1\alpha}$$
$$-w_\alpha\sum_{\alpha=1}^{n}\delta(\xi_1-\xi_{1\alpha})\delta'(\xi_2-\xi_{2\alpha})b_{2\alpha}=S(\xi_1,\xi_2;t) \quad (3.25)$$

If we define m_{kl} as the mixed moment of order k and l with respect to the first and second internal coordinates respectively, after the application of the quadrature approximation we obtain:

$$m_{kl} = \int_0^\infty \int_0^\infty n(\xi_1, \xi_2) \xi_1^k \xi_2^l \mathrm{d}\xi_1 \mathrm{d}\xi_2 \approx \sum_{\alpha=1}^n w_\alpha \xi_{1\alpha}^k \xi_{2\alpha}^l , \tag{3.26}$$

and after applying the moment transform to Eq. (3.25) the population balance equation is transformed into this set of linear equations:

$$\sum_{\alpha=1}^n \left[(1-k-l) \xi_{1\alpha}^k \xi_{2\alpha}^l a_\alpha + k \xi_{1\alpha}^{k-1} \xi_{2\alpha}^l b_{1\alpha} + l \xi_{1\alpha}^k \xi_{2\alpha}^{l-1} b_{2\alpha} \right] = \overline{S}_{kl}^{(n)} , \tag{3.27}$$

and where the source term[4] is:

$$\overline{S}_{kl} = \int_0^{+\infty} \int_0^{+\infty} S(\xi_1, \xi_2) \xi_1^k \xi_2^l d\xi_1 d\xi_2 . \tag{3.28}$$

If a quadrature approximation of order n is tracked, then $3n$ transport equations must be solved, n for the weights, n for the nodes with respect to the first internal coordinate, and n for the nodes with respect to the second internal coordinate. In order to calculate the $3n$ source terms ($a_1, ..., a_n$, $b_{11}, ..., b_{1n}$, $b_{21}, ..., b_{2n}$) the linear system reported in Eq. (3.27) must be solved for $3n$ values of k and l.

For example, if only one node is used ($n = 1$) then the NDF is represented by one Dirac function. This is equivalent to the representation of the population of particles with only one class of particles with number density w_1 and with characteristics values of the internal coordinates $\xi_{1\alpha}$ and $\xi_{2\alpha}$. The evolution of this unique node is tracked by solving three transport equations and by calculating three source terms through the solution of the linear system determined by imposing three values of k and l. For example, a possible choice could be $(k,l) = (0,0);(1,0);(0,1)$ resulting in the following transport equations:

$$\frac{dw_1}{\mathrm{d}t} = a_1 = \overline{S}_{00} ; \frac{\mathrm{d}(w_1 \xi_{11})}{\mathrm{d}t} = b_{11} = \overline{S}_{10} ; \frac{\mathrm{d}(w_2 \xi_{21})}{\mathrm{d}t} = b_{21} = \overline{S}_{01} . \tag{3.29}$$

It is interesting to highlight here that with this choice we are tracking the quadrature approximation guaranteeing that three mixed moments are correctly predicted, namely m_{00}, m_{10}, m_{01}.

If two nodes are chosen ($n = 2$) then the population of particles is represented by two Dirac functions, and therefore the population of particles is represented by two classes of particles, with number densities w_1 and w_2 and with characteristic internal coordinate values ξ_{11} and ξ_{12} for the first one, and ξ_{21} and ξ_{22} for the second one. In order to solve the transport equations for these six variables a linear system with six equations must be constructed and therefore six values of k and l must be chosen. As already said by choosing these six values we are implicitly choosing six moments to be correctly described by the quadrature approximation. For example a possible choice could be (k,l)=(0,0);(1,0);(0,1);(2,0);(0,2);(3,0) resulting in the following linear system:

---

[4] It is interesting to highlight here that the source term is indicated with the superscript $^{(n)}$ to remind readers that it is evaluated and closed through a quadrature approximation of order n.

$$\begin{pmatrix} 1 & 1 & 0 & 0 & 0 & 0 \\ 0 & 0 & 1 & 1 & 0 & 0 \\ 0 & 0 & 0 & 0 & 1 & 1 \\ -\xi_{11}^2 & -\xi_{12}^2 & 2\xi_{11} & 2\xi_{12} & 0 & 0 \\ -\xi_{21}^2 & -\xi_{22}^2 & 0 & 0 & 2\xi_{21} & 2\xi_{22} \\ -2\xi_{11}^3 & -2\xi_{12}^3 & 3\xi_{11}^2 & 3\xi_{12}^2 & 0 & 0 \end{pmatrix} \begin{pmatrix} a_1 \\ a_2 \\ b_{11} \\ b_{12} \\ b_{21} \\ b_{22} \end{pmatrix} = \begin{pmatrix} \overline{S}_{00} \\ \overline{S}_{10} \\ \overline{S}_{01} \\ \overline{S}_{20} \\ \overline{S}_{02} \\ \overline{S}_{30} \end{pmatrix}. \quad (3.30)$$

The set of six moments chosen to build the linear system has a strong impact on the stability of the algorithm as well as on the quality of the prediction. In fact, as already pointed out the moments chosen to build the linear system are the ones that are going to be accurately predicted, and therefore one must include in this set the most important moments. For example, low order moments, such as m_{00}, m_{10} and m_{01} that represent important properties of the population of particles[5], must always be included. The other moments to complete the set can be chosen following other criteria, as it will become clearer in the following paragraphs.

Another problem related to the choice of the moments is that the set must lead to a non-singular system, or in other words the determinant of the matrix of the linear system must not be null. In fact, some choices of the set of moments result in singular linear systems, and for example for two nodes ($n = 2$) the most common choices of six moments to avoid are the ones highlighted in blue (keeping the first three moments equal to m_{00}, m_{10} and m_{01}) and reported in Figure 1. Any other choice of the moments will lead to a non-singular system. If three nodes are used ($n = 3$) then nine moments must be chosen and the sets to avoid are the ones reported in Figure 2.

An interesting alternative is the use of fractional moments. For example if two nodes are used ($n = 2$) the following set of fractional moments can be used: (k,l) = (0,0); (1/3,0); (0,1/3); (2/3,0); (1,0); (0,1), whereas for three nodes the following nine moments can be chosen (k,l) = (0,0); (1/3,0); (0,1/3); (2/3,0); (1/3,1/3); (0,2/3); (1,0); (0,1); (4/3,0). Use of fractional moments also gives the advantage of reducing the number of possible singular matrices. Moreover, with fractional moments it is possible to work with a higher number of nodes without including very high order moments, that in some cases make the matrix of the linear system nearly-singular.

[5] As already pointed out m_{00} represents the total number particle density (i.e., the number of particles per unit volume), whereas if for example ξ_1 is particle mass and ξ_2 is particle surface area, m_{10} and m_{01} represent the total particle volume and the total particle area, respectively.

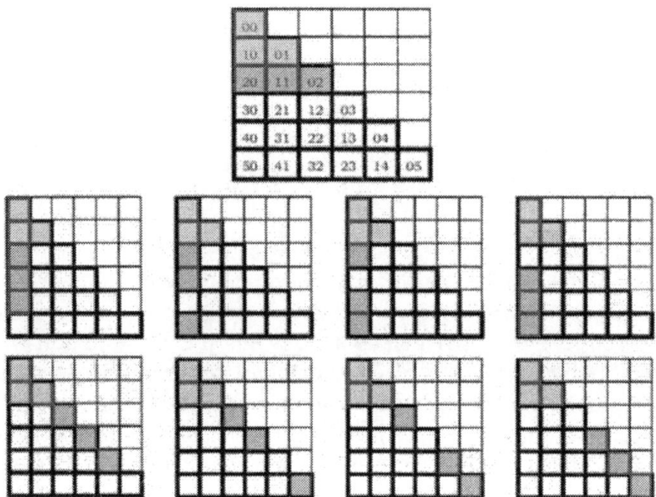

Figure 1. Sets of moments leading to singular matrices for two internal coordinates and $n = 2$ [taken from Zucca (2006)].

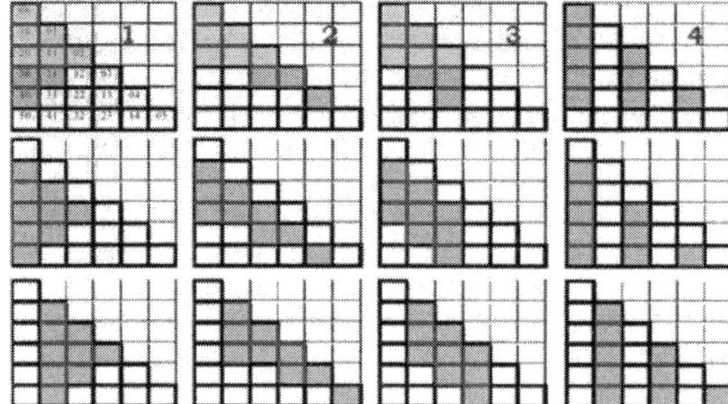

Figure 2. Set of moments leading to singular matrices for two internal coordinates and $n = 3$ [taken from Zucca (2006)].

3.4 Extension to spatially heterogeneous problems

Also for the bi-variate case the extension to the spatially heterogeneous problem is straightforward. The transport equations that have to be solved for tracking the evolution of the quadrature approximation are now the following ones:

$$\begin{aligned}
&\frac{\partial w_\alpha}{\partial t}+\frac{\partial}{\partial x_i}\left[U_{\mathrm{p}\alpha,i}\, w_\alpha\right]-\frac{\partial}{\partial x_i}\left[\Gamma\frac{\partial w_\alpha}{\partial x_i}\right]=a_\alpha\\
&\frac{\partial(w_\alpha\xi_{1\alpha})}{\partial t}+\frac{\partial}{\partial x_i}\left[U_{\mathrm{p}\alpha,i}\,(w_\alpha\xi_{1\alpha})\right]-\frac{\partial}{\partial x_i}\left[\Gamma\frac{\partial(w_\alpha\xi_{1\alpha})}{\partial x_i}\right]=b_{1\alpha}\quad,\\
&\frac{\partial(w_\alpha\xi_{2\alpha})}{\partial t}+\frac{\partial}{\partial x_i}\left[U_{\mathrm{p}\alpha,i}\,(w_\alpha\xi_{2\alpha})\right]-\frac{\partial}{\partial x_i}\left[\Gamma\frac{\partial(w_\alpha\xi_{2\alpha})}{\partial x_i}\right]=b_{2\alpha}
\end{aligned}\tag{3.31}$$

where the source terms are calculated by solving a linear system created by the following equation evaluated for different values of k and l:

$$\sum_{\alpha=1}^{n}\left[(1-k-l)\xi_{1\alpha}^{k}\xi_{2\alpha}^{l}a_\alpha+k\xi_{1\alpha}^{k-1}\xi_{2\alpha}^{l}b_{1\alpha}+l\xi_{1\alpha}^{k}\xi_{2\alpha}^{l-1}b_{2\alpha}\right]=\overline{C}_{kl}+\overline{S}_{kl}^{(n)},\tag{3.32}$$

where as before the introduction of spatial gradients created a correction term:

$$\overline{C_{kl}}=\Gamma_x\sum_{\alpha=1}^{N}w_\alpha\left[k(k-1)\xi_{1\alpha}^{k-2}\xi_{2\alpha}^{l}\frac{\partial\xi_{1\alpha}}{\partial x_i}\frac{\partial\xi_{2\alpha}}{\partial x_i}+2kl\xi_{1\alpha}^{k-1}\xi_{2\alpha}^{l-1}\frac{\partial\xi_{1\alpha}}{\partial x_i}\frac{\partial\xi_{2\alpha}}{\partial x_i}+l(l-1)\xi_{1\alpha}^{k}\xi_{2\alpha}^{l-2}\frac{\partial\xi_{2\alpha}}{\partial x_i}\frac{\partial\xi_{2\alpha}}{\partial x_i}\right]\tag{3.33}$$

that is non zero only when the global order of the moment is equal to or greater than two ($k+l\geq 2$). Similarly to what previously explained for the mono-variate case, in the limit of small Stokes number flows, the particles move with the same velocity of the fluid, and in the case of turbulent flows the transport equations for the Reynolds-averaged nodes become:

$$\begin{aligned}
&\frac{\partial\langle w_\alpha\rangle}{\partial t}+\frac{\partial}{\partial x_i}\left[\langle U_{\mathrm{f},i}\rangle\langle w_\alpha\rangle\right]-\frac{\partial}{\partial x_i}\left[(\Gamma+\Gamma_{\mathrm{t}})\frac{\partial\langle w_\alpha\rangle}{\partial x_i}\right]=a_\alpha\\
&\frac{\partial\langle w_\alpha\xi_{1\alpha}\rangle}{\partial t}+\frac{\partial}{\partial x_i}\left[\langle U_{\mathrm{f},i}\rangle\langle w_\alpha\xi_{1\alpha}\rangle\right]-\frac{\partial}{\partial x_i}\left[(\Gamma+\Gamma_{\mathrm{t}})\frac{\partial\langle w_\alpha\xi_{1\alpha}\rangle}{\partial x_i}\right]=b_{1\alpha}\quad,\\
&\frac{\partial\langle w_\alpha\xi_{2\alpha}\rangle}{\partial t}+\frac{\partial}{\partial x_i}\left[\langle U_{\mathrm{f},i}\rangle\langle w_\alpha\xi_{2\alpha}\rangle\right]-\frac{\partial}{\partial x_i}\left[(\Gamma+\Gamma_{\mathrm{t}})\frac{\partial\langle w_\alpha\xi_{2\alpha}\rangle}{\partial x_i}\right]=b_{2\alpha}
\end{aligned}\tag{3.34}$$

where as already stated the quantities between brackets are Reynolds-averaged values, particles are convected with the fluid velocity, and where the spatial diffusion coefficient is the summation of the molecular coefficient (i.e., particle diffusion due to Brownian motions) and of the turbulent diffusion coefficient (i.e., particle diffusion due to turbulent fluctuations). An interesting applications of this approach to the simulation of formation and evolution of nano-particles in turbulent non-premixed flames can be found in our previous works (Zucca et al., 2006; Zucca et al., 2007).

3.5 Extension to multi-phase problems

When the Stokes number is much higher than one, the particle relaxation time is longer than the fluid characteristic time, and therefore the population of particles moving from point to point do not have enough time to relax to the characteristic local fluid velocity. In this case particles move with a velocity that is different from the velocity of the fluid and that is different from particle to particle. In this case, as already explained the NDF is function of the internal coordinates ξ_1 and ξ_2 and of particle velocity $\mathbf{U}_p$. The use of the quadrature approximation is equivalent to an assumption on the functional form of the NDF as follows:

$$n\left(\mathbf{U}_p,\xi;\mathbf{x},t\right)=\sum_{\alpha=1}^{n} w_\alpha\left(\mathbf{x},t\right)\delta\left[\xi_1-\xi_{1\alpha}\left(\mathbf{x},t\right)\right]\delta\left[\xi_2-\xi_{2\alpha}\left(\mathbf{x},t\right)\right]\delta\left[\mathbf{U}_p-\mathbf{U}_{p\alpha}\left(\mathbf{x},t\right)\right], \tag{3.35}$$

where $\mathbf{U}_{p\alpha}$ is the characteristic velocity of the particles with internal coordinate values equal to $\xi_{1\alpha}$ and $\xi_{2\alpha}$, and number density equal to w_α. It is clear that now a quadrature approximation can be tracked if n transport equations are solved for the number densities w_α, $2n$ transport equations are solved for the first ($\xi_{1\alpha}$) and the second ($\xi_{2\alpha}$) internal coordinate values (see Eq. 3.36), and eventually n transport equations are solved for particle velocity $\mathbf{U}_{p\alpha}$.

Usually these transport equations are solved within the multi-fluid model framework (Anderson and Jackson, 1967) where the poly-disperse system is described as constituted by n particle classes. In this framework the transport equations for weights and nodes of the quadrature approximation must be solved (see Eq. 3.31) convecting now each node of the quadrature approximation with its own velocity $\mathbf{U}_{p\alpha}$ along with the momentum balance equations for the n particle classes[6] as explained by Fan, Marchisio and Fox (2004):

$$\begin{aligned}\frac{\partial}{\partial t}\left[\varepsilon_{p\alpha}U_{p\alpha,i}\right]+\frac{\partial}{\partial x_j}\left[U_{p\alpha,j}\left(\varepsilon_{p\alpha}U_{p\alpha,i}\right)\right]=\frac{\varepsilon_{p\alpha}}{\rho_p}\frac{\partial p}{\partial x_i}+f\left(C_D\right)\left(U_{p\alpha,i}-U_{f,i}\right)+\varepsilon_{p\alpha}g_i\\+\int H\left(\mathbf{U}_p,\xi;\mathbf{x},t\right)\mathbf{U}_p\mathrm{d}\mathbf{U}_p\mathrm{d}\xi\end{aligned}, \tag{3.36}$$

where $\varepsilon_{p\alpha}$ is the volume fraction of the particle class corresponding to the node α and characterized by the number density w_α and by some relevant properties (i.e., particle size, volume, fractal dimension) identified by the internal coordinates. It is also clear that the summation of the volume fractions of all dispersed phases equals the total solid particle volume fraction introduced in Eq. (1.2):

$$\varepsilon_p\left(\mathbf{x},t\right)=\sum_{\alpha=1}^{n}\varepsilon_{p\alpha}\left(\mathbf{x},t\right). \tag{3.37}$$

The terms on the left-hand side of Eq. (3.36) instead represents transport in space and time of momentum, whereas the first term on the right-hand side represents the continuous change in momentum imprinted by the pressure gradients of the continuous phase, the second term represents the continuous change in momentum due to fluid-particle interactions (only the drag force contribution) that is derived from Eq. (2.32), whereas the last term represents the rate of change in momentum due to discrete events (such as particle-particle collision). For example, in

[6] It is interesting to point out here that the momentum balance equation for the n particle classes can be derived from Eq. (1.3) by substituting the functional form of the NDF reported in Eq. (3.2) and by applying the moment transform of order zero with respect to both internal coordinates and of order one with respect to particle velocity.

the case of dense solid-fluid systems this last term is closed by resorting to the kinetic theory for granular flows as explained by Gidaspow (1994).

4 Application to Solid-Liquid Turbulent Systems

In what follows an application of QMOM and DQMOM will be presented and discussed. The multi-phase system presented here is constituted by some solid particles dispersed into a liquid. The solid particles are primary particles of uniform size with radius equal to R_p. Two types of primary particles exist: primary particles of type A (white) and primary particles of type B (black), that tend to aggregate and form binary fractal aggregates, as sketched in Figure 3. These aggregates can be fully characterized by the total number of primary particles, ξ, and by the number of primary particles of type A, ζ included in one single aggregate. The size of a single aggregate is related to the total number of primary particles ξ through the following relationship:

$$\xi = k_f \left(\frac{R_g}{R_p} \right)^{d_f}, \tag{4.1}$$

where d_f is the fractal dimension and where k_f is usually close to unity (Sorensen, 2001). In this description two aggregates constituted by the same number of primary particles, and therefore characterized by the same dimensionless mass ξ can have different radius of gyration. When the fractal dimension is close to three the aggregates are very compact and the radius of gyration is smaller compared to that of aggregates characterized by very open and loose structures and by smaller values of the fractal dimension (e.g., $d_f = 1.8$).

These primary particles due to Brownian motions and turbulent fluctuations tend to aggregate increasing their characteristic size and tend to form fractal objects. The typical evolution of these aggregates is sketched in Figure 4. The aggregation kernel in fully destabilized systems can be expressed as the summation of two contributions, one due to Brownian motions, known as peri-kinetic aggregation, and the second one due to velocity gradients, known as ortho-kinetic aggregation (Waldner et al., 2005):

$$\beta(\gamma;\xi,\xi') = \frac{2k_B T}{3\mu_f}\frac{1}{W}\left(\xi^{1/d_f} + \xi'^{1/d_f}\right)\left(\xi^{-1/d_f} + \xi'^{-1/d_f}\right) + \alpha(\xi,\xi')\gamma R_p \left(\xi^{1/d_f} + \xi'^{1/d_f}\right)^3, \tag{4.2}$$

where k_B is the Boltzman constant, T is the absolute temperature, μ_f is the fluid viscosity, W is the stability ratio, α is the aggregation pre-factor, γ is the shear rate, and as already reported R_p is the radius of primary particles.

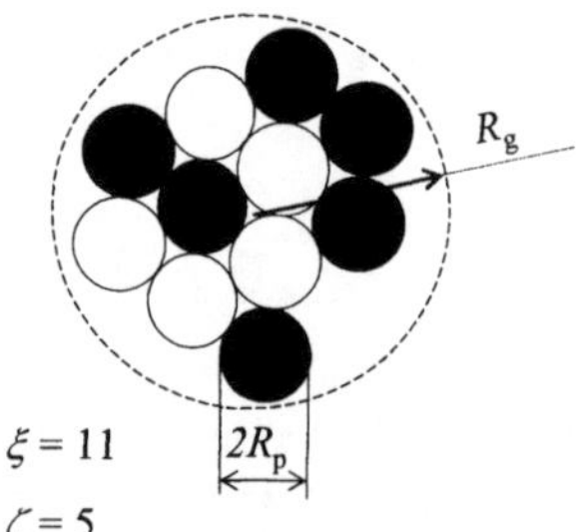

Figure 3. Sketch of the typical aggregate.

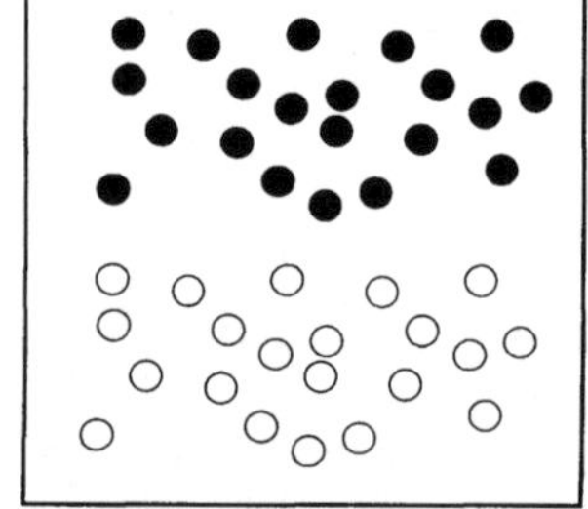

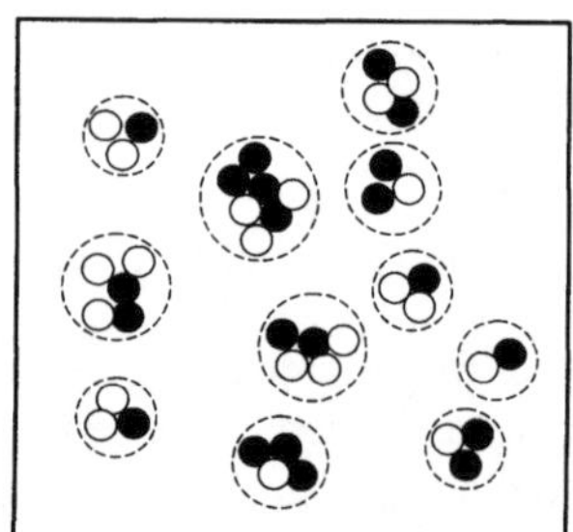

Figure 4. Typical aggregation process of a binary mixture of primary particles.

The shear rate quantifies the local velocity gradients that are the driving force for ortho-kinetic aggregation. For turbulent systems the shear rate is defined as the characteristic velocity gradient inside one eddy (Saffman and Turner, 1956).:

$$\gamma(\mathbf{x}) = \sqrt{\frac{\varepsilon}{\nu_{\mathrm{f}}}}, \tag{4.3}$$

where ε is the turbulent dissipation rate, and where ν_{f} is the kinematic viscosity of the continuous phase.

As clusters get bigger and bigger due to aggregation they become more fragile and start to break up forming smaller ones. Breakage is also caused by the shear rate γ, that produces hydrodynamic stresses which break the aggregate when they overcome cohesive forces. Several expressions are available in literature; some of them are based on turbulence theory and are characterized by an exponential dependence of the breakage rate upon dimensionless aggregate mass. One of the most popular is the so-called power law kernel expressed by:

$$a(\gamma;\xi) = P_1\gamma^{P_2}\left(R_{\mathrm{p}}\xi^{1/d_{\mathrm{f}}}\right)^{P_3}, \tag{4.4}$$

where P_1, P_2 and P_3 are fitting parameters usually derived by comparison with experimental data.

The daughter distribution function $b(\xi,\zeta|\xi',\zeta')$ contains information about the size of the fragments characterized by ξ and ζ, produced by breaking up an aggregate of size ξ' and ζ'. Various expressions can be used corresponding to symmetric fragmentation, erosion, uniform distribution, as explained in previous works [see for example Marchisio, Vigil, and Fox (2003a)]. Experimental evidences, concerning breakage of solid aggregates of primary particles, support the hypothesis that very often only binary symmetric breakage occurs and the relative daughter distribution function is as follows:

$$b(\xi,\zeta|\xi',\zeta') = 2\delta\left[\xi - \xi'/2\right]\delta\left[\zeta - \zeta'/2\right]. \tag{4.5}$$

The description of the evolution of this multi-phase system can be based on a NDF expressed in terms of particle velocity $\mathbf{U}_{\mathrm{p}}$, number of total primary particles ξ, number of primary particles of type A ζ, space $\mathbf{x}$ and time t.

As previously mentioned it is sometimes convenient to integrate out particle velocity and define a new NDF as follows:

$$n(\xi,\zeta;\mathbf{x},t) = \int_{\Omega_{U_p}} n(\mathbf{U}_p,\xi,\zeta;\mathbf{x},t)\mathrm{d}\mathbf{U}_p, \tag{4.6}$$

only function of the two internal coordinates and of space and time. From this NDF we can define some integral properties of the population of particles through mixed moments. Based on this definition the total number of aggregates per unit volume can be calculated as follows:

$$m_{00}(\mathbf{x},t) = \int_0^\infty\int_0^\infty n(\xi,\zeta;\mathbf{x},t)\mathrm{d}\xi\mathrm{d}\zeta\,. \tag{4.7}$$

The total number of primary particles of type A and B (a quantity that remains constant during particle aggregation and breakage) corresponds to:

$$m_{10}(\mathbf{x},t) = \int_0^\infty\int_0^\infty n(\xi,\zeta;\mathbf{x},t)\xi\mathrm{d}\xi\mathrm{d}\zeta, \tag{4.8}$$

whereas the total number of primary particles of type A (also constant during particle aggregation and breakage) is defined as:

$$m_{01}(\mathbf{x},t) = \int_0^\infty\int_0^\infty n(\xi,\zeta;\mathbf{x},t)\zeta\mathrm{d}\xi\mathrm{d}\zeta\,. \tag{4.9}$$

From the ratio of these quantities it is possible to define the average number of primary particles per aggregate as follows:

$$\langle\xi\rangle(\mathbf{x},t) = \frac{\int_0^\infty\int_0^\infty n(\xi,\zeta;\mathbf{x},t)\xi\mathrm{d}\xi\mathrm{d}\zeta}{\int_0^\infty\int_0^\infty n(\xi,\zeta;\mathbf{x},t)\mathrm{d}\xi\mathrm{d}\zeta} = \frac{m_{10}(\mathbf{x},t)}{m_{00}(\mathbf{x},t)}, \tag{4.10}$$

whilst the average number of primary particles of type A per cluster is calculated as follows:

$$\langle\zeta\rangle(\mathbf{x},t) = \frac{\int_0^\infty\int_0^\infty n(\xi,\zeta;\mathbf{x},t)\zeta\mathrm{d}\xi\mathrm{d}\zeta}{\int_0^\infty\int_0^\infty n(\xi,\zeta;\mathbf{x},t)\mathrm{d}\xi\mathrm{d}\zeta} = \frac{m_{01}(\mathbf{x},t)}{m_{00}(\mathbf{x},t)}. \tag{4.11}$$

Moreover from the NDF it is possible to calculate an average dimensionless aggregate size, also called mean radius of gyration, defined as follows:

$$\frac{\langle R_\mathrm{g}\rangle}{R_\mathrm{P}} = \sqrt{\frac{\int_0^\infty\int_0^\infty n(\xi,\zeta;\mathbf{x},t)\xi^{2(1+1/d_\mathrm{f})}\mathrm{d}\xi\mathrm{d}\zeta}{\int_0^\infty\int_0^\infty n(\xi,\zeta;\mathbf{x},t)\xi^2\mathrm{d}\xi\mathrm{d}\zeta}}\,. \tag{4.12}$$

The equation that governs the evolution of the NDF is:

$$\frac{\partial n(\xi,\zeta;\mathbf{x},t)}{\partial t}+\frac{\partial}{\partial x_i}\left[\langle U_{\mathrm{p},i}\rangle n(\xi,\zeta;\mathbf{x},t)\right]-\frac{\partial}{\partial x_i}\left[(\Gamma+\Gamma_\mathrm{t})\frac{\partial n(\xi,\zeta;\mathbf{x},t)}{\partial x_i}\right]=$$
$$\frac{1}{2}\int_0^{\zeta}\int_0^{\xi}\beta(\gamma;\xi-\xi',\xi')n(\xi-\xi',\zeta-\zeta';\mathbf{x},t)n(\xi',\zeta';\mathbf{x},t)\mathrm{d}\xi'\mathrm{d}\zeta'-n(\xi,\zeta;\mathbf{x},t)\int_0^{\infty}\int_0^{\infty}\beta(\gamma;\xi,\xi')n(\xi',\zeta';\mathbf{x},t)\mathrm{d}\xi'\mathrm{d}\zeta' \tag{4.13}$$
$$+\int_{\zeta}^{\infty}\int_{\xi}^{\infty}a(\gamma;\xi')b(\xi,\zeta|\xi',\zeta')n(\xi',\zeta';\mathbf{x},t)\mathrm{d}\xi'\mathrm{d}\zeta'-a(\gamma;\xi,\zeta)n(\xi,\zeta;\mathbf{x},t)$$

where Eq. (4.13) was derived knowing that during aggregation and breakage events both internal coordinates are additive. The problem can be tackled with several approaches. For example, it could be transformed into a mono-variate population balance equation, integrating out the second internal coordinate (i.e., the number of primary particle of type A per aggregate ζ) and then solving the resulting equation using QMOM as described in the previous paragraphs and as reported in Marchisio, Vigil and Fox (2003b) and Marchisio et al. (2006a,b).

An interesting alternative is to directly solve Eq. (4.13) by resorting to DQMOM. If the moment transform of order k and l with respect to the two internal coordinates is applied Eq. (4.13) the following equation is obtained:

$$\frac{\partial m_{kl}(\mathbf{x},t)}{\partial t}+\frac{\partial}{\partial x_i}\left[\langle U_{\mathrm{p},i}\rangle m_{kl}(\mathbf{x},t)\right]-\frac{\partial}{\partial x_i}\left[(\Gamma+\Gamma_\mathrm{t})\frac{\partial m_{kl}(\mathbf{x},t)}{\partial x_i}\right]=\overline{S}_{kl}(\mathbf{x},t)=$$
$$\frac{1}{2}\int_0^{+\infty}\int_0^{+\infty}\int_0^{+\infty}\int_0^{+\infty}\left[(\xi+\xi')^k(\zeta+\zeta')^l-\xi^k\zeta^l-\xi'^k\zeta'^l\right]\beta(\xi',\zeta';\xi,\zeta)n(\xi',\zeta';\mathbf{x},t)n(\xi,\zeta;\mathbf{x},t)\mathrm{d}\xi'\mathrm{d}\zeta'\mathrm{d}\xi\mathrm{d}\zeta \tag{4.14}$$
$$+\int_0^{+\infty}\int_0^{+\infty}\int_0^{+\infty}\int_0^{+\infty}a(\xi',\zeta')b(\xi,\zeta|\xi',\zeta')\xi^k\zeta^l n(\xi',\zeta';\mathbf{x},t)\mathrm{d}\xi'\mathrm{d}\zeta'\mathrm{d}\xi\mathrm{d}\zeta-\int_0^{+\infty}\int_0^{+\infty}a(\xi,\zeta)n(\xi,\zeta;\mathbf{x},t)\xi^k\zeta^l\mathrm{d}\xi\mathrm{d}\zeta$$

and after applying the quadrature approximation the source term becomes:

$$\overline{S}_{kl}(\mathbf{x},t)=\frac{1}{2}\sum_{\alpha=1}^{n}\sum_{\gamma=1}^{n}\left[(\xi_\alpha+\xi_\gamma)^k(\zeta_\alpha+\zeta_\gamma)^l-\xi_\alpha^k\zeta_\alpha^l-\xi_\gamma^k\zeta_\gamma^l\right]w_\alpha w_\gamma\beta_{\alpha,\gamma}+\sum_{\alpha=1}^{n}a_\alpha w_\alpha\xi_\alpha^k\zeta_\alpha^l\left(2^{1-k-l}-1\right) \tag{4.15}$$

As previously explained using DQMOM a quadrature approximation of order n is directly tracked by solving $3n$ transport equations: n for the weights w_α, and $2n$ for the nodes ξ_α and ζ_α with $\alpha = 1,\ldots,n$. The source terms of these $3n$ equations are calculated by solving a linear system, obtained by choosing some mixed moments of the NDF. In Figure 5 the typical evolution of some moments is reported. As it is seen m_{00} decreases because of aggregation and then when breakage starts playing an important role a steady state solution is reached, whereas m_{10} and m_{01} are conserved during the process. The average total number of primary particles per cluster and the average number of primary particles of type A per aggregate, as well as the mean radius of gyration increase until the balance between aggregation and breakage is achieved.

The predictions obtained with increasing the number of nodes of the quadrature approximation resulted in almost identical low-order moments predictions, showing that as long as the low order moments (i.e., m_{00}, m_{01}, m_{10}, ...) are included into the set of moments used for the construction of the linear system, DQMOM is able to track low-order moments themselves with good accuracy.

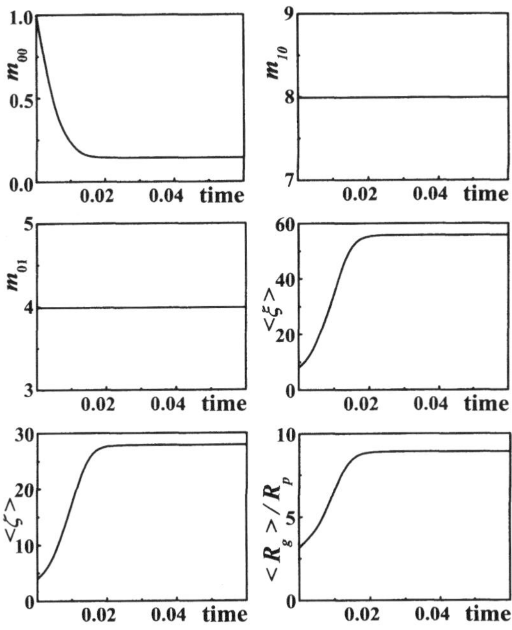

Figure 5. Evolution versus dimensionless time of some lower-order moments for realistic aggregation and breakage kernels.

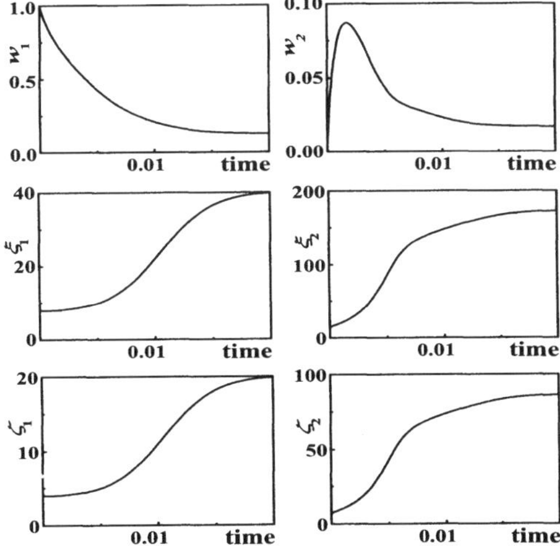

Figure 6. Evolution versus dimensionless time of the quadrature approximation for $n = 2$ in the case of realistic kernels.

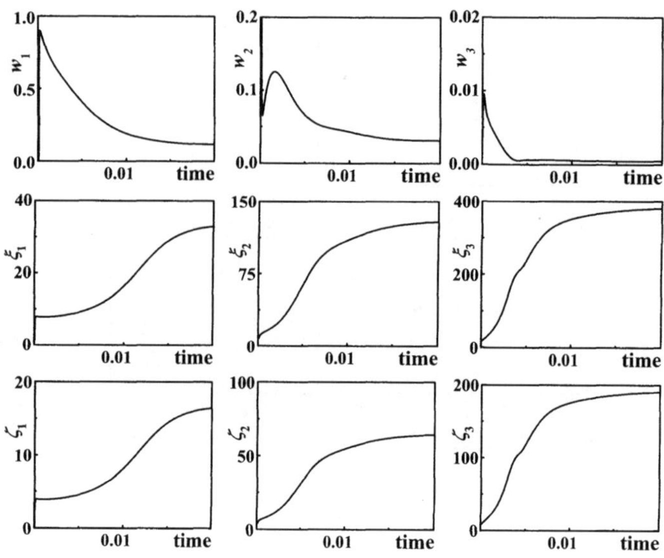

Figure 7. Evolution versus dimensionless time of the quadrature approximation for $n = 3$ in the case of realistic kernels.

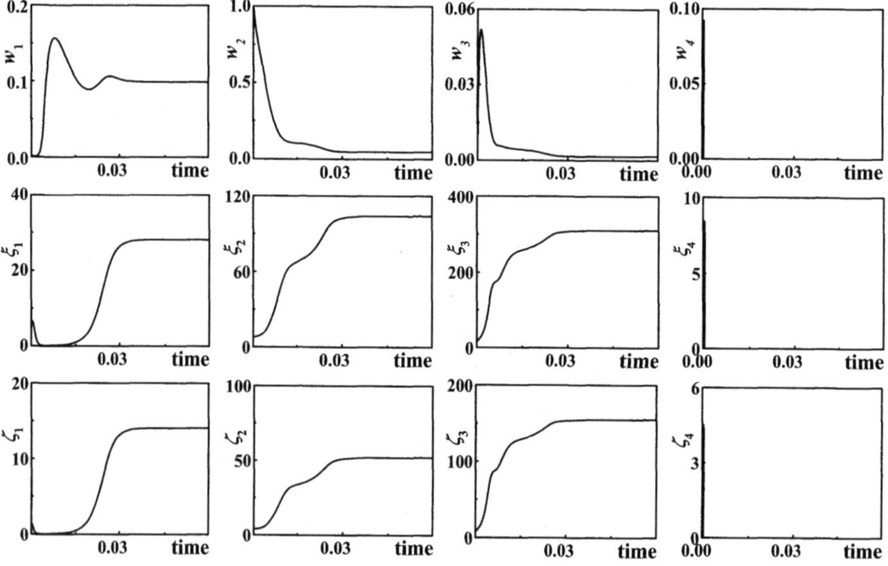

Figure 8. Evolution versus dimensionless time of the quadrature approximation for $N = 4$ in the case of realistic kernels.

In Figures 6, 7 and 8 the evolution of the quadrature approximation with two nodes (n = 2; tracked moments: m_{00}; m_{10}; m_{01}; $m_{1/3,0}$; $m_{0,1/3}$; $m_{0,2/3}$), three nodes (n = 3; tracked moments: m_{00}; m_{10}; m_{01}; $m_{1/3,0}$; $m_{0,1/3}$; $m_{2/3,0}$; $m_{2,2}$; $m_{0,3}$; $m_{3,0}$), and four nodes (n = 4; tracked moments: m_{00}; m_{10}; m_{01}; $m_{1/3,0}$; $m_{0,1/3}$; $m_{2/3,0}$; $m_{1/3,1/3}$; $m_{0,2/3}$; $m_{4/3,0}$; $m_{0,4/3}$; $m_{2,0}$; $m_{0,2}$) is reported[7].

Also in this case the stability of the algorithm is strongly affected by the choice of the moments included in the set for the construction of the linear system. Only including some higher-order moments (e.g., m_{20}; m_{02}; m_{30}; m_{03}) the tail of the distribution, represented by some extreme node, is well tracked and numerical gelation is avoided. However, if the order of these higher-order moments exceeds a certain value, than the code becomes unstable again, due this time to the fact that the matrix of the linear system is nearly singular.

A crucial role is played by the total solid concentration. The solid concentration is expressed in terms of the solid particle volume fraction α_p. When the solid volume fraction is very small the final aggregate size is quite small too and the multi-phase system is very dilute. As a consequence the velocity of all solid particles equals that of the fluid (i.e., $\mathbf{U}_\mathrm{p} = \mathbf{U}_\mathrm{f}$). Moreover, at very low solid volume fractions the aggregation and breakage processes are very slow with respect to particle mixing. In order to quantify these concepts it is useful to define, based on the aggregation and breakage kernels, some characteristic aggregation and breakage timescales (Marchisio et al., 2006a):

$$\tau_\mathrm{A} = \frac{1}{a(\bar{\gamma};\langle\xi\rangle,\langle\xi\rangle)m_{00}}; \quad \tau_\mathrm{B} = \frac{1}{b(\bar{\gamma};\langle\xi\rangle)}, \tag{4.16}$$

where τ_A and τ_B represent the time interval between two aggregation and breakage events, and where $\bar{\gamma}$ is the volume-averaged shear rate. These two characteristic timescales must be compared with the particle mixing timescale τ_M. This latter expresses the time required to perfectly mix the solid particles inside the vessel, in order to smooth down all the spatial gradients in particle properties. As it has been reported above, when the system is dilute usually $\tau_\mathrm{M} \ll \tau_\mathrm{A}, \tau_\mathrm{B}$ meaning that the time required for the solid particles to mix is much smaller than the time required for the solid particles to aggregate and break. In this situation not only particles move with the fluid velocity, but they move so fast (in comparison to aggregation and breakage) that they are perfectly mixed throughout the vessel. In this situation the NDF defining the state of the population of particles does not depend on space and particle velocity.

If the solid volume fraction increases beyond a certain critical value then aggregation and breakage speed up and their characteristic timescales get closer and closer to the mixing timescale. When this happens mixing is not fast enough to smooth down all the gradients in particle properties and therefore the NDF does depend on spatial coordinates. It is important to remind here that usually this situation occurs when the solid concentration is too low to have an effect on the flow and turbulent field of the continuous phase, moreover the final aggregate size is small enough to satisfy the small Stokes number limit. In fact, under these operating conditions the solid-liquid system can be modelled as a pseudo single-phase system.

A further increase in solid volume fraction causes a further reduction of the aggregation and breakage timescales making them smaller than the particle mixing timescales and comparable

[7] Simulations reported here were run under operating conditions very similar to those reported in Marchisio et al. (2006a,b) and readers interested in the details are remanded to those works.

with turbulent fluctuations[8]. Under these operating conditions the final aggregate size is quite large and therefore particles will be characterized by their own velocity. Moreover particle loading is so high that the continuous phase is strongly affected by the presence of the disperse phase, generating what is known as two-way coupling.

The threshold values of solid volume fractions where the transitions between the three different regimes are located will depend upon the densities of the continuous and disperse phases, the viscosity of the continuous phase, and the aggregation and breakage kernels. For a system characterized by latex primary particles of diameter equal to 2.17 μm dispersed in water under turbulent stirring in a range of shear rate $\gamma = 75\text{-}150\ s^{-1}$ as in the experimental work of Oles (1992) these transitions are positioned according to the plot reported in Figure 9 (taken from Marchisio et al., 2006a).

In this figure the characteristic timescales for aggregation, breakage and mixing are reported versus the solid volume fraction and as it is possible to see three regimes are individuated. Summarizing, in the first one (I) the solid-liquid system is very dilute, the final aggregate size is quite small and particles adapt to the fluid velocity. Moreover particles move in the vessel faster than they aggregate and break up, therefore the system can be modelled as a pseudo single-phase spatially homogeneous system. In the second regime (II) the characteristic timescales for particle aggregation and breakage are comparable to that of particle mixing, the solid-liquid system is now characterized by strong gradients in particle properties and therefore is spatially heterogeneous but the final aggregate size is still rather small and the system is dilute enough to be described as pseudo single-phase. Eventually at very high solid volume fractions (III) the solid-liquid system must be described with a multi-phase approach taking into account spatial heterogeneities.

In Figure 10 the contour plots of the mean radius of gyration at steady state in a stirrer tank (Marchisio et al., 2006b) is reported for two different values of the solid volume fractions. As it is seen at low volume fractions (regime I) the steady state values of the mean radius of gyration are very uniform within the vessel, whereas at higher volume fractions (regime II) some evident gradients are present, confirming what reported above.

5 Conclusions

Several approaches for modelling multi-phase systems have been presented and discussed. In particular the use of QMOM and DQMOM for describing the evolution of mono- and bi-variate Number Density Functions (NDFs) has been discussed in details. Moreover the possibility to describe the multi-phase systems at different level of details, neglecting spatial heterogeneities, or as a pseudo-single phase system or with a multi-phase model has been discussed. Eventually the treatment of a practical case, namely binary aggregation and breakage of primary particles has been described.

[8] Under these operating conditions the approximation reported in Eq. (2.38) is not valid anymore and a model to take into account the interaction between turbulent fluctuations and particle aggregation and breakage (i.e., micro-mixing model) must be included.

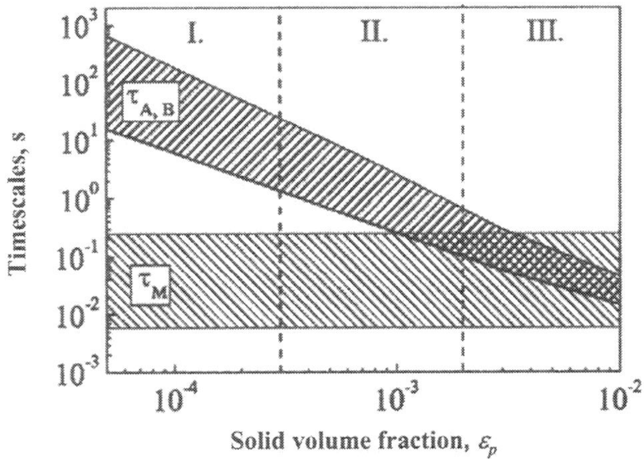

Figure 9. Particles aggregation, breakage, and mixing timescales versus the solid volume fraction.

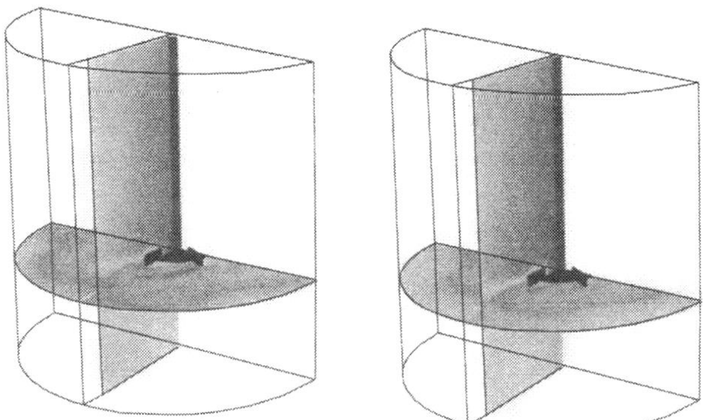

Figure 10. Contour plot of $\langle R_g \rangle / R_p$ at steady-state for $\gamma = 110\ \mathrm{s}^{-1}$ for the stirrer tank; left: low solid concentration $\varepsilon_p = 5.0 \times 10^{-5}$ (min: 5.4 - max: 5.9); right: high solid concentration $\varepsilon_p = 1.0 \times 10^{-3}$ (min: 5.96 - max: 16.4).

Acknowledgements

The theoretical and practical findings presented in this work are the results of research efforts that involved many people at Iowa State University (USA) and Politecnico of Torino (Italy). The contributions of these people have been meticulously mentioned in the manuscript citing their original works. However, the author feels particularly indebted to Rodney Fox, Antonello Barresi, Marco Vanni, Giancarlo Baldi, Miroslav Soos, Jan Sefcik, and Massimo Morbidelli. Particularly important is the contribution of Alessandro Zucca for the work carried out for the solution of bi-variate population balance equations. Last but not least, the support of Donata Mostardini during Christmas break 2006 for making the deadline for this manuscript is gratefully acknowledged. Grazie Donata!

References

T. B. Anderson and R. Jackson (1967). A fluid mechanical description of fluidized beds. *Industrial & Engineering Chemistry Fundamentals,* 6: 527-534.

R. Andersson and B. Andersson (2006). On the breakup of fluid particles in turbulent flows. *AIChE Journal,* 52: 2020-2030.

J. Baldyga and W. Orciuch (2001). Some hydrodynamic aspects of precipitation. *Powder Technology,* 121: 9-19.

J. C. Barrett and N. A. Webb (1998) A comparison of some approximate methods for solving the aerosol general dynamic equation. *Journal of Aerosol Science,* 29: 31-39.

C. T. Crowe, M. Sommerfeld, and Y. Tsuji (1998). *Multiphase flows with droplets and particles.* CRC Press, Boca Raton.

H. Dette and W. J. Studden (1997). *The Theory of Canonical Moments with Applications in Statistics, Probability, and Analysis.* Wiley, New York, NY.

R. B. Diemer and J. H. Olson (2002). A moment methodology for coagulation and breakage problems: Part 3 - Generalized daughter distribution functions. *Chemical Engineering Science,* 57: 4187-4198.

J. A. Dirksen and T. A. Ring (1991). Fundamentals of crystallization: kinetics effect of particle size distribution and morphology. *Chemical Engineering Science,* 46:2389-2427.

R. Fan, D. L. Marchisio, R. O. Fox (2004). Application of the direct quadrature method of moments to polydisperse gas-solid fluidized beds. *Powder Techonology* 139: 7-20.

Fluent Inc. (2004) *Fluent 6.2 User's Guide.* Fluent Inc.: Lebanon, NH.

E. Gavi, L. Rivautella, D. L. Marchisio, M. Vanni, A. A. Barresi, and G. Baldi (2007). CFD modelling of nano-particle precipitation in Confined Impinging Jet Reactors. *Chemical Engineering Research & Design,* in press.

D. Gidaspow (1994). *Multiphase Flow and Fluidization,* Academic Press, Boston, MA.

R. G. Gordon (1968). Error bounds in equilibrium statistical mechanics. *Journal of Mathematical Physics,* 9: 655-667.

H. M. Hulburt and S. Katz (1964). Some problems in particle technology. *Chemical Engineering Science,* 19: 555-574.

Z. Jaworski, P. Pianko-Oprych, D. L. Marchisio, and A. W. Nienow (2007). CFD modelling of turbulent drop breakage in a Kenics static mixer and comparison with experimental data. *Chemical Engineering Research & Design,* in press.

C. M. Jones and M. A. Larson (1999). Characterizing growth-rate dispersion of NaNO3 secondary nuclei. *AIChE Journal,* 45: 2128-2135.

M. Manninen, V. Taivassalo, and S. Kallio (1996). On the mixture model for multiphase flow. *VTT Publications* 288, Technical Research Centre of Finland.

D. L. Marchisio, A. A. Barresi, and R. O. Fox (2001). Simulation of turbulent precipitation in a semi-batch Taylor-Couette reactor using CFD. *AIChE Journal,* 47: 664-676.

D. L. Marchisio, A. A. Barresi, and M. Garbero (2002). Nucleation, growth, and agglomeration in barium sulfate turbulent precipitation. *AIChE Journal,* 48: 2039-2050.

D. L. Marchisio, J. T. Pikturna, R. O. Fox, R. D. Vigil, and A. A. Barresi (2003). Quadrature method of moments for population balances. *AIChE Journal,* 49: 1266-1276.

D. L. Marchisio, R. D. Vigil, R.D., and R. O. Fox (2003a). Quadrature method of moments for aggregation-breakage processes. *Journal of Colloid and Interface Science,* 258: 322-334.

D. L. Marchisio, R. D. Vigil, and R. O. Fox (2003b). Implementation of the quadrature method of moments in CFD codes for aggregation–breakage problems. *Chemical Engineering Science,* 58: 3337 – 3351.

D. L. Marchisio and R. O. Fox (2005). Solution of population balance equations using the direct quadrature method of moments. *Journal of Aerosol Science,* 36: 43-73.

D. L. Marchisio, L. Rivautella, and A. A. Barresi (2006). Design and scale-Up of chemical reactors for nanoparticle precipitation. *AIChE Journal,* 52: 1877-1887.

D. L. Marchisio, M. Soos, J. Sefcik, J., and M. Morbidelli (2006a). Role of turbulent shear rate distribution in aggregation and breakage processes. *AIChE Journal,* 52: 158-173.

D. L. Marchisio, M. Soos, J. Sefcik, M. Morbidelli, A. A. Barresi, and G. Baldi (2006b). Effect of fluid dynamics on particle size distribution in particulate processes. *Chemical Engineering & Technology,* 29: 1-9.

R. McGraw (1997). Description of aerosol dynamics by the quadrature method of moments. *Aerosol Science Technology,* 27: 255-267.

V. Oles (1992). Shear-induced aggregation and breakup of polystyrene latex particles. *Journal of Colloid and Interface Science,* 154: 351-358.

S. B. Pope (2000). *Turbulent Flows.* Cambridge Universiy Press, Cambridge.

W. H. Press, S. A. Teukolsky, W. T. Vetterling, and B. P. Flannery (1992). *Numerical Recipes in Fortran 77: The Art of Scientific Computing.* Cambridge University Press, Cambridge, UK.

D. Ramkrishna (2000). *Population balances: theory and applications to particulate systems in engineering.* Academic Press. San Diego.

D. E. Rosner, and S. Yu (2001). MC simulation of aerosol aggregation and simultaneous spheroidization. *AIChE Journal,* 47: 545-561.

P. G. Saffman, and J. S. Turner (1956). On the collision drops in turbulent clouds. *Journal of Fluid Mechanics,* 1: 16-31.

J. Sanyal, D. L. Marchisio, R. O. Fox, and D. Dhanasekharan (2005). On the comparison between population balance models for CFD simulation of bubble columns. *Industrial & Engineering Chemistry Research,* 44: 5063-5072.

C. M. Sorensen (2001). Light scattering by fractal aggregates: A Review. *Aerosol Science and Technology,* 35: 648-687.

S. Valerio, M. Vanni, A. A. Barresi, and G. Baldi (1998). Engineering modelling of turbulent flows in chemical engineering applications. *Trends in Chemical Engineering,* 5: 1-44.

M. H. Waldner, J. Sefcik, M. Soos, and M. Morbidelli (2005). Initial growth kinetics of aggregates in turbulent coagulator. *Powder Technology,* 156: 226-237.

L. Wang, D. L. Marchisio, R. D. Vigil, and R. O. Fox (2005). CFD simulation of aggregation and breakage processes in laminar Taylor-Couette flow. *Journal of Colloid and Interface Science,* 282, 380-396.

A. Zucca (2006) *Modelling of turbulence-chemistry interaction and soot formation in turbulent flames.* PhD Dissertation. Politecnico di Torino, Torino, IT.

A. Zucca, D. L. Marchisio, A. A. Barresi, and R. O. Fox (2006). Implementation of the population balance equation in CFD codes for modelling soot formation in turbulent flames. *Chemical Engineering Science,* 61: 87 – 95.

A. Zucca, D. L. Marchisio, M. Vanni, and A. A. Barresi (2007). Validation of the bivariate DQMOM for nano-particle processes simulation. *AIChE Journal,* in press.

Eulerian Multi-Fluid Models for Polydisperse Evaporating Sprays

Marc Massot*

Laboratoire EM2C - CNRS UPR 288, Ecole Centrale Paris, France

Abstract In this contribution we propose a presentation of Eulerian multi-fluid models for polydisperse evaporating sprays. The purpose of such a model is to obtain a Eulerian-type description with three main criteria: to take into account accurately the polydispersity of the spray as well as size-conditioned dynamics and evaporation; to keep a rigorous link with the Williams spray equation at the kinetic, also called mesoscopic, level of description, where elementary phenomena such as coalescence can be described properly; to have an extension to take into account non-resolved but modeled fluctuating quantities in turbulent flows. We aim at presenting the fundamentals of the model, the associated precise set of related assumptions as well as its implication on the mathematical structure of solutions, robust numerical methods able to cope with the potential presence of singularities and finally a set of validations showing the efficiency and the limits of the model.

1 Introduction

In many industrial combustion applications such as Diesel engines, fuel is stocked in condensed form and burned as a disperse liquid phase carried by a gaseous flow. Two-phase effects as well as the polydisperse character of the droplet size distribution (since the droplet dynamics depend on their inertia and are conditioned by size) can significantly influence flame structure. Size distribution effects are also encountered in a crucial way in solid propellant rocket boosters, where the cloud of alumina particles experiences coalescence and become polydisperse in size, thus determining their global dynamical behavior (Hylkema, 1999; Hylkema and Villedieu, 1998). The coupling of dynamics, conditioned on particle size, with coalescence or aggregation as well as with evaporation can also be found in the study of fluidized beds (Tsuji et al., 1998) and planet formation in solar nebulae (Bracco et al., 1999; Chavanis, 2000). Consequently, it is important to have reliable models and numerical methods in order to be able to describe precisely the physics of two-phase flows where the disperse phase is constituted of a cloud of particles of various sizes that can evaporate, coalesce or aggregate, break-up and also have their own inertia and size-conditioned dynamics. Since our main area of interest is combustion, we will work with sprays throughout this paper, keeping in mind the broad application fields related to this study.

By spray, we denote a disperse liquid phase constituted of droplets carried by a gaseous phase. Even with this seemingly precise definition, two approaches corresponding to two levels of description can be distinguished. The first, associated with a full direct numerical simulation

*marc.massot@em2c.ecp.fr

(DNS) of the process, provides a model for the dynamics of the interface between the gas and liquid, as well as the exchanges of heat and mass between the two phases using various techniques such as the Volume of Fluids (VOF) or Level Set methods (see, for example, Aulisa et al., 2003; Herrmann, 2005; Josserand et al., 2005; Tanguy and Berlemont, 2005). This "microscopic" point of view is very rich in information on the detailed properties at the single-droplet level concerning, for example, the resulting drag exerted on a droplet depending on its surroundings flow or the details of one event of droplet break-up following the geometry of the interface between the phases. The second approach, based on a more global point of view (thus called "mesoscopic") describes the droplets as a cloud of point particles for which the exchanges of mass, momentum and heat are described using a statistical point of view, with eventual correlations, and the details of the interface behavior, angular momentum of droplets, detailed internal temperature distribution inside the droplet, etc., are not predicted. Instead, a finite set of global properties such as size of spherical droplets, velocity of the center of mass, and temperature are modeled. Because it is the only one for which numerical simulations at the scale of a combustion chamber or in a free jet can be conducted, this "mesoscopic" point of view will be adopted in the present work.

The principal physical processes that must be accounted for are (1) transport in real space, (2) droplet heating and evaporation, (3) acceleration of droplets due to drag, and (4) coalescence and break-up of droplets leading to polydispersity. Spray models (where a spray is understood as a disperse phase of liquid droplets, i.e. where the liquid volume fraction is much smaller than one) have a common basis at the mesoscopic level (also called "the kinetic level" by analogy to the kinetic theory of gases), under the form of a number density function (NDF) satisfying a Boltzmann type equation, the so-called Williams (1958, 1985) equation. This equation is also called a Population Balance Equation (PBE) in the chemical engineering community. The Williams equation implies several underlying assumptions, the major one being associated with the possibility of closing the equation in the "one-particle" probability density function (PDF) from the Liouville equation in the multi-particle joint PDF (Cercignani et al., 1994). The reader who is not familiar with the kinetic theory of gases approach will find a precise adapted version to the present subject in Fox (2007). The internal variables characterizing one droplet are the size, velocity, and temperature, so that the total phase space is usually high-dimensional. Such a transport equation describes the evolution of the NDF of the spray due to evaporation, to the drag force of the gaseous phase, to the heating of the droplets by the gas and finally to the droplet-droplet interactions, such as coalescence and break-up phenomena (Dukowicz, 1980; O'Rourke, 1981; Amsden et al., 1989; Raviart and Sainsaulieu, 1995; Domelevo and Sainsaulieu, 1997; Hylkema, 1999; Laurent and Massot, 2001). The spray transport equation is then coupled to the gas phase equations. The two-way coupling of the phases occurs first in the spray transport equations through the rate of evaporation, drag force and heating rate, which are functions of the gas-phase variables, and second through exchange terms in the gas-phase equations.

There are several strategies in order to solve the liquid phase and the major challenge in numerical simulations is to account for the strong coupling between all the involved processes. A first choice is to approximate the NDF by a sample of discrete numerical parcels of particles of various sizes through a Lagrangian–Monte-Carlo approach (Dukowicz, 1980; O'Rourke, 1981; Amsden et al., 1989; Hylkema, 1999; Rüger et al., 2000). It is called Direct Simulation Monte-Carlo method (DSMC) by Bird (1994) and is generally considered to be the most accurate for solving Williams equation; it is especially suited for DNS since it does not introduce any numerical diffusion, the particle trajectories being exactly resolved. This approach has been widely used and has been shown to be efficient in a number of cases. Its main drawback, that has shown recently to be a major one with the development of new combustion chambers leading to combustion instabilities (Lean Premixed Prevaporized combustor with spray injection), is the coupling of a Eulerian description for the gaseous phase to a Lagrangian description of the disperse phase; thus offering limited possibilities of vectorization/parallelization and use of implicit schemes. Besides, it brings another difficulty associated with the repartition of the evaporated mass at the droplet location onto the Eulerian grid for the gas description. Moreover for unsteady computations of polydisperse sprays, a large number of parcels in each cell of the computational domain is generally needed, thus yielding large memory requirement and CPU cost.

This drawback makes the use of an Eulerian formulation for the description of the disperse phase attractive, at least as a complementary tool for Lagrangian solvers, and leads to the use of moments methods since the high dimension of the phase space prevents the use of DNS on the NDF equation with deterministic numerical methods like finite volumes. The use of moments methods leads to the loss of some information but the cost of such methods is usually much lower than the Lagrangian ones for two reasons: the first one is related to the fact that the number of unknowns we solve for is limited; the second one is related to the high level of optimization one can reach when the two phases are both described by a Eulerian model. Let us consider first the droplet size and velocity as the internal variables (the temperature and other variables lead to the same kind of reasoning). Basically, the velocity variable, for particles having their own inertia, has to be treated by the use of global moment methods, i.e. moments on the whole real line. In this context, the key question is then how to treat the size variable as well as its correlation with the velocity variable.

There are basically two classes of methods. Let us present them in the case where the only the size variable is present. For submicron particles like aerosols (see Ramkrishna and Fredrickson (2000) and references therein) and soot (see Hall et al., 1997; Zucca et al., 2006), the numerical solution of the PBE for the particle size distribution describing the various phenomena (aggregation, breakage, growth, oxidation, etc.) can be done either considering a relevant set of global moments in size (quadrature method of moments (QMOM) by Wright et al. (2001); McGraw (1997) or direct quadrature method of moments (DQMOM) by Marchisio et al. (2003)) or by discretizing the size phase space into "sections" or "classes" (usually called a multi-fluid model), thus leading to a finite volume-like numerical method (Greenberg et al., 1993; McEnally et al., 1998). Let us underline that it is not possible in such problems to consider a presumed NDF approach since the structure of the particle size distribution is strongly influenced by both the drift velocity in the size variable due to growth and oxidation as well as by the aggregation-breakage phenomena.

In the context of inertial droplets, for which the inertia determines the dynamical behavior, the two methods can be used combined with moment methods in the velocity variable and making use of the large literature in the field of kinetic theory of gases (Domelevo and Sainsaulieu, 1997; Laurent and Massot, 2001; Reveillon et al., 2002) or in two-fluid type models used for separated two-phase flows (Drew and Passman, 1999; Chanteperdrix et al., 2002; Murrone and Guillard, 2005). Thus, either the set of global moments in both size and velocity can be considered (a degenerate form of this is to be found in the classical two-fluid models with the transport of surface area density and mass density of the liquid phase) as in Fox et al. (2006); however, the question of the choice of moments is still open. Or one can consider velocity moments conditioned by size and include this into the formalism of the multi-fluid model.

A first attempt at deriving a fully Eulerian model for sprays polydisperse in size in laminar configurations with droplets having their own inertia and which was able to capture the dynamics of droplets with various sizes, whatever the profile of the NDF, was developed by Greenberg et al. (1993) with the so-called sectional method. The idea was to consider the disperse phase as a set of continuous media: "fluids", each "fluid" corresponding to a statistical average between two fixed droplet sizes, the section. The sprays were then described by a set of conservation equations for each "fluid". Greenberg et al. (1993) noticed that such a model has also its origin at the kinetic level, trying to make the link with the Williams spray equation. However, they only provided a partial justification, the complete derivation for the conservation of mass and number of droplets, the momentum and energy equations being out of the scope of their paper since the set of rigorous closure assumptions of the conservation equations was not provided. Finally, their coalescence model did not take into account the relative velocity of colliding particles, thus making the model only suited for very small particles like soot such as in Hall et al. (1997).

The purpose of the present contribution is to present the class of methods called the Eulerian multi-fluid approach after Laurent and Massot (2001), which was inspired by the sectional approach and extended to coalescence in Massot and Villedieu (2001); Laurent et al. (2004a), to break-up in Dufour et al. (2003); Dufour (2005), to turbulent flows in Reveillon et al. (2002); Massot et al. (2004) and to multi-dimensional configurations in de Chaisemartin et al. (2006). The basis idea behind such a class of models is to build a model meeting the following criteria: (1) An Eulerian description of the disperse liquid phase. (2) Preservation of the link with the kinetic level of description. (3) Ability to capture the polydispersity in size of evaporating sprays. (4) Ability to capture the dynamics of the evaporating droplets conditioned by size. The key ingredient in order to obtain such a model is to preserve the droplet size as an internal variable, but to use a moment approach as far as velocity and eventually temperature are concerned. For the sake of legibility, the derivation of the various level of modeling will be conducted within the framework of simplified droplet models, since it is straightforward to extend it to richer droplet modeling for which we provide precise references. Even if it is important for combustion application (see Section 5), we will neglect the heating of the droplets and illustrate the proposed model in the case where the internal variables are only the size and the droplet velocity.

This first step is denoted the semi-kinetic model. The purpose of the first section is to present the set of assumptions as well as the resulting semi-kinetic set of conservation equations. It is equivalent to the original Williams equation under the assumption that there exists a single mean characteristic velocity for a given droplet size, at a given space location and for a given time, around which a generalized Maxwellian velocity distribution is to be found thus leading to a possible closure of the set of equations. In the laminar case or in the framework of DNS, which will be treated first, it corresponds to the so-called monokinetic velocity distribution, i.e. no dispersion around the mean, which is a generalization of the usual concept of Maxwellian distribution (a common assumption of zero temperature used in astrophysics by Zel'dovich (1970) for example). Even if the whole droplet size range is covered, the support of the droplet number distribution in velocity phase space is restricted to a one-dimensional sub-manifold of $\mathbb{R}^d$, parameterized by droplet size (Laurent and Massot, 2001), d being the spatial dimension. In the turbulent case for which some scales are not resolved but modeled, it corresponds to the usual Maxwellian distribution for which the dispersion becomes a variable of the problem: the droplet agitation "internal energy". We will successively present the semi-kinetic model for polydisperse dense evaporating sprays experiencing break-up and coalescence and then for turbulent sprays.

The set of assumptions we have identified allows one to relate the obtained model in the laminar case to the pressure-less gas dynamics studied elsewhere (Zel′dovich, 1970; Bouchut, 1994). Consequently, the mathematical structure of the resulting conservation equations leads to identified singularities. The second section is devoted to an illustration of typical singularity formation and we characterize the critical Stokes number associated with it. It will be very helpful in understanding the physics behind such mathematical behavior, as well as in designing optimal and robust numerical methods that can reproduce such singularities. In the turbulent case, we also provide some very interesting insight about singularity formation in a highly compressible medium that explains the difficulties which can be encountered in the numerical simulation performed in more complex configurations when not enough numerical diffusion is added. From the knowledge of the semi-kinetic model and of its structural behavior, we design the numerical tools in order to simulate such a model in a third section. This semi-kinetic model is then discretized in the size variable using a finite volume-like technique (Laurent, 2006). It yields the multi-fluid model for which we can preserve some level of information about the size distribution with a reasonable and adaptive computational cost. If a good level of precision is required about the size distribution, the computational cost is going to be lower but comparable to the Lagrangian one (however the optimization of the solver through the fully Eulerian description of the two phases leads to a substantial gain in CPU time). The method is still able to capture the behavior of the spray with a coarse discretization in the size phase space (Laurent et al., 2004b) and thus a low computational cost, a definite advantage in comparison with the Lagrangian methods. The question of the computational efficiency of such Eulerian approaches is a key question since these methods are intended to be used in more realistic unsteady configurations as an alternative to the too costly Lagrangian methods for polydisperse sprays. We have already studied this question in Laurent et al. (2004a); de Chaisemartin et al. (2006) where the Eulerian multi-fluid approach was shown to offer good precision at relatively low cost.

Finally the fourth section is devoted to the numerical illustration and validation of the multi-fluid model in several test-cases. We briefly recall the results from Massot et al. (1998); Laurent et al. (2004b) for heptane counter-flow spray diffusion flames for which quantitative comparisons of numerical simulation with complex chemistry and detailed transport against experimental measurements lead to the validation of the model. We then show some comparisons between a Lagrangian Monte-Carlo approach and the multi-fluid model for dense polydisperse evaporating sprays that experience coalescence in the framework of a conical nozzle. The two-dimensional Taylor-Green configuration is a very interesting benchmark which allows us to prove the computational efficiency and the good behavior of the multi-fluid model for multi-dimensional configurations (de Chaisemartin et al., 2006). We end-up with the configuration of spatially decaying turbulence, which is a "simple" turbulent flow that is non-homogeneous in one direction, and we perform a detailed comparison study with a Euler/Lagrange DNS. This leads us to conclude on the efficiency and capability of the proposed method and we indicate further research direction in connection with the other methods presented in the CISM Lecture, i.e. the DQMOM method.

2 The Starting Point: A semi-kinetic model conditioned by size

2.1 Modeling fundamentals in a simplified framework

For the purpose legibility, we will present the derivation of the Eulerian multi-fluid model for polydisperse evaporating sprays that experience coalescence and secondary break-up from a simplified Williams equation at the kinetic level of description.

Williams transport equation Let us define the number distribution function f^α of the spray, where $f^\alpha(t, \mathbf{x}, \phi, \mathbf{u})\, \mathrm{d}\mathbf{x}\, \mathrm{d}\phi\, \mathrm{d}\mathbf{u}$ denotes the average number of droplets (in a statistical sense), at time t, in a volume of size $\mathrm{d}\mathbf{x}$ around $\mathbf{x}$, with a velocity in a $\mathrm{d}\mathbf{u}$-neighborhood of $\mathbf{u}$ and with a size in a $\mathrm{d}\phi$-neighborhood of ϕ. The droplets are considered to be spherical and characterized by $\phi = ar^\beta$, where r is the radius of the droplet; ϕ can be the radius ($a = 1$ and $\beta = 1$; $\phi = r$), the surface, ($a = 4\pi$ and $\beta = 2$; $\phi = s$) or the volume ($a = 4\pi/3$ and $\beta = 3$; $\phi = v$). The associated evaporation rate will be denoted R_ϕ. We will work mainly with the volume and the surface in the following. Within this section, we will use the volume and f^3 will be noted f. For the sake of simplicity and for the purpose of this paper, we are going to consider that the evaporation process is described by a d^2 law without convective corrections, that the drag force is given by a Stokes law, and finally that the unstationary heating of the droplets does not need to be modeled so that the evaporation law coefficient does not depend on the heating status of the droplet. We refer to Abramzon and Sirignano (1989), Hylkema (1999) and Laurent and Massot (2001) for more detailed droplet models for which the derivation can be easily extended. The evolution of the spray is then described by the Williams transport equation:

$$\partial_t f + \mathbf{u} \cdot \partial_\mathbf{x} f + \partial_v (R_v f) + \partial_\mathbf{u} \cdot (\mathbf{F} f) = \Gamma_{\mathrm{co}} + \Gamma_{\mathrm{bu}}, \tag{2.1}$$

where R_v denotes the d^2-law evaporation rate, $\mathbf{F}$ the Stokes drag force due to the velocity difference with the gaseous phase; Γ_{co} is the collision operator leading to coalescence and Γ_{bu} the operator modeling the secondary break-up. These quantities have the following dependence $(t, \mathbf{x}, \mathbf{u}, v)$ (except for Γ_{co} and Γ_{bu}, which are integral operators depending on f); they depend on the local gas composition, velocity and temperature and this dependence is implicitly written in the $(t, \mathbf{x})$ dependence.

It has to be noticed that the refinement of the drag, evaporation and heating models, can not go beyond a given limit in the context of a kinetic description. The added mass effect in the aerodynamical forces can be described only through the addition of a new variable in the phase space, which is the time derivative of the velocity of the particle; in the applications we are considering, it is going to be negligible. Actually, the history terms such as the Basset forces or the inner temperature distribution of a droplet in the effective conductivity model by Abramzon and Sirignano (1989) can not be modeled in the context of a kinetic description of the spray, as already discussed in Laurent and Massot (2001). As a conclusion, the derivation presented in the following on a simplified model can be extended to more refined droplet models as long as they do not include history terms.

Modeling of coalescence The kinetic model for the collision operator leading to coalescence is taken from Villedieu and Hylkema (1997) and we neglect the influence of the impact parameter on the probability of rebound of two collisional partners:

[Co1] We only take into account binary collisions (small volume fraction of the liquid phase).

[Co2] The mean collision time is very small compared to the intercollision time.

[Co3] Every collision leads to coalescence of the partners.

[Co4] During coalescence, mass and momentum are conserved.

Thus $\Gamma_{co} = Q^-_{coll} + Q^+_{coll}$, where Q^-_{coll} and Q^+_{coll} respectively correspond to the quadratic integral operators associated with creation and destruction of droplets due to coalescence:

$$Q^-_{coll} = -\int_{v^*}\int_{\mathbf{u}^*} f(t,\mathbf{x},v,\mathbf{u})f(t,\mathbf{x},v^*,\mathbf{u}^*)B(|\mathbf{u}-\mathbf{u}^*|,v,v^*)\,dv^*d\mathbf{u}^*, \tag{2.2}$$

$$Q^+_{coll} = \frac{1}{2}\int_{v^*\in[0,v]}\int_{\mathbf{u}^*} f(t,\mathbf{x},v^\diamond(v,v^*),\mathbf{u}^\diamond(v,v^*,\mathbf{u}))f(t,\mathbf{x},v^*,\mathbf{u}^*)B(|\mathbf{u}^\diamond-\mathbf{u}^*|,v^\diamond,v^*)J\,dv^*d\mathbf{u}^*, \tag{2.3}$$

where $v^\diamond$ and $\mathbf{u}^\diamond$ are the pre-collisional parameters, $v^\diamond(v,v^*) = v - v^*$ and J is the Jacobian of the transform $(v,\mathbf{u}) \to (v^\diamond,\mathbf{u}^\diamond)$, at fixed $(v^*,\mathbf{u}^*)$: $J = (v/v^\diamond)^d$, with d the dimension of the velocity phase space (Villedieu and Hylkema, 1997) and where $B(|\mathbf{u}-\mathbf{u}^*|,v,v^*) = \beta(v,v^*)|\mathbf{u}-\mathbf{u}^*|$ with

$$\beta(v,v^*) = \pi\left(r(v)+r(v^*)\right)^2, \qquad r(v) = \left(\frac{3v}{4\pi}\right)^{1/3} \qquad \mathbf{u}^\diamond = \frac{v\mathbf{u}-v^*\mathbf{u}^*}{v-v^*}, \tag{2.4}$$

where $v^\diamond$ and $\mathbf{u}^\diamond$ are the pre-collisional parameters, $v^\diamond(v,v^*) = v - v^*$ and J is the Jacobian of the transform $(v,\mathbf{u}) \to (v^\diamond,\mathbf{u}^\diamond)$, at fixed $(v^*,\mathbf{u}^*)$: $J = (v/v^\diamond)^d$ (Hylkema, 1999; Villedieu and Hylkema, 1997).

Remark 2.1. The assumption that all collisions lead to the coalescence of the partners is not realistic, especially if the colliding droplets have comparable sizes (Brazier-Smith et al., 1972; Ashgriz and Poo, 1990). In such situations the probability E_{coal}, that coalescence really occurs from the collision of two droplets has to be taken into account; the expression for B then becomes $B(|\mathbf{u}-\mathbf{u}^*|,v,v^*) = E_{coal}(|\mathbf{u}-\mathbf{u}^*|,v,v^*)\beta(v,v^*)|\mathbf{u}-\mathbf{u}^*|$. It is possible to treat the coalescence efficiency factor within the proposed model and the reader is referred to Laurent et al. (2004a) for the details. Nevertheless, in the following, for the sake of simplicity, we will assume $E_{coal}(|\mathbf{u}-\mathbf{u}^*|,v,v^*) = 1$, which is equivalent to assumption [Co3].

Modeling of secondary break-up Similarly the modeling of secondary break-up relies on a statistical approach and can be taken from Hylkema (1999); Dufour (2005):

$$\Gamma_{bu}(f)(v,\mathbf{u}) = -\nu^{bup} f + \int_{v^*>v}\int_{\mathbf{u}} \nu^{bup}(v^*,\mathbf{u}^*)h(v,v^*,\mathbf{u},\mathbf{u}^*)f^*dv^*d\mathbf{u}^*. \tag{2.5}$$

ν^{bup} is the mean break-up frequency of the droplets, which is defined through experimental correlations on the mean break-up time of a droplet from Achim (1999); Nigmatulin (1991). Following Hsiang and Faeth (1993) (a thorough discussion of this issue can be found in Dufour (2005)), for Weber numbers beyond the critical one, we take

$$\nu^{bup}(v,\mathbf{u}) = \frac{\|\mathbf{U}-\mathbf{u}\|}{10r(v)}\left(\frac{\rho_d}{\rho_g}\right)^{1/2}, \tag{2.6}$$

where $\mathbf{U}$ is the gas-phase velocity. However, the mean break-up frequency can be also a function of both the Weber and the Ohnsorge numbers as in Nigmatulin (1991). $h(.,.,v^*,\mathbf{u}^*)$ denotes the daughter distribution in terms of size and velocity $(v^*, \mathbf{u}^*)$. This distribution must satisfy the mass-conservation constraint:

$$\int_u \int_{v=0}^{v^*} v\, h(v, \mathbf{u}, v^*, \mathbf{u}^*)\, dv\, d\mathbf{u} = v^*, \tag{2.7}$$

and must be further defined. We take $h(v, \mathbf{u}, v^*, \mathbf{u}^*) = g_u(v, \mathbf{u}, v^*, \mathbf{u}^*) g_v(v, v^*, \mathbf{u}^*)$, where $g_u(v, v^*, \mathbf{u}, \mathbf{u}^*) = \delta(\mathbf{u} - \mathbf{u}^{\text{bup}}(v, v^*, \mathbf{u}^*))$ with

$$\mathbf{u}^{\text{bup}}(v, v^*, \mathbf{u}^*) = \mathbf{U} + \frac{\mathbf{u}^* - \mathbf{U}}{1 + C_{\text{We}} \left(\frac{\rho_g}{\rho_d}\right)^{1/3} \left(\frac{r(v^*)}{r(v)}\right)^{2/3}} \tag{2.8}$$

from Hsiang and Faeth (1993) and $g_v(v, v^*, \mathbf{u}^*)$ is the droplet size distribution after break-up.

There is not a single and standard way of modeling the statistics of the daughter droplets. O'Rourke and Amsden (1987) for example take an exponentially decreasing function of droplet radius with the use of a correlation for the Sauter mean radius r_{smd} in terms of Weber number: the Wert (1995) correlation. Since most of the time, for evaporating sprays, the number density function as a function of droplet radius linearly approaches zero at zero droplet radius, Dufour (2005) rather introduced, using

$$r_{\text{smd}} = \left(\frac{3}{4\pi}\right)^{1/3} \frac{\int_0^{v^*} v\, g_v\, dv}{\int_0^{v^*} v^{2/3} g_v\, dv}, \qquad (v^*)^3 = \int_0^{v^*} v^3 g_v\, dv, \tag{2.9}$$

a profile of the droplet size distribution, which he assumed exponentially decreasing as a function of droplet surface:

$$g_s(s, v^*, \mathbf{u}^*) = \frac{\alpha}{8\pi} \exp\left(-\frac{\gamma}{4\pi} s\right), \quad g_s(s, v^*, \mathbf{u}^*)\, ds = g_v(v, v^*, \mathbf{u}^*)\, dv, \tag{2.10}$$

where the two functions $\alpha(v^*, \mathbf{u}^*)$ and $\gamma(v^*, \mathbf{u}^*)$ are determined from the above equalities. The reader will find the details of the calculation of these two coefficients in Dufour (2005).

Eulerian semi-kinetic model In this section, we will recall the first step of the formalism and the associated assumptions introduced in Laurent and Massot (2001) in order to derive the Eulerian multi-fluid method and explain how this formalism can be extended in order to treat the coalescence phenomenon between droplets having their own inertia governed by their size. It is worth noticing that we do take into account the mean velocity difference of the droplets in the coalescence process as opposed to the model proposed in Greenberg et al. (1993), which is mainly suited for very small particles such as soot. The key idea is to reduce the size of the phase space and to consider only the moments of order zero and one in the velocity variable at a given time, a given position and for a given droplet size. The obtained conservation equations, called the semi-kinetic model for the two fields $n(t, \mathbf{x}, v) = \int f\, d\mathbf{u}$ and $\overline{\mathbf{u}}(t, \mathbf{x}, v) = \int f \mathbf{u}\, d\mathbf{u} / n(t, \mathbf{x}, v)$, are only in a closed form under a precise assumption on the support of the original NDF in the whole phase space: the velocity distribution at a given time, given location and for a given droplet size is a Dirac delta function (Laurent and Massot, 2001).

However, this assumption is not directly compatible with the coalescence phenomenon, since there is no reason for a droplet created by the coalescence of two droplets of various sizes, which is deduced from momentum conservation, to exactly match the velocity corresponding to its new size. We first relax the assumption of zero dispersion and assume Gaussian velocity dispersion and handle the whole positive real line for the possible sizes so that all the collisions can be described by the model: $(v, \mathbf{u}) \in (0, +\infty) \times \mathbb{R}^d$. The semi-kinetic system of conservation laws is then obtained by taking the limit of zero dispersion in the source terms coming from coalescence, uniformly in $(t, \mathbf{x}, v)$. Summarizing, we obtain the semi-kinetic model in the limit of zero dispersion of a more general problem where dispersion is allowed and "project" the original NDF at the kinetic level onto a one-dimensional sub-manifold of velocity phase space parameterized by droplet size (Laurent and Massot, 2001).

Proposition 2.2. *Let us make the following assumptions on the spray distribution function:*

[H1] *For a given droplet size, at a given point $(t, \mathbf{x})$, there is only one characteristic average velocity $\overline{\mathbf{u}}(t, \mathbf{x}, v)$.*

[H2] *The velocity dispersion around the average velocity $\overline{\mathbf{u}}(t, \mathbf{x}, v)$ is zero in each direction, whatever the point $(t, \mathbf{x}, v)$.*

[H3] *The droplet number density $n(t, \mathbf{x}, v)$ is exponentially decreasing at infinity as a function of v uniformly in $(t, \mathbf{x})$.*

Assumptions [H1] *and* [H2] *define the structure of f: $f(t, \mathbf{x}, v, \mathbf{u}) = n(t, \mathbf{x}, v)\delta(\mathbf{u} - \overline{\mathbf{u}}(t, \mathbf{x}, v))$ and the semi-kinetic model is given by two partial differential equations in the variables $n(t, \mathbf{x}, v)$ and $\overline{\mathbf{u}}(t, \mathbf{x}, v)$, which express respectively, the conservation of the number density of droplets and their momentum, at a given location $\mathbf{x}$ and for a given size v:*

$$\partial_t n + \partial_{\mathbf{x}} \cdot (n\overline{\mathbf{u}}) + \partial_v (n\overline{R}_v) - \mathcal{G}_{\mathrm{co}}^n + \mathcal{G}_{\mathrm{bu}}^n \tag{2.11}$$

$$\partial_t (n\overline{\mathbf{u}}) + \partial_{\mathbf{x}} \cdot (n\overline{\mathbf{u}} \otimes \overline{\mathbf{u}}) + \partial_v (n\overline{R}_v\overline{\mathbf{u}}) - n\overline{\mathbf{F}} = \mathcal{G}_{\mathrm{co}}^{nu} + \mathcal{G}_{\mathrm{bu}}^{nu}. \tag{2.12}$$

where $\overline{\mathbf{F}}(t, \mathbf{x}, v)$ and $\overline{R}_v(t, \mathbf{x}, v)$ are the Stokes's drag force and the evaporation rate, respectively, taken at $\mathbf{u} = \overline{\mathbf{u}}$, and with

$$\mathcal{G}_{\mathrm{co}}^n = -n(v) \int_{v^* \in [0,+\infty)} n(v^*)\beta(v, v^*)I_n^- \mathrm{d}v^* + \frac{1}{2} \int_{v^* \in [0,v]} n(v^\diamond(v, v^*))n(v^*)\beta(v^\diamond(v, v^*), v^*)I_n^+ \mathrm{d}v^*, \tag{2.13}$$

$$\mathcal{G}_{\mathrm{co}}^{nu} = -n(v) \int_{v^* \in [0,+\infty)} n(v^*)\beta(v, v^*)I_u^- \mathrm{d}v^* + \frac{1}{2} \int_{v^* \in [0,v]} n(v^\diamond(v, v^*))n(v)\beta(v^\diamond(v, v^*), v^*)I_u^+ \mathrm{d}v^*, \tag{2.14}$$

where the partial collisional integrals I_n^-, I_n^+, I_u^- and I_u^+ are functions of $(t, \mathbf{x}, v, v^*)$ and take the following expressions:

$$I_n^- = |\overline{\mathbf{u}}(v) - \overline{\mathbf{u}}(v^*)|, \quad I_u^- = \overline{\mathbf{u}}(v)|\overline{\mathbf{u}}(v) - \overline{\mathbf{u}}(v^*)|, \tag{2.15}$$

$$I_n^+ = |\overline{\mathbf{u}}(v^*) - \overline{\mathbf{u}}(v - v^*)|, \quad v\, I_u^+ = \big((v - v^*)\overline{\mathbf{u}}(v - v^*) + v^*\overline{\mathbf{u}}(v^*)\big)\, |\overline{\mathbf{u}}(v^*) - \overline{\mathbf{u}}(v - v^*)|. \tag{2.16}$$

The source terms associated with secondary break-up read, noting $\overline{\nu}^{\mathrm{bup}}(v) = \nu^{\mathrm{bup}}(v, \overline{\mathbf{u}}(v))$:

$$\mathcal{G}_{\mathrm{bu}}^n = -\overline{\nu}^{\mathrm{bup}}(v)n(v) + \int_{v^* > v} \overline{\nu}^{\mathrm{bup}}(v^*)g_v(v, v^*, \overline{\mathbf{u}}(v^*))n(v^*)\, \mathrm{d}v^*, \tag{2.17}$$

$$\mathcal{G}^{nu}_{\text{bu}} = -\overline{\nu}^{\text{bup}}(v)n(v)\overline{\mathbf{u}}(v)) + \int_{v^*>v} \overline{\nu}^{\text{bup}}(v^*)g_v(v, v^*, \overline{\mathbf{u}}(v^*))n(v^*)\overline{\mathbf{u}}(v^*)\,dv^*. \tag{2.18}$$

We refer the reader to the thesis of Dufour (2005) for the extension of the presented coalescence and break-up operators to sprays with the temperature as another internal variable which describes the unsteady heating of the droplets by the gaseous carrier phase. The proof of the preceding proposition can be found in Laurent et al. (2004a) and Dufour (2005). The validity of the assumption on the velocity distribution conditioned by size will be discussed in a more general framework in Section 3, whereas its applicability for polydisperse sprays with evaporation and coalescence to a nozzle test case will be treated in Section 5.

Remark 2.3. It has to be noted that the coalescence source terms obtained in the equations does not conserve the global kinetic energy of the droplets. However, for the velocity and the temperature range we are interested in, the change in the temperature of the droplets due to the dissipation of the kinetic energy lost in the coalescence process is totally negligible. Besides, due to the statistical treatment of break-up, mass is conserved by the corresponding source terms, however, momentum is not.

2.2 Extension to turbulent flows

We have considered so far two-phase flows where, the model being given through the Williams equation for the spray, the disperse liquid phase is coupled to the Navier-Stokes equations for the gaseous carrier flow and we intend to resolve the whole range of temporal and spatial scales of the resulting model (either with a Lagrangian or an Eulerian approach for the liquid and with an Eulerian method for the gas). However, in a number of configurations, the gaseous flow is turbulent and we are only interested in the behavior of the system "in the mean". Some average quantities will be resolved and fluctuations or small scales are modeled. The modeling of turbulent flows for the gas has received considerable attention and we will focus on the modeling of the liquid phase. Our strategy is to start at the kinetic level with a realization of the Williams equation and to derive a kinetic equation in the average where the effect of the fluctuating scales of the turbulent gas phase are modeled. The idea is to obtain a closed equation at the mesoscopic level first, since it is easier, and then, from there, at the fluid level, but not directly at the fluid level. This approach as well as its main interests are described originally in Reeks (1991) and used in Reveillon et al. (2002); Massot et al. (2004).

The derivation of a transport equation for the NDF of droplets in turbulent flows has been the subject of interest of many researchers in the past two decades. Some excellent reviews have appeared recently on the subject, of which the most relevant are Mashayek and Pandya (2003); Minier and Peirano (2001); Kaufmann (2004). A detailed description of the various works is redundant and not in the scope of this work and only some relevant results will be repeated for completeness. For the purpose of the presentation, we restrict ourselves to a d^2 evaporation law, to a Stokes drag and neglect the heat exchange between the droplets and the gas. We introduce a characteristic gas velocity A, a characteristic length L and the corresponding time $\tau_{\text{gas}} = L/A$; s^0 is a characteristic droplet surface, $\tau_{\text{ev}} = s^0/R_s$, a characteristic evaporation time and we use the two non-dimensional numbers $\text{St} = \tau_{\text{p}}/\tau_{\text{gas}}$ the ratio of the typical droplet dynamical response time τ_{p}, based on s^0, over the gas characteristic time, called the Stokes number, as well as $\text{K} = \tau_{\text{gas}}/\tau_{\text{ev}}$, the ratio of the gas time over the evaporation time based on s^0.

The non-dimensional evolution equations for an individual droplet then read

$$\mathrm{d}_t \mathbf{x}_\mathrm{p} = \mathbf{u}_\mathrm{p}, \qquad \mathrm{d}_t \mathbf{u}_\mathrm{p} = \frac{\mathbf{U}(t, \mathbf{x}_\mathrm{p}) - \mathbf{u}}{\mathrm{St}\; s_\mathrm{p}}, \qquad \mathrm{d}_t s_\mathrm{p} = -\mathrm{K}, \tag{2.19}$$

where $\mathbf{x}_\mathrm{p}$ is the position of the droplet at time t, $\mathbf{v}_\mathrm{p}$ its velocity and s_p its surface. The system of equations (2.19) can also be described through what is called the fine-grained phase-space distribution function, defined by

$$W(t, \mathbf{x}, \mathbf{u}, s) = \delta(\mathbf{x} - \mathbf{u}_\mathrm{p}(t))\delta(\mathbf{u} - \mathbf{u}_\mathrm{p}(t))\delta(s - s_\mathrm{p}(t)), \tag{2.20}$$

which is therefore an Eulerian quantity, taking different values from one realization to another. The single particle PDF, P, is then defined as the ensemble average of the fine-grained (see Pope, 2000) distribution function $P(t, \mathbf{x}, \mathbf{v}, s) = \langle W \rangle$, which satisfies, as well as $f(t, \mathbf{x}, \mathbf{u}, s)$, the corresponding NDF (the summation over all single particle PDF's), the kinetic Williams equation in non-dimensional form:

$$\partial_t f + \mathbf{u} \cdot \partial_\mathbf{x} f + \partial_\mathbf{u} \cdot \left(\frac{\mathbf{U} - \mathbf{u}}{\mathrm{St} s} f \right) - \mathrm{K}\, \partial_s f = 0. \tag{2.21}$$

The initial conditions on the NDF are deduced from the distribution function of particles at time $t = 0$. If no other randomness is introduced in the system, that is, for example, if $\mathbf{U}(t, \mathbf{x}_\mathrm{p})$ is deterministically determined, the ensemble average mentioned above is only the expectation over the probability space associated with the initial distribution in the phase or sample space. This has two implications. First it allows a particle-based definition of the NDF and provide a derivation of the Williams equation when the forces acting on the particle are localized in phase space. It will be useful in order to make the link between Eulerian and stochastic Lagrangian descriptions of the liquid phase as in de Chaisemartin et al. (2006). Second, we can then envision the difficulties associated with non predictable gaseous flows for which two random variables are introduced: one associated with the initial condition and one associated with the fluctuations of the gaseous field around its mean or with the small scales.

NDF transport equation "in the mean" In a series of papers, Reeks (1991, 1992, 1993) obtained rigorously the NDF equation for inhomogeneous flows by using Kraichnan's Lagrangian history direct interaction (LHDI) approximation. The same result was later obtained by Hyland et al. (1999) through the application of the Furutsu-Novikov-Donsker formula. The first step is to split the carrier velocity into a mean and fluctuation part, according to $\mathbf{U}(t, \mathbf{x}) = \mathbf{U}'(t, \mathbf{x}) + \overline{\mathbf{U}}(t, \mathbf{x})$, where $\overline{\mathbf{U}} = \langle \mathbf{U} \rangle$ is an abbreviation of the ensemble averaged gas velocity and $\langle \cdot \rangle$ denotes the ensemble averaging operator with respect to the randomness introduced through the fluctuating quantity. Similarly, the drag force introduced in the momentum equation (2.19) for an individual droplet may be decomposed in a mean and fluctuating part.

The derivation of a kinetic equation starts from the fine-grained phase-space distribution function (2.20). The one-particle PDF is then defined as the ensemble average of the fine-grained distribution function: $P(t, \mathbf{x}, \mathbf{u}, s) = \langle W \rangle$. The kinetic equation "in the mean" for the associated NDF $\bar{f}$ then reads

$$\partial_t \bar{f} + \mathbf{u} \cdot \partial_\mathbf{x}(\bar{f}) + \partial_\mathbf{u} \cdot \left(\frac{\overline{\mathbf{U}} - \mathbf{u}}{\mathrm{St} s} \bar{f} \right) - \mathrm{K}\, \partial_s \bar{f} = -\partial_\mathbf{u} \cdot \left(\frac{\langle \mathbf{U}' f \rangle}{\mathrm{St} s} \right), \tag{2.22}$$

where we have isolated the effect of the fluctuating part of the velocity. It has to be underlined that the phase-space diffusion current $\langle \mathbf{U}' f \rangle/(\mathrm{St}\, s)$ in not closed since we introduce some randomness in the fluctuating quantity $\mathbf{U}'$, which makes f a stochastic variable. It corresponds to the mean effect of the gaseous flow fluctuations along the trajectory of the droplets and there is no natural closure of this non-linear product. The closure proposed by Reeks (1992) reads

$$\frac{\langle \mathbf{U}' f \rangle}{\mathrm{St} s} = -\partial_{\mathbf{x}} \cdot (\lambda f) - \partial_{\mathbf{u}} \cdot (\mu f) - \gamma f \tag{2.23}$$

where $\lambda(t, \mathbf{x}, \mathbf{u}, s)$ and $\mu(t, \mathbf{x}, \mathbf{u}, s)$ are the diffusivity tensors and γ, the velocity drift vector. In homogeneous mean flows, simplified expressions for these coefficients may be derived (indeed, when $\partial_{\mathbf{x}} \overline{\mathbf{U}} = 0$, the generalized response function (see Reeks, 1992) $G_{kj}(t'|t)$ can explicitly be calculated as $G_{kj}(t'|t) = [1 + (t - t')/(\mathrm{St} s \alpha)]^{-\alpha} \delta_{kj}$ where $\alpha = \tau_{\mathrm{ev}}/\tau_{\mathrm{p}} = 1/(\mathrm{St K})$ and $1 + (t - t')/(\mathrm{St} s \alpha) = 1 + \mathrm{K}\ (t - t')/s$). Their expression can be shown to lead to

$$\mu_{ij} = \frac{1}{(\mathrm{St} s)^2} \int_0^t \langle U_i'(t, \mathbf{x}) U_j'(t, \mathbf{x}, \mathbf{u}, t|t') \rangle \left(1 + \frac{t - t'}{\mathrm{St} s \alpha}\right)^{-\alpha} \mathrm{d}t', \tag{2.24}$$

$$\lambda_{ij} = \frac{1}{\mathrm{St} s} \int_0^t \langle U_i'(t, \mathbf{x}) U_j'(t, \mathbf{x}, \mathbf{u}, s|t') \rangle \frac{\alpha}{\alpha - 1} \left[1 - \left(1 + \frac{t - t'}{\mathrm{St}\, s\, \alpha}\right)^{-\alpha + 1}\right] \mathrm{d}t', \tag{2.25}$$

where $\mathbf{U}'(t, \mathbf{x}, \mathbf{u}, s|t')$ represents the value of $\mathbf{U}'$ measured at time t' along the particle trajectory that passes through $\mathbf{x}$, $\mathbf{u}$ and s at time t. A similar expression can be derived for γ_i following Reeks (1992). In order to make a clear link to these studies, let us note that in the limit of infinitely slowly evaporating droplets, i.e. $\mathrm{K} \to 0$, $\alpha \to +\infty$ and $[1 + (t - t')/(\mathrm{St} s \alpha)]^{-\alpha} \to \exp(-(t - t')/(\mathrm{St} s))$, one recovers the usual expressions associated with solid particles for $s = 1$ in the non-dimensional setting. A simple modeling approach to these expressions can be obtained by using the extension of the usual type of correlation defined by two parameters: k the turbulent kinetic energy of the gas and T_{L} is the gas Lagrangian time scale. We refer to Minier and Peirano (2001) for a detailed study in the non-evaporating framework, which can be extended easily to the present case.

Remark 2.4. Other models can be considered for the underlying kinetic level of description. In particular, the velocity of the gas seen by the droplets along their trajectories can be considered to be a stochastic variable itself and Langevin type of stochastic ODEs can be obtained in this framework as well as Fokker-Planck-like diffusion equations. We refer to Minier and Peirano (2001) for more information about this approach and the link with our point of view.

Eulerian semi-kinetic model Starting from the modeling at the kinetic level, we introduce, in the same way as before, an equilibrium assumption on the velocity distribution conditioned by size, which will allow us to derive a closed set of macroscopic conservation equations on the velocity moments of order 0, 1 and 2, the droplet size being kept as a phase-space variable. We assume that the velocity distribution function conditioned by droplet size is an isotropic Maxwellian distribution. Such an assumption corresponds to the previous "monokinetic" assumption when all the scales are resolved.

Transport equations for the velocity moments of the NDF are obtained by multiplying the ensemble averaged kinetic equation and the model for the source term with 1, $\mathbf{u}$ and $\mathbf{u}^2/2$ and integrating over the velocity space at fixed size. The zeroth-order moment of the NDF is thus defined by $\overline{n} = \int f \, d\mathbf{u}$ and for any other moment, a density-weighted quantity $\overline{Q}$ can be defined as: $\overline{Q} = \int Q f \, d\mathbf{u}/\overline{n}$. Besides we define $Q' = Q - \overline{Q}$. It is convenient at this point to introduce the droplet pressure defined by

$$p_{\mathrm{d}} = \frac{\overline{n}}{d}\left(\overline{u_i' u_i'} + \overline{\lambda}_{ii}\right) \tag{2.26}$$

where d is the spatial dimension of the problem and where we use the Einstein summation rule. Furthermore, we define the droplet total energy by $E_{\mathrm{d}} = \overline{n}\,\overline{u}_i\overline{u}_i/2$. The system of moment equations can now be written in conservative form:

$$\begin{aligned}
\mathcal{D}_t\overline{n} + \partial_{\mathbf{x}} \cdot (\overline{n}\,\overline{\mathbf{u}}) &= 0, \\
\mathcal{D}_t\,(\overline{n}\,\overline{\mathbf{u}}) + \partial_{\mathbf{x}} \cdot (\overline{n}\,\overline{\mathbf{u}} \otimes \overline{\mathbf{u}} + p_{\mathrm{d}}) &= \overline{n}\frac{\overline{\mathbf{U}} - \overline{\mathbf{u}}}{\mathrm{St}s} + \overline{n}\,\overline{\boldsymbol{\gamma}}, \\
\mathcal{D}_t E_{\mathrm{d}} + \partial_{\mathbf{x}} \cdot [\overline{\mathbf{u}}(E_{\mathrm{d}} + p_{\mathrm{d}})] &= \frac{\overline{n}\,\overline{\mathbf{u}} \cdot \overline{\mathbf{U}} - 2E_{\mathrm{d}}}{\mathrm{St}s} + \overline{n}\,\mathrm{Tr}(\overline{\mu}) + \overline{n}\,\overline{\mathbf{u}} \cdot \overline{\boldsymbol{\gamma}},
\end{aligned} \tag{2.27}$$

where $\mathcal{D}_t Q = \partial_t Q - \mathrm{K}\,\partial_s Q$ denotes a special type of material derivative associated with the size phase space and where $\mathrm{Tr}(\overline{\mu}) = \overline{\mu}_{ii}$. Furthermore, $\sigma_{ij} = \overline{n}(\overline{u_i' u_j'} + \overline{\lambda}_{ij}) - p_{\mathrm{d}}\delta_{ij}$ are the anisotropic components of the droplet turbulent stress tensor (they are assumed to be zero through the semi-kinetic closure assumption), and $\overline{\gamma}$ the velocity drift. The correlations $\overline{u_i' \lambda_{ij}'}$ and $\overline{u_i' \gamma_i'}$ have been neglected in the equations.

Remark 2.5. The present study could be extended to the case of a non-isotropic velocity distribution but at the cost of very heavy notations. Since we will use it in an isotropic framework in Sections 3 and 5, we have chosen to restrict ourselves to this simplified configuration.

3 Mathematical Issues – Singularities

In this section, we illustrate some of the mathematical difficulties associated with the semi-kinetic model and the related assumption for both laminar and turbulent modeling: an equilibrium velocity distribution around the mean velocity conditioned by droplet size. By illustration, we aim at giving a few relevant examples that provide the right insight about the potential singularities as well as their physical interpretation. The link with the related mathematical theories will be given through references. The conclusions to be drawn from this section are also relevant to the global multi-fluid model as well as to the DQMOM approach, since the transport term in the final set of conservation equations are the same pressure-less gas dynamics, as explained in Fox et al. (2006), where both approaches are compared in detail.

3.1 Laminar case – pressure-less gas dynamics

Let us underline that the semi-kinetic model contains a transport term similar to the pressure-less gas system studied for example in Bouchut (1994). Indeed, it looks like the Euler gas system for the conservation of mass and momentum, but without the pressure term in this momentum equation. This system can be found for example in astrophysics, when describing the formation of large-scale structures in the universe, in the modeling of sticky particles (Zel′dovich, 1970). It has the peculiarity to be weakly hyperbolic and can generate " δ-shocks" (that is a discontinuity in velocity that leads to Dirac concentration in density) or create vacuum zones. Note that it is important to be able to cope efficiently with vacuum zones since they represent areas of the flow where no droplets are to be found and are commonly encountered in most applications. In fact the equation on the velocity itself decouples from the conservation of mass and takes the form of the Burger's system of equations. A shock may then arise, leading to the concentration of density at its interface.

In the following we will consider the non-dimensional variables introduced previously, where the reference time is τ_{gas}, the one of the gas, and where the non-dimensional spray equation on $f(t, x, v)$ satisfies equation (2.21) with $\mathrm{K} = 0$ since we consider a monodisperse non-evaporating spray for the sake of simplicity. The droplet size only comes into the picture through the Stokes number that is proportional to the droplet surface area s^0. We consider also the initial distribution f^0 to be monokinetic $f^0 = n^0(\mathbf{x})\delta(\mathbf{u} - \overline{\mathbf{u}}^0(\mathbf{x}))$ in order to be compatible with the semi-kinetic approach. As long as the velocity field $\overline{\mathbf{u}}$ remains a regular field, there is an equivalence between the original equation (2.21) and the semi-kinetic formulation:

$$\begin{aligned} &\partial_t n + \partial_{\mathbf{x}} \cdot (n\overline{\mathbf{u}}) = 0, \\ &\partial_t (n\overline{\mathbf{u}}) + \partial_{\mathbf{x}} \cdot \left(\frac{(n\overline{\mathbf{u}})^2}{n} \right) = n \frac{\mathbf{U} - \overline{\mathbf{u}}}{\mathrm{St}}, \end{aligned} \tag{3.1}$$

for which the smooth velocity field satisfies the Burgers equation, independently of the density:

$$\partial_t \overline{\mathbf{u}} + \overline{\mathbf{u}} \cdot \partial_{\mathbf{x}} \overline{\mathbf{u}} = \frac{\mathbf{U} - \overline{\mathbf{u}}}{\mathrm{St}}. \tag{3.2}$$

1-D model problem and critical Stokes number The purpose of the present section is to identify the critical point for the appearance of "δ-shocks" (Bouchut, 1994; Bouchut et al., 2003), that is the eventual concentration up to infinity of the density field related to the creation of a discontinuous velocity field. Such events correspond to the crossing of characteristic curves in physical space (LeVeque, 2002) and to the limit of the monokinetic character of the NDF at the kinetic level, i.e. the velocity distribution at a given location becomes multi-valued. These characteristic curves are defined, for both the kinetic equation (2.21) and the system of conservation laws (3.1), by a set of ordinary differential equations (ODE) and initial conditions:

$$\begin{aligned} &\mathrm{d}_t X_{\mathrm{p}} = V_{\mathrm{p}}, & & X_{\mathrm{p}}(0) = X_{\mathrm{p}}^0, \\ &\mathrm{d}_t V_{\mathrm{p}} = \frac{U(t, X_{\mathrm{p}}) - V_{\mathrm{p}}}{\mathrm{St}}, & \text{with} \quad & V_{\mathrm{p}}(0) = V_{\mathrm{p}}^0 = u^0(X_{\mathrm{p}}^0), \end{aligned} \tag{3.3}$$

because the initial distribution is monokinetic. It should be noticed that the non-linear coupling between the two fields is contained in the fact that the gas velocity field U is only sampled by the droplet trajectory at X_{p}. Thus the characteristic curves are the integral curves of the vector field defined by (3.3), parameterized by the initial spatial coordinate so that we will adopt the notation $(X_{\mathrm{p}}, V_{\mathrm{p}})^t(t, X_{\mathrm{p}}^0)$. Under some standard conditions on the regularity of the field U, the characteristic curves exist and are well-defined for all time and spatial initial conditions. However, as soon as some characteristic curves cross in the only spatial projection of the characteristic diagram (x, t), the distribution ceases to be monokinetic and the equivalence between the macroscopic and kinetic descriptions is not valid any more. In fact the characteristics never cross in the

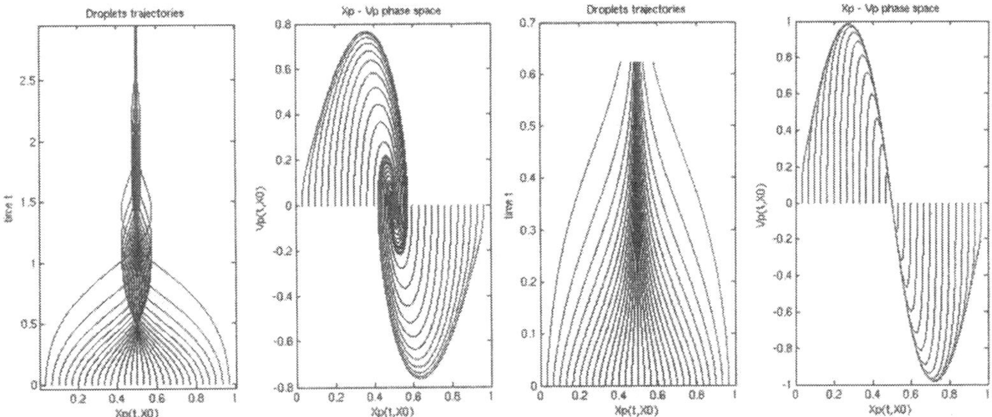

Figure 1. Left: St $= 0.3$, characteristics crossing in the (x, t) plane and phase-plane dynamics (x, v) for 30 initial conditions. Right: St $= 0.03$, no characteristics crossing in the (x, t) plane and phase-plane dynamics (x, v) for 30 initial conditions.

(x, v) phase space so that the distribution admits multiple velocities at a point where the spatial projection of the characteristics cross.

For the sake of simplicity the generic example of steady gas flow field is given by a spatial harmonic oscillation $U(x) = \sin(2\pi x)$, with periodic boundary conditions on the spatial interval $[0, 1]$. The initial condition for the spray is a uniform zero velocity distribution $u^0 = 0$, as well as a constant density distribution $n^0 = 1$. The characteristic first crossing point can be shown to be at $x = 1/2$ since this is the point of maximal strain. In order to characterize the limit we linearize the original system of ODEs (3.3) at $x = 1/2$, for which the eigenvalues of the associated matrix are real, if and only if $8\pi\mathrm{St} \leq 1$. The limiting value of is then $1/(8\pi) \approx 0.0398$. Taking a look at Figure 1 will provide the reader with the intuitive picture of two cases where there is or there is not characteristics crossing. Besides, as shown in Dufour (2005) in the 1-D case, the droplet velocity field is rapidly attracted, within a few Stokes numbers, to an invariant velocity manifold, which is smooth only if the non-dimensional Stokes number is below the critical limit and becomes discontinuous beyond this threshold; thus allowing the droplets to go from one half of the domain to the other (see Figure 1 for St $= 0.3$). This manifold is easily observed in the $(X_\mathrm{p}, V_\mathrm{p})$ phase plane in Figure 1 for St $= 0.03 < 1/(8\pi)$.

For Stokes numbers beyond the critical limit, let us underline the fact that, even for the Williams equation at the kinetic level, there is a singularity at the time when the characteristics are crossing in the (x, t) diagram. At this exact time, the zeroth-order moment of the NDF, that is the number density of droplets, becomes infinite at $x = 1/2$ and the original modeling ceases to be valid since either a "granular-pressure" term limiting the amount of liquid present in this neighborhood, or a collision term should be added. For the Eulerian model, below the critical limit, we observe a progressive concentration of droplets at $x = 1/2$ which tends to infinity with infinite time, whereas, beyond the critical limit, the concentration up to infinity of the density occurs in finite time through a δ-shock. This will prove to be symptomatic of what happen in

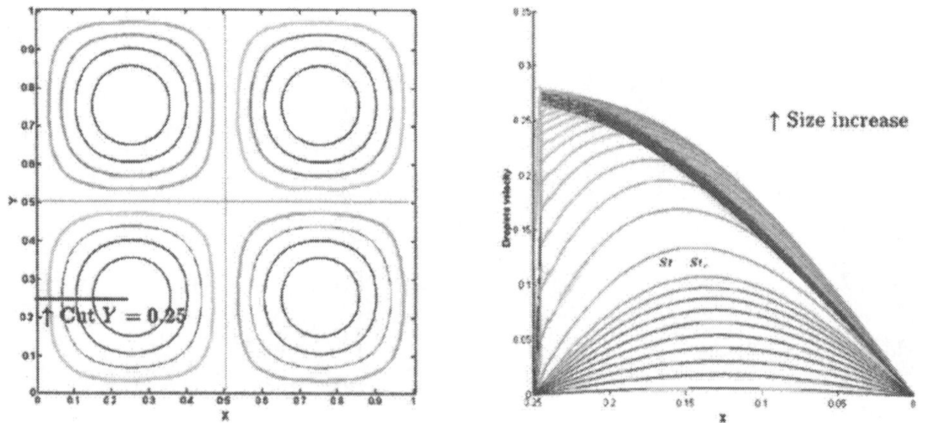

Figure 2. Left: Stream lines of the Taylor-Green vortices and position of the cut for the study of normal ejection velocity. Right: Structure of the invariant droplet velocity manifold versus Stokes number in term of the velocity component orthogonal to the gas flow.

multi-dimensional configurations with complex flows.

Taylor-Green vortices In the two-dimensional version of the steady spatially oscillating gas flow field, we consider a steady solution of the incompressible Euler equations with periodic boundary conditions, which reads in the non-dimensional setting $U = \sin(2\pi x)\cos(2\pi y)$ for the horizontal velocity, and $V = -\cos(2\pi x)\sin(2\pi y)$ for the vertical one, $(x, y) \in [0, 1]\times[0, 1]$. The structure of the flow field is presented in Figure 2 through the stream lines. In order to analytically characterize the critical Stokes number, we focus on the behavior of the system around the central point $(1/2, 1/2)$. The characteristics in their non-dimensional form are then linearized at this point and it yields

$$\begin{aligned} \mathrm{d}_t X_\mathrm{p} &= U_\mathrm{p}, & \mathrm{d}_t Y_\mathrm{p} &= V_\mathrm{p}, \\ \mathrm{d}_t U_\mathrm{p} &= \frac{2\pi X_\mathrm{p} - U_\mathrm{p}}{\mathrm{St}}, & \mathrm{d}_t V_\mathrm{p} &= \frac{-2\pi Y_\mathrm{p} - V_\mathrm{p}}{\mathrm{St}}. \end{aligned} \tag{3.4}$$

The systems in the two directions then decouples and the eigenvalues in the x direction are always real, whereas one recovers the same analysis in the y direction as in the 1-D case, with the same critical value of the Stokes number St. It can be shown that, for the considered initial monokinetic velocity distribution, the first point of characteristics crossing is at the center: $(1/2, 1/2)$ which is again the point of maximum rate of strain. Besides, as shown in de Chaisemartin et al. (2006), the droplet velocity field is rapidly attracted, within a few Stokes number, to an invariant velocity manifold that is smooth only if the non-dimensional Stokes number is below its critical value and becomes discontinuous beyond this threshold, thus allowing the droplets to go from one vortex to its neighbors. This phenomenon is presented in Figure 2 where the invariant droplet velocity normal to the stream lines of the gaseous velocity field is plotted. Beyond the obtained Stokes critical value, the droplets are ejected from the vortices and encounter the droplets coming from the other vortices, since the original number density of droplets is symmetric. Consequently, for Stokes numbers below this critical value, we are sure that the multi-fluid assumptions will be valid in the sense that the kinetic modeling and the fluid modeling provide identical descriptions. We will then be able to conduct thorough comparisons with a Lagrangian solver as in de Chaisemartin et al. (2006). For Stokes numbers beyond this critical value, the multi-fluid model and the kinetic model are not coherent any more. The Williams spray equation also encounters a singularity if no collision model or granular interaction is added. The Eulerian semi-kinetic and multi-fluid models both lead to infinite density concentrations and discontinuous velocity fields as presented in Figure 2.

Relation to mathematical studies In fact, in the non-dimensional gas velocity variables, the maximal value of the strain for both the 1-D and 2-D cases is 2π at the symmetry point. From Jabin (2002) it can be shown that system (3.1) is equivalent to the kinetic Williams equation for monokinetic initial velocity distributions under two conditions on both the initial velocity field $\overline{\mathbf{u}}^0$ and on the maximum $a^{\max}$ of the derivative of the steady gas velocity field. The variable $a^{\max}$ denotes the maximal rate of strain of the gas flow field. In non-dimensional variables, the conditions can be written $a^{\max}\mathrm{St} < 1/(4d)$ and $|\partial_{\mathbf{x}}\overline{\mathbf{u}}^0|_\infty\mathrm{St} < 1/(2d)$. Since in the preceeding case, $a^{\max} = 2\pi$, one recovers the obtained condition on the critical Stokes number in the one-dimensional setting. These two conditions thus insure, from a mathematical point of view that the kinetic NDF will remain monokinetic, if it was originally so, for all times. In this context this provides a rigorous basis in order to insure the validity of the semi-kinetic and multi-fluid model.

3.2 Turbulent case – shock wave formation

In the 1-D configuration, with the same sinusoidal gas velocity field in the dimensional setting with periodic boundary conditions, we introduce a simple model for the Williams equation "in the mean":

$$\partial_t \bar{f} + u\,\partial_x \bar{f} + \partial_u\left(\frac{\overline{U}(x) - u}{\tau_\mathrm{p}}\bar{f}\right) = \partial_u\left[\partial_u\left(\mu^0 \bar{f}\right) + \partial_x\left(\lambda^0 \bar{f}\right)\right] \tag{3.5}$$

where eventually, the droplet dynamical time τ_p could evolve due to evaporation, but the spray is here assumed monodisperse and non-evaporating, and where μ^0 and λ^0 are the terms associated to a local Brownian-like motion due to the gas weak turbulence. The latter is taken of small amplitude and, in the limit of delta correlation in time, it can be related to the study of the previous section. In fact, assuming that $k = \langle U'U'\rangle/2$ is the turbulent kinetic energy of the gas and that T_L, the gas Lagrangian time scale becomes asymptotically small as compared to τ_p, i.e. the fluctuating gas velocity becomes delta-correlated in time, it can be shown through stochastic calculus that $kT_\mathrm{L}/\tau_\mathrm{p}$ converges to a constant (Minier and Peirano, 2001), so that we finally get for the diffusion coefficients $\mu^0 = \overline{\mu}^0 A^2/\tau_\mathrm{p}$ and $\lambda^0 = 0$. Thus, using the same dimensional references, adding the non-dimensional sound speed $\mathcal{C} = (\mu^0\tau_\mathrm{p})^{1/2}/A = (\overline{\mu}^0)^{1/2}$, we obtain the following non-dimensional equation:

$$\partial_t \bar{f} + u\,\partial_x \bar{f} + \frac{1}{\mathrm{St}}\partial_u\Big(\left[\sin(2\pi x) - u\right]\bar{f} - \mathcal{C}^2\partial_u \bar{f}\Big) = 0. \tag{3.6}$$

Let us emphasize that AC represents the velocity dispersion around zero mean velocity of the *global* equilibrium associated with the previous Fokker-Planck equation, that is to say, the velocity dispersion generated by the delta-correlated in time weak turbulence we have considered. Besides, the final density profile associated to this global equilibrium is

$$n^{\mathrm{eq}}(x) = \tilde{n}^0 \exp\left(\frac{1}{\mathrm{St}\mathcal{C}^2} \int_0^x \sin(2\pi x')\,\mathrm{d}x'\right) \tag{3.7}$$

with $\tilde{n}^0$ defined by $\int_0^1 n^{\mathrm{eq}}(x')\,\mathrm{d}x' = n^0$.

It is particularly interesting to note that the corresponding non-dimensional semi-kinetic system associated to a *local* equilibrium Maxwellian velocity distribution around the mean leads to the gas dynamics-like system of conservation equations:

$$\begin{aligned}
\partial_t \overline{n} + \partial_x \left(\overline{n}\,\overline{u}\right) &= 0, \\
\partial_t (\overline{n}\,\overline{u}) + \partial_x \left(\overline{n}\,\overline{u}^2 + 2\overline{n}\mathcal{E}\right) &= \frac{\overline{n}}{\mathrm{St}} \left(\overline{U} - \overline{u}\right), \\
\partial_t E_{\mathrm{d}} + \partial_x \left(E_{\mathrm{d}}\overline{u} + 2\overline{n}\mathcal{E}\overline{u}\right) &= \frac{\overline{n}}{\mathrm{St}} \left[\overline{u}(\overline{U} - \overline{u}) - 2\mathcal{E} + \mathcal{C}^2\right],
\end{aligned} \tag{3.8}$$

where $E_{\mathrm{d}} = \overline{n}(\mathcal{E} + \overline{u}^2/2)$ and $\mathcal{E} = \overline{u'u'}/2$. Considering the initial condition for which $\overline{n} = 1$, $\overline{u} = 0$ and $\mathcal{E} = \mathcal{C}^2/2$, it is easy to see that at the global equilibrium, we have $\overline{u} = 0$, $\overline{n} = \overline{n}^{\mathrm{eq}}$ and $\mathcal{E} = \mathcal{C}^2/2$. In the context of small $\mathcal{C}$, and for Stokes values below the previously derived critical limit, one could think that no singularity will occur, as in the pressure-less case. However, even if no δ-shock are to be expected, we encounter shock formation. We present an example with the same value 0.03 for the Stokes number as before with a sound velocity of $\mathcal{C} = 0.1$ in Figure 3. These initial distributions in density, velocity and agitation internal energy, as well as six profiles spanning the time interval $[0, 1.8]$ have been plotted both with a global point of view in order to get the whole dynamics and with a zoom in order to capture the singularity formation in velocity and internal energy. The numerical schemes used for this simulation are the classical WENO schemes and we refer to Shu (1998) and the subsequent publications in the field. In a time of order one, the equilibrium density profile is reached and we observe a shock formation making the link between the *global* equilibrium solution and the invariant manifold from the pressure-less case (see the $\overline{u}$ velocity field, which is almost invariant and close to the pressure-less profile in Figure 3). However, it can be shown from a mathematical point of view that the moments of the kinetic diffusion equation (3.6) in velocity are smooth. Consequently, the singularity formation only appears in time of order 1 due to the "equilibrium" assumption adopted for the derivation of the semi-kinetic model. This mathematical issue is treated in details in a coming paper (Dufour et al., 2007).

3.3 Conclusion

The few examples we have presented in this section allow us to understand that the semi-kinetic model for spray modeling, through the projection of the kinetic velocity distribution functions onto local equilibrium, not only constitutes a highly compressible model as expected, but generates various types of singularities. It is then important to design and use dedicated numerical methods in order to be able to cope with the associated numerical difficulties: either the

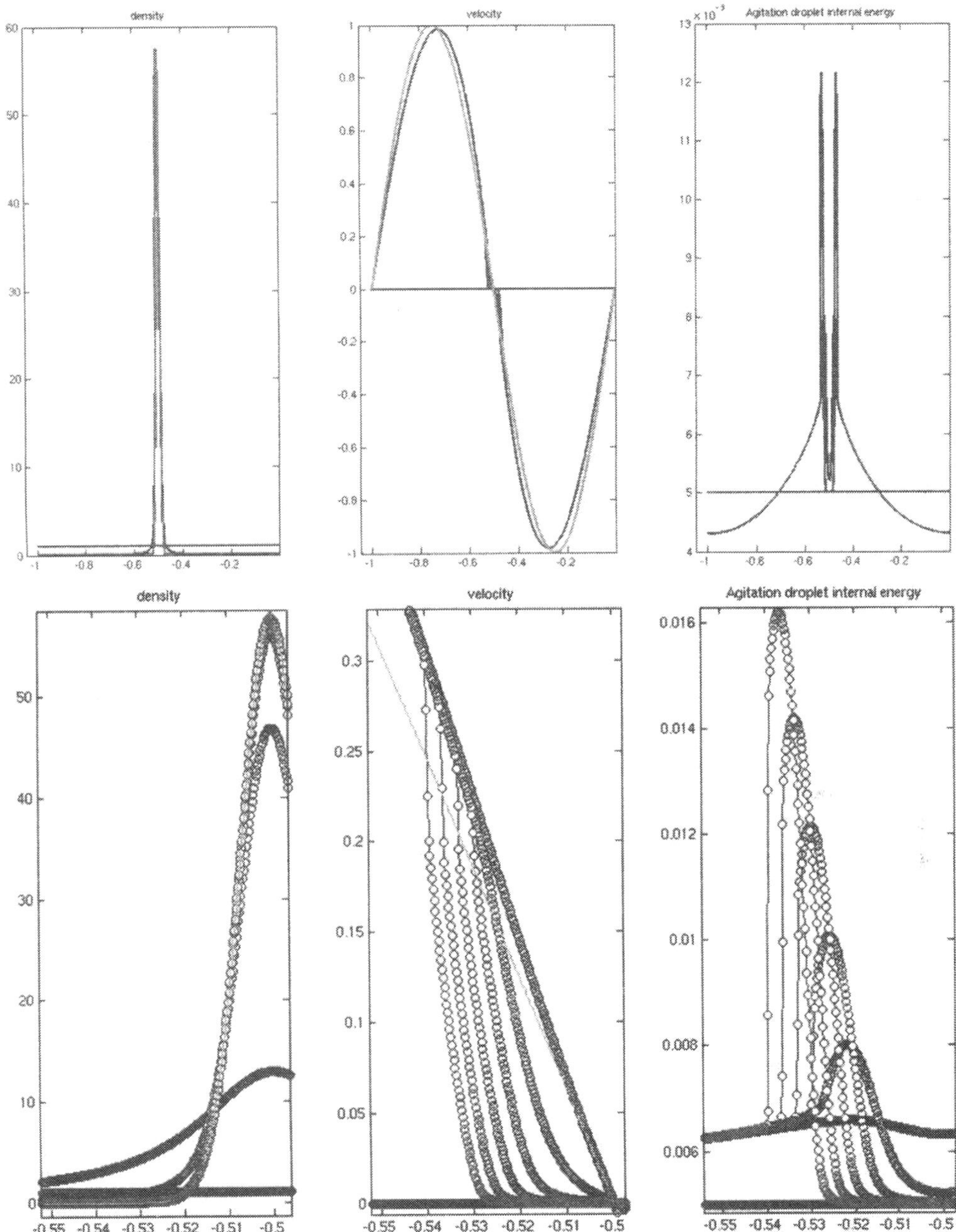

Figure 3. Shock wave formation for St $= 0.03$ below the critical value and $\mathcal{C} = 0.1$ with initial droplet agitation internal energy $\mathcal{E} = 0.005$, over the time $[0, 1.8]$ with six successive profiles. Top: Global behavior. Bottom: Zoom of the same figures as at top with details of the shock formation. Left: Density profile. Middle: Velocity field with gaseous velocity profile. Right: Internal energy.

presence of no droplets in some spatial areas (so-called vacuum regions) or the possibility of very high density concentration in other areas and discontinuous velocity fields. This is the purpose of the next section.

4 Multi-Fluid Model and Related Numerical Methods

In Section 2 we introduced the semi-kinetic model as well as the set of associated assumptions. We start from this model and now derive the multi-fluid model in the present section and present the related numerical methods.

4.1 Discretization in the size phase space

We choose a level of discretization for the droplet size phase space and we average the obtained semi-kinetic system of conservation equations over some fixed size intervals, each "fluid" corresponding to the set of droplets in each size interval. It can be interpreted as a finite volume discretization in the size variable, the order of which has been studied in Laurent (2006) as far as the evaporation is concerned. The coalescence phenomenon, when taken into account, results in quadratic source terms in the Eulerian multi-fluid conservation equations for the mass and momentum of each "fluid". The coefficients involved in these source terms are collisional integrals; they do not depend on t, $\mathbf{x}$, nor on the droplet size but only on the given droplet size discretization. Consequently, they can be pre-calculated before the resolution of the system of conservation laws is conducted.

Original multi-fluid approach The Eulerian multi-fluid model is based on the reduction of a continuous semi-kinetic equation (2.11–2.12) as a function of size, to a finite number of degrees of freedom. This reduction is performed by averaging, in fixed size intervals: the sections (the jth section being defined by $v^{(j-1)} \leq v < v^{(j)}$) of the semi-kinetic model. As a fundamental assumption, the form of n as a function of the droplet geometry is supposed to be independent of t and x in a given section. Thus the evolution of the mass concentration of droplets in a section $m^{(j)}$ is decoupled from the repartition in terms of sizes $\kappa^{(j)}(v)$ inside the section (Greenberg et al., 1993):

$$n(t, \mathbf{x}, v) = m^{(j)}(t, \mathbf{x})\kappa^{(j)}(v), \qquad \int_{v^{(j-1)}}^{v^{(j)}} \rho_l v \kappa^{(j)}(v)\,\mathrm{d}v = 1. \tag{4.1}$$

The sections have fixed sizes, which is a major difference compared to a sampling method; however, they are not independent from each other, they exchange mass and momentum. The choice of the discretization points $v^{(j)}$, $j \in [1, N]$ has been studied in Laurent et al. (2004b); consequently we choose the $(N+1)^{\text{th}}$ section to be $[v^{(N)}, +\infty)$ in order to be able to describe the whole size spectrum. The final model is then obtained in the next proposition.

Proposition 4.1. *Besides assumptions* [H1]–[H3] *associated to the derivation of the semi-kinetic model, we make the following assumption on the velocity distribution inside a section:*

[H4] *In each section, the averaged velocity* $\overline{\mathbf{u}}(t, \mathbf{x}, v)$ *does not depend on* v, $\overline{\mathbf{u}}(t, \mathbf{x}, v) = \overline{\mathbf{u}}^{(j)}(t, \mathbf{x})$, $v^{(j-1)} \leq v < v^{(j)}$.

We obtain the multi-fluid system of $2(N+1)$ *conservation equations:*

$$\partial_t m^{(j)} + \partial_{\mathbf{x}} \cdot \left(m^{(j)}\overline{\mathbf{u}}^{(j)}\right) = -\left(E_1^{(j)} + E_2^{(j)}\right) m^{(j)} + E_1^{(j+1)} m^{(j+1)} + C_{\mathrm{co}}^{\mathrm{m}(j)} + C_{\mathrm{bu}}^{\mathrm{m}(j)}, \quad (4.2)$$

$$\partial_t \left(m^{(j)}\overline{\mathbf{u}}^{(j)}\right) + \partial_{\mathbf{x}} \cdot \left(m^{(j)}\overline{\mathbf{u}}^{(j)} \otimes \overline{\mathbf{u}}^{(j)}\right) = -\left(E_1^{(j)} + E_2^{(j)}\right) m^{(j)}\overline{\mathbf{u}}^{(j)} + E_1^{(j+1)} m^{(j+1)}\overline{\mathbf{u}}^{(j+1)} + m^{(j)}\overline{\mathbf{F}}^{(j)} + C_{\mathrm{co}}^{\mathrm{mu}(j)} + C_{\mathrm{bu}}^{\mathrm{mu}(j)} \quad (4.3)$$

where $E_1^{(j)}$ *and* $E_2^{(j)}$ *are the "classical" pre-calculated evaporation coefficients (Greenberg et al., 1993; Laurent and Massot, 2001):*

$$E_1^{(j)} = -\rho_l v^{(j)} R_v(t, \mathbf{x}, v^{(j)}) \kappa^{(j)}(v^{(j)}), \quad E_2^{(j)} = -\int_{v^{(j-1)}}^{v^{(j)}} \rho_l R_v(t, \mathbf{x}, v) \kappa^{(j)}(v)\, dv, \quad (4.4)$$

where $\overline{\mathbf{F}}^{(j)}$ *is the mean drag force:* $\overline{\mathbf{F}}^{(j)}(t,\mathbf{x}) = \int_{v^{(j+1)}}^{v^{(j)}} \rho_l v \overline{\mathbf{F}}(t,\mathbf{x},v)\kappa^{(j)}(v)\,dv$*, and where the source terms associated with the coalescence phenomenon ,* $C_{\mathrm{co}}^{\mathrm{m}(j)}$ *and* $C_{\mathrm{co}}^{\mathrm{mu}\,(j)}$ *in the mass and momentum equation respectively of the* j^{th} *section are quadratic in the variables* $(m^{(j)})_{j\in[1,N+1]}$ *and include the velocity difference between droplet of various sections* $V_{jk} = |\overline{\mathbf{u}}^{(j)} - \overline{\mathbf{u}}^{(k)}|$ *as well as collisional integrals* $\mathcal{Q}_{jk}$*,* $j = [1, N+1], k = [1, N+1], k \neq j$*,* $\mathcal{Q}_{ji}^{\diamond}$*,* $\mathcal{Q}_{ji}^{*}$*,* $j = [2, N+1], i = [1, I^{(j)}]$*, the expression of which can be found in Massot and Villedieu (2001); Laurent et al. (2004a). The source terms associated with secondary break-up are to be found from the semi-kinetic model in Dufour (2005); they are linear in the variables* $(m^{(j)})_{j\in[1,N+1]}$*. The source terms due to coalescence conserve mass and momentum, whereas the source term due to break-up only conserve mass, since the daughter droplets all relax to a single after-break-up velocity:*

$$\sum_{j=1}^{N+1} C_{\mathrm{co}}^{\mathrm{m}(j)} = 0, \qquad \sum_{j=1}^{N+1} C_{\mathrm{co}}^{\mathrm{m}(j)} = 0, \qquad \sum_{j=1}^{N+1} C_{\mathrm{bu}}^{\mathrm{m}(j)} = 0 \quad (4.5)$$

Let us mention that this model does not predict any coalescence between droplets of the same section since they have the same velocity and their probability of colliding is zero. Finally, the coefficients used in the model, either for the evaporation process or the drag force $E_1^{(j)}$, $E_2^{(j)}$ and $\overline{\mathbf{F}}^{(j)}$, $j = [1, N+1]$ in (4.2–4.3), or the collisional integrals for evaluating the coalescence can be pre-evaluated from the choice of the droplet size discretization and from the choice of $\kappa^{(j)}$ since the collision integrals do not depend on time nor space. The algorithms for the evaluation of these coefficients are provided in Laurent et al. (2004a) for coalescence and in Dufour (2005) for break-up.

Extension to higher order In the previous subsection, we have presented the original multi-fluid approach as it was originally derived in Laurent and Massot (2001) for the sake of coherence of the present contribution. However, since then various extension of the model have been conducted. Laurent (2006) and Dufour and Villedieu (2005) have proposed two versions of a second-order extension of the method since it has shown to be first order in the droplet size variable discretization step in Laurent (2006). We refer to these publications for the details of such extensions. However, the field is in evolution at the present time and new ideas are to come up, as will be discussed in the conclusion. Consequently we decided only to refer to the relevant literature, and not to present such improvements in details, and to focus on the global treatment of the subject.

4.2 Treatment of transport in physical space

Let us recall that the transport term in the semi-kinetic model is the pressure-less gas system (Bouchut, 1994):

$$\begin{aligned} \partial_t \rho + \partial_{\mathbf{x}} \cdot (\rho \overline{\mathbf{u}}) &= 0, \\ \partial_t (\rho \overline{\mathbf{u}}) + \partial_{\mathbf{x}} \cdot (\rho \overline{\mathbf{u}} \otimes \overline{\mathbf{u}}) &= 0. \end{aligned} \tag{4.6}$$

A specific numerical method is needed in order to cope with discontinuous velocity fields, "δ-shocks" as well as with a vacuum. We need a method that strictly preserves the positivity of the density field, allows the concentration of all the density in one cell without leading to an instability of the method, and guarantees the maximum principle on the velocity. This is quite a constraint; however, we still at least require a second-order scheme in order to limit the numerical diffusion as well as the level of discretization required for a precise calculation. Bouchut et al. (2003) developed second-order kinetic schemes, which are finite-volume schemes based on the equivalence between a macroscopic and a microscopic level of description for the pressure-less gas equations. For the purpose of the exposition, we will work in one dimension:

$$\partial_t f + u \partial_x f = 0 \iff \left\{ \begin{aligned} \partial_t \rho + \partial_x (\rho \overline{u}) &= 0 \\ \partial_t (\rho \overline{u}) + \partial_x (\rho \overline{u}^2) &= 0 \end{aligned} \right. \tag{4.7}$$

with

$$f(t, x, u) = \rho(t, x) \delta(u - \overline{u}(t, x)). \tag{4.8}$$

The values of ρ and $\overline{u}$ are then recovered from f by the formula:

$$\begin{pmatrix} \rho \\ \rho \overline{u} \end{pmatrix} (t, x) = \int_{\mathbb{R}} \begin{pmatrix} 1 \\ u \end{pmatrix} f(t, x, u) du. \tag{4.9}$$

This is precisely the same framework as the one we encountered for the semi-kinetic model and will apply to the multi-fluid model.

The principle of such a method is illustrated on Figure 4 for the 1-D case. It is a finite-volume method giving approximations ρ_j^n and $q_j^n = \rho_j^n u_j^n$ of the following average values on each cell $[x_{j-1/2}, x_{j+1/2}]$ of ρ and $\rho \overline{u}$ at each discrete time t^n:

$$\rho_j^n \simeq \frac{1}{\Delta x} \int_{x_{j-1/2}}^{x_{j+1/2}} \rho(t^n, x) \, \mathrm{d}x, \qquad q_j^n = \rho_j^n u_j^n \simeq \frac{1}{\Delta x} \int_{x_{j-1/2}}^{x_{j+1/2}} \rho(t^n, x) \overline{u}(t^n, x) \, \mathrm{d}x. \tag{4.10}$$

First, at time $t = t^n$, a distribution function $f^n(x, u)$ is reconstructed from the averaged values ρ_j^n and q_j^n. This comes from Eq. (4.8) and, for example, a piecewise linear reconstruction of $\rho(t^n, x)$ and $\overline{u}(t^n, x)$ with adequate slope limiters will work. Second, the kinetic equation is solved analytically between t^n and t^{n+1}: $f(t, x, u) = f^n(x - u(t - t^n), u)$. Finally, a projection

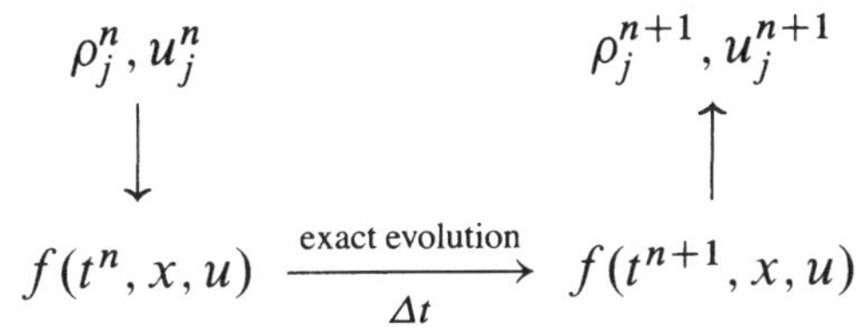

$$\begin{array}{ccc} \rho_j^n, u_j^n & & \rho_j^{n+1}, u_j^{n+1} \\ \downarrow & & \uparrow \\ f(t^n, x, u) & \xrightarrow[\Delta t]{\text{exact evolution}} & f(t^{n+1}, x, u) \end{array}$$

Figure 4. Main steps of the kinetic-based transport scheme.

of $f(t^{n+1-}, x, u)$ is conducted in order to find ρ_j^{n+1} and q_j^{n+1}, which corresponds to the average on each cell of (4.9) at $t = t^{n+1-}$. This leads to the following scheme:

$$\rho_j^{n+1} = \rho_j^n - \frac{\Delta t}{\Delta x}\left(F_{j+1/2}^{(1)} - F_{j-1/2}^{(1)}\right), \tag{4.11}$$

$$q_j^{n+1} = q_j^n - \frac{\Delta t}{\Delta x}\left(F_{j+1/2}^{(2)} - F_{j-1/2}^{(2)}\right), \tag{4.12}$$

with the fluxes

$$F_{j+1/2} = \begin{pmatrix} F_{j+1/2}^{(1)} \\ F_{j+1/2}^{(2)} \end{pmatrix} = \frac{1}{\Delta t}\int_{t^n}^{t^{n+1}} \int_{\mathbb{R}} \begin{pmatrix} 1 \\ u \end{pmatrix} u f^n(x_{j+1/2} - u(t - t^n), u)\, \mathrm{d}u\, \mathrm{d}t. \tag{4.13}$$

The obtained fluxes rely, through Eq. (4.8), on the reconstructions of $\rho^n(x) = \rho(t^n, x)$ and $\overline{u}^n(x) = \overline{u}(t^n, x)$ from the discrete values ρ_j^n and u_j^n. Different type of reconstructions are proposed in Bouchut et al. (2003). We choose the one which gives good results without being too complex, a piecewise linear reconstruction. The detailed expressions of the fluxes from the reconstructed fields can be found in Bouchut et al. (2003) and the adapted version of the scheme to our case in de Chaisemartin et al. (2006).

For the 2-D case we are dealing with, in the framework of structured meshes, the same type of numerical method is written in Bouchut et al. (2003). However, it is hard to find a good slope limiter that does not induce too large a numerical diffusion. That is why we rather use a dimensional splitting of the 1D scheme previously described, with a Strang type splitting to preserve the second-order nature of the method as indicated in LeVeque (2002). Besides, such an approach is very competitive as far as computational cost is concerned. The corresponding scheme then offers the ability to treat the delta-shocks and vacuum. It guarantees a maximum principle on the velocity as well as the positivity of the density. Moreover, it is second-order accurate in space and in time.

4.3 Evaporation and drag

As presented in the assumptions to derive the multi-fluid model, we have to choose a form $\kappa^{(j)}(s)$ of the distribution in the jth section (Hypothesis [H4]). Here, this function is chosen constant since it is the optimal choice, as shown in Laurent (2006) for the evaporation process. We obtain then the multi-fluid system for the only drag and evaporation:

$$d_t m^{(j)} = -(E_1^{(j)} + E_2^{(j)}) m^{(j)} + E_1^{(j+1)} m^{(j+1)}, \tag{4.14}$$

$$d_t (m^{(j)} \mathbf{u}^{(j)}) = -(E_1^{(j)} + E_2^{(j)}) m^{(j)} \mathbf{u}^{(j)} + E_1^{(j+1)} m^{(j+1)} \mathbf{u}^{(j+1)} + m^{(j)} \overline{\mathbf{F}}^j. \tag{4.15}$$

The expression of $E_1^{(j)}$ and $E_2^{(j)}$ are to be found in Laurent (2006). The finite-volume discretization recalled here is proved to be a first-order method in Laurent (2006). We need then to perform a time discretization in order to solve the multi-fluid system of equations: we have chosen here a θ-scheme with $\theta = 1/2$. This scheme offers unconditional stability, preserves the positivity of the mass density in each section and is second-order accurate in time. It is a semi-implicit scheme and can be written $(\mathbb{I} + \Delta t C/2) W^{n+1} = (\mathbb{I} - \Delta t C/2) W^n$ with W^n the vector of the mass densities and momentum of the sections at time t_n, with $\mathbb{I}$ the identity matrix and with C a matrix independent of n associated to the previous linear system of ODEs.

4.4 Time integration using splitting methods

We have presented separately the different contributions: transport (evolution in space), evaporation and drag (evolution in the chosen internal variables), because they are treated separately in our computation through a splitting algorithm. We choose a Strang splitting algorithm that has the following structure:

- Evaporation and drag force during $\Delta t/2$.
- Transport during Δt.
- Evaporation and drag force during $\Delta t/2$.

There are two key issues in this choice of a numerical scheme; the first one is related to the properties satisfied by separate parts contained in the system of conservation equations, which will no longer be satisfied if the two parts are coupled together. It is then crucial to be able to use dedicated numerical schemes on each part of the system and couple them afterwards. Moreover, there exists well-behaved splitting algorithms, even in the presence of fast scales in the problem and we refer to Strang (1968); Descombes and Massot (2004). The second key feature of such a method is the fact that the evolution of the internal variables is treated separately from the transport step; thus leading to a set of ODEs parameterized by space that can be parallelized easily and very efficiently as in de Chaisemartin et al. (2006). This approach has the great advantage to preserve the properties of the different schemes we use for the different contributions, as for example maximum principle or positivity. If we assume that the involved phenomena evolve at roughly the same time scales this Strang splitting algorithm guarantees a second-order accuracy in time provided that each of the elementary schemes has a second-order time accuracy. So, finally, our numerical method is second-order accurate in time and in space and first-order accurate in droplet surface.

5 Numerical Validation

5.1 Counterflow spray diffusion flames

In most industrial combustion applications the fuel is stored in condensed form and injected as a spray into a gaseous stream. Rather than tackling the impossible task of dealing with such practical systems, we focus on laminar spray flames, which provides rich physical scenarios amenable to detailed modeling. They are intermediate in complexity between practical spray combustion systems, which are difficult to interpret, and the classic experiments and modeling on single droplet burning that have been the subject of a vast literature over the past fifty years. They provide an ideal test-case with some of the ingredients of practical flames, but without turbulence complications. In particular, disperse-phase dynamics, heating and vaporization for sprays with a distribution of droplet sizes can be systematically investigated. Within the category of laminar spray flames, counterflow diffusion flames are perhaps the simplest. They have been the focus of various studies in the past (see Massot et al., 1998, and the references therein) in view of the fact that, besides their natural cylindrical symmetry, under a similarity assumption, they can be treated as a one-dimensional flow field along the centerline (see Figure 5.1). We restrict ourselves to dilute sprays for which the volume fraction occupied by the liquid is small compared to that of the gas mixture. Although the region of high chemical activity is usually located far enough from the injection and atomization area, two-phase effects can significantly influence flame structure, even for relatively dilute sprays (Massot et al., 1998).

Most atomizers produce droplets of various sizes, which experience different dynamics, vaporization and heating histories. To predict the flame structure accurately, it is then necessary to model the polydisperse character of the spray in a multicomponent reactive gas flow. For unconfined flames, governing equations under the constant pressure, low Mach Number approximation apply (Giovangigli, 1999; Smooke et al., 1990). The coupling of the phases occurs through additional source terms describing mass, momentum and heat exchanges between the two phases. In the present contribution, we recall the main results from Massot et al. (1998) and Laurent et al. (2004b) where quantitative comparisons are provided between numerical simulations (with complex chemistry and detailed transport properties for the gaseous phase and both Lagrangian and Eulerian multi-fluid model for the polydisperse spray) and experimental measurements. The compared quantities concern both the gaseous flame properties such as temperature and velocity fields in order to show that the flame structure is well-captured, as well as droplet properties such as mass density, droplet velocities conditioned by size at various locations in order to prove that the multi-fluid model is able to capture correctly the dynamics, heating and evaporation of the droplets of various sizes.

In Laurent et al. (2004b) the key issue of the tail of the droplet size distribution is tackled from both a modeling and experimental point of view. We conduct a detailed study in order to optimize the number of sections for an accurate prediction of flame structure. We show that within the framework of a few sections, we are able to describe accurately the spray evolution. In this subsection, we only wanted to underline that the original multi-fluid model had been validated against experimental measurements. Even if there was still the possibility for a lot of improvement, especially from a numerical point of view, as was shown since then, it is able to describe properly the evolution of polydisperse sprays in a combustion environment. In the following, we will focus on idealized flows in order to conduct more numerical oriented studies and compare the multi-fluid approach to its Lagrangian counterpart.

5.2 Coalescence for dense sprays in a conical nozzle

In this subsection, the break-up of the droplets is not taken into account. We want to first validate the multi-fluid model with a reference Lagrangian solver that uses an efficient and already

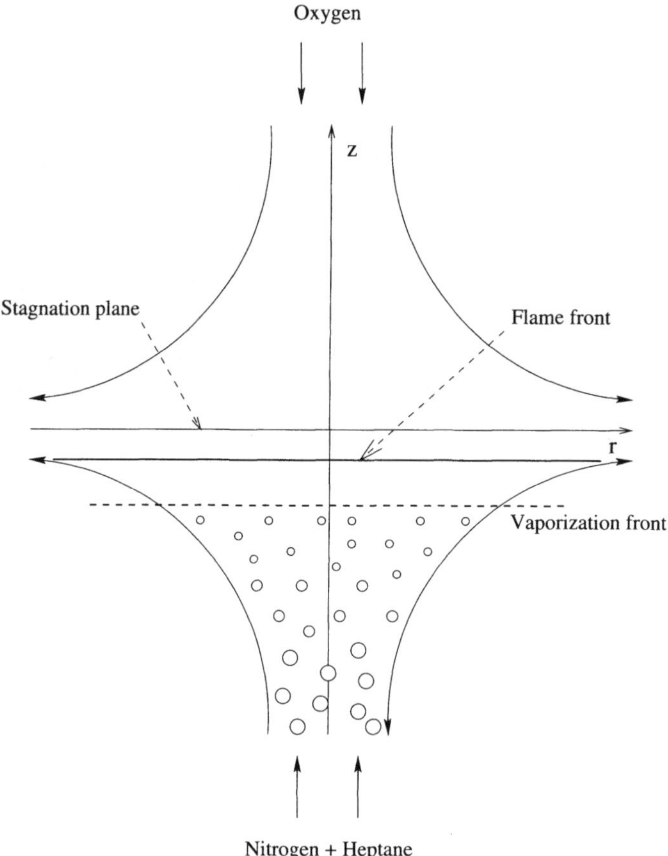

Figure 5. Counterflow spray diffusion flame. Top: Sketch of the configuration. Bottom: Laser light showing the trajectories of methanol droplet towards the flame front (Source: A. Gomez, Center for Combustion Studies, Yale University)

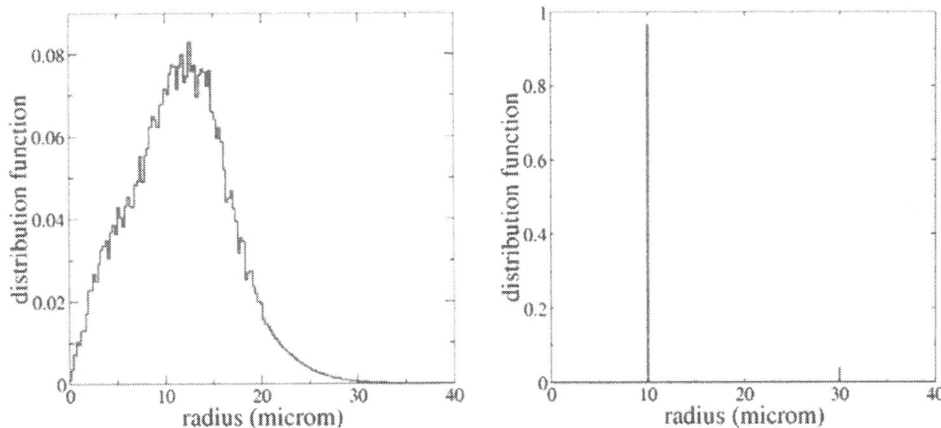

Figure 6. Left: Monomodal distribution function from experimental measurements. Right: Bimodal distribution function.

validated stochastic algorithm for the description of droplet coalescence (Hylkema and Villedieu, 1998). We need a well-defined configuration, both of stationary and unstationary laminar flows where evaporation, coalescence and the dynamics of droplets of various sizes are coupled together. Secondly, we want to evaluate the validity of the assumption underlying the model and the computational efficiency of the multi-fluid approach as compared to the Lagrangian solver. The chosen test-case is a decelerating self-similar 2-D axisymmetrical nozzle. The deceleration generates a velocity difference between droplets of various sizes and induces coalescence. The temperature of the gas is taken high enough in order to couple the evaporation process to the coalescence one. The configuration is stationary 2-D axisymmetrical in space and 1-D in droplet size. It is described in detail, along with the Lagrangian solver, in Laurent et al. (2004a). Hence, only its essential characteristics are given here. The influence of the evaporation process on the gas characteristics is not taken into account in our one-way coupled calculation. It is clear that the evaporation process is going to change the composition of the gas phase and then of the evaporation itself. However, we do not attempt to achieve a fully coupled calculation, but only to compare two ways of evaluating the coupling of the dynamics, evaporation and coalescence of the droplets. It has to be emphasized that it is not restrictive in the framework of this study, which is focused on the numerical validation of Eulerian solvers for the liquid phase.

Let us finally consider two droplet distribution functions. The first one, called monomodal, is composed of droplets with radii between 0 and 35 microns with a Sauter mean radius of 15.6 microns. It is represented in Figure 6 and is typical of the experimental condition reported in Laurent et al. (2004b). The droplets are only constituted of liquid heptane, their initial velocity is the one of the gas, their initial temperature, fixed at the equilibrium temperature 325.4 K (corresponding to an infinite conductivity model), does not change along the trajectories. The second distribution is called bimodal since it involves only two groups of radii respectively 10 microns and 30 microns with equivalent mass density. This bimodal distribution function is typical of alu-

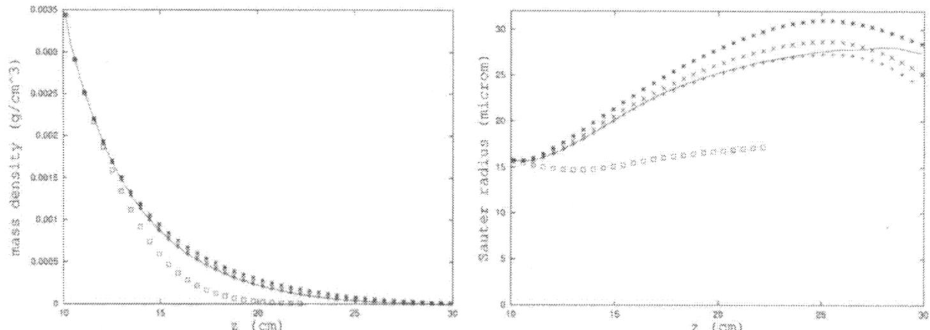

Figure 7. Left: Evolution of the mass density of liquid for the Lagrangian reference solution (solid line) and for various discretizations with the Eulerian multi-fluid model (+: 90 sections, ×: 30 sections, ∗: 15 sections, □: 90 sections without collisions). Right: Sauter mean radius for the Lagrangian reference solution (solid line) and for various discretizations with the Eulerian multi-fluid model (+: 90 sections, ×: 30 sections, ∗: 15 sections, □: 90 sections without collisions).

mina particles in solid propergol rocket boosters (Hylkema, 1999). It is represented in Figure 6 and is probably the most difficult test case for a Eulerian description of the size phase space. The initial injected mass density is then taken at $m_0 = 3.609$ mg cm^{-3} so that the volume fraction occupied by the liquid phase is 0.57% in the stationary case. Because of the deceleration of the gas flow in the conical nozzle, droplets are going to also decelerate; however at a different rate depending on their size and inertia. This will induce coalescence.

For the problem to be one-dimensional in space, conditions for straight trajectories are used and are compatible with the assumption of an incompressible gas flow. This leads to the following expression for the gaseous axial velocity U_z and the reduced radial velocity U_r/r:

$$U_z = V(z) = \frac{z_0^2 V(z_0)}{z^2}, \quad \frac{U_r}{r} = \frac{V(z)}{z} = \frac{z_0^2 V(z_0)}{z^3}, \quad \text{for } z \geq z_0, \tag{5.1}$$

where $z_0 > 0$ is the coordinate of the nozzle entrance and the axial velocity $V(z_0)$ at the entrance is fixed. The trajectories of the droplets are also assumed straight since their injection velocity is co-linear to the one of the gas. This assumption is only valid when no coalescence occurs. However, even in the presence of coalescence, it is valid in the neighborhood of the centerline. The deceleration at the entrance of the nozzle is taken at $a(z_0) = -2V(z_0)/z_0$; it is chosen large enough so that the velocity difference developed by the various sizes of droplet is important. In the stationary configuration, we have chosen a very large value as well as a strong deceleration leading to extreme cases: $V(z_0) = 5$ m/s, $z_0 = 10$ cm for the monomodal case and $V(z_0) = 5$ m/s, $z_0 = 5$ cm for the bimodal case. It generates a very strong coupling of coalescence and dynamics and induces an important effect on the evaporation process.

As mentioned previously, the strength of such multi-fluid models, is to be able to reproduce the global behavior of the spray with a limited number of sections. Consequently, calculations with various numbers of sections were performed: 90, 60, 30 and 15. For completeness, we have

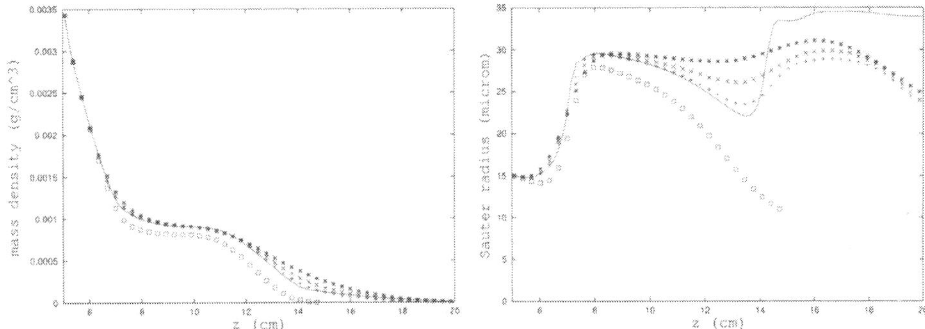

Figure 8. Left: Evolution of the mass density of liquid for the Lagrangian reference solution (solid line) and for various discretizations with the Eulerian multi-fluid model (+: 100 sections, ×: 50 sections, ∗: 25 sections, □: 100 sections without collisions). Right: Sauter mean radius for the Lagrangian reference solution (solid line) and for various discretizations with the Eulerian multi-fluid model (+: 100 sections, ×: 50 sections, ∗: 25 sections, □: 100 sections without collisions).

also represented in Figure 7 the solution without coalescence calculated with 90 sections. The conclusion to be drawn from these figures is that the multi-fluid model is able to predict fairly well, even in the case of 15 sections for which the computational cost is very reasonable, the global coupling of the various phenomena occurring in the nozzle. One crucial point is related to the localization of the evaporation front for pollutant formation purposes and even with 15 sections, the evaporation front is precisely computed. It is particularly interesting to note that the average dynamics are correctly reproduced with 90 sections. Concerning the Sauter mean radius of the distribution, it is extremely well-predicted by the 90-section solution, fairly well-reproduced by the 30-section one and the difference gets bigger when we use 15 sections, even if the mass difference does not grow beyond two percent of the initial one.

If the monomodal distribution is well-suited for the Eulerian multi-fluid approach, the bimodal one can be considered as the most difficult task; the method can be shown to be of first order in the size discretization step (Laurent, 2006). In such a situation, the numerical diffusion is introducing some artificial coupling at the dynamical level since only one velocity is prescribed per section. However, the results presented in Figure 8 show that the mass evolution is very well captured, the difference with the reference solution stays below 1%. The evolution of the Sauter mean radius in Figure 8 is also well-captured. However, once 93% of the initial mass has evaporated, there is a little difference in the mass density decrease, which seems to be due to a difficulty to correctly reproduce the dynamics and size distribution of the spray.

In order to have a more precise idea of what is happening, we have observed the mass distribution function at the point $z = 10.53$ cm as well as the velocity distribution as a function of the droplet size at this point. It appears very clearly that the numerical diffusion, if too high, can smooth out the mass distribution function and consequently the velocity distribution function because of the assumption, the Eulerian multi-fluid model relies on. In the 25-section case, the peaks of the distribution have disappeared and the velocity distribution function has become monotone. This example allows one to understand what will be the limits of such an approach. However, the simulation with 100 sections allows one to predict very accurately the various peaks of the mass distribution function, as well as their dynamics, except for the very "big" droplets, the velocity of which is becoming higher thus causing the difference to be observed in Figure 8 on the Sauter mean radius. This discrepancy can be attributed to the numerical diffusion (Laurent, 2006), which acts on a size distribution function which is very singular and remains so through the coalescence phenomenon.

The computational efficiency of the model is presented in details in Laurent et al. (2004a) where a series of unsteady computations are conducted in a challenging configuration for both Eulerian and Lagrangian methods. If a good level of accuracy in the size phase space is required, the cost is lower than the Lagrangian method but not much lower thus showing that, in such a case, the improvement through the use of the Eulerian model is going to be achieved with the optimization of the solver coupling the two Eulerian description. However, the Eulerian model is going to allow the user to perform several levels of accuracy for the size phase space discretization without having any trouble with the smoothness of the calculated solution (an essential point for combustion applications), a feature which is not present with the Lagrangian solver. We show that a coarse discretization allows us to obtain a good qualitative description of the phenomenon; it proves to be very computationally efficient compared to the reference Lagrangian solution and still allows us to take into account the polydisperse character of the spray. The level of code optimization that can be obtained from a global Eulerian description has already been demonstrated in the case of two-fluid models, but the detailed study in the framework of Eulerian multi-fluid models is in progress and beyond the scope of the present paper. Finally, the assumption that the velocity dispersion around its mean at a given time, for a given space location and a given droplet size is zero, the Eulerian multi-fluid model relies on, is investigated by considering the results from the Lagrangian solver and we prove that this assumption is fulfilled. We then have validated the Eulerian multi-fluid model by showing the good correspondence with the reference solution when the size phase space is finely discretized. It is also shown that the behavior of the spray is correctly captured even if a limited number of "fluids" is used for the Eulerian model corresponding to a coarser discretization in the size phase space.

5.3 Taylor-Green configuration

After focusing on droplet interactions in a configuration reduced to a monodimensional setting, we do not consider droplets interactions any more but focus on a multi-dimensional problem with the Taylor-Green configuration. The equations for both the kinetic and multi-fluid levels of descriptions are the one derived in Sections 2 and 4; they are taken in non-dimensional form. The configuration as well as the reference quantities have already been presented in Section 3.

Initial conditions We have used in our computations two different initial repartitions in space for the droplets. The first one is a simple uniform distribution, the droplets being equally distributed in the computational domain. To realize computations respecting the monokinetic condition of the multi-fluid method we have in this case to make sure that the maximal Stokes number of the droplets is under the critical value: $\mathrm{St}_c = 1/8\pi$ previously introduced. Otherwise, the

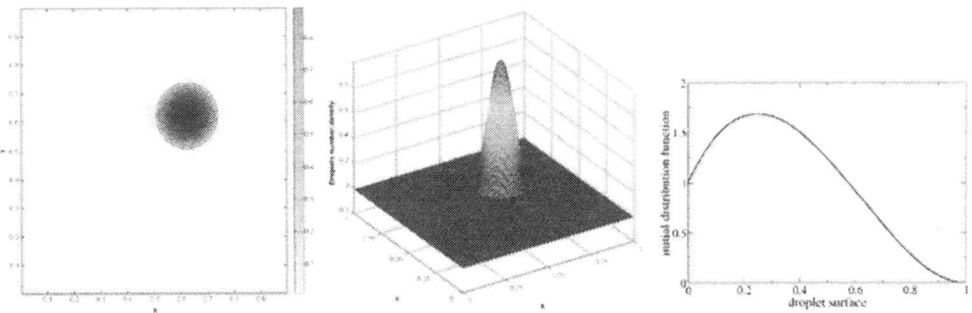

Figure 9. Initial conditions. Left/Center: Initial droplets spatial distribution, non-uniform repartition in the domain. Right: Non-dimensional initial polydisperse size distribution.

droplets would leave their initial vortices and would cross each-other, leading to contradiction with the monokinetic assumption. We have also defined another non-uniform distribution, offering some advantages compared to the uniform one. This repartition is defined thanks to the function $\psi(x) = \sin(x)/x$, for x belonging to the interval $[-\pi, \pi]$. A representation of this repartition can be done representing iso-surfaces of the number density of the droplets as in Figure 9.

This repartition presents different advantages. First of all, this distribution will be interesting to analyze the numerical diffusion of the method, and to determine the spatial refinements needed. Indeed the gradients of this distribution are important as we can see in Figure 9 where we realized a three-dimensional plot to represent the droplets number density in the domain. Besides, this non-uniform distribution offers a second advantage in the ability to perform computations with droplets having a Stokes number greater than the critical value of $1/8\pi$ within the limits of the multi-fluid model's assumptions. Indeed the exit of the vortex does not lead in this case frontal crossing of sprays originated from different vortices. This distribution will then allow us to study the behavior of more inertial droplets. Concerning the initial distribution in sizes, we want to reproduce the distribution $f^0(r)$ represented in Figure 9. This droplet size distribution is uniform in the spatial coordinate.

Density concentration for uniform initial distributions The first point we make is related to the ability to capture high concentrations of droplets due to the ejection of the spray from one vortex. Two successive times $t = 0.33$ and $t = 1$ in the non-dimensional setting are presented for droplets of sizes corresponding to 0.9 times the critical Stokes number with a spatial resolution of 100×100 cells in Figure 10. The numerical schemes does not encounter any problem even if the main part of the mass is concentrated in only a few cells. Moreover, if the Stokes number is increased beyond its critical value, the dynamics is correctly reproduced until the droplets are ejected from one vortex to the other. After this time, an averaging phenomena already discussed in Laurent and Massot (2001) leads to a high concentration on the edges of the lattice where the normal velocity is zero by symmetry and the spray remains there. A more detailed study can be found in de Chaisemartin et al. (2006), the key information being the robustness of the proposed

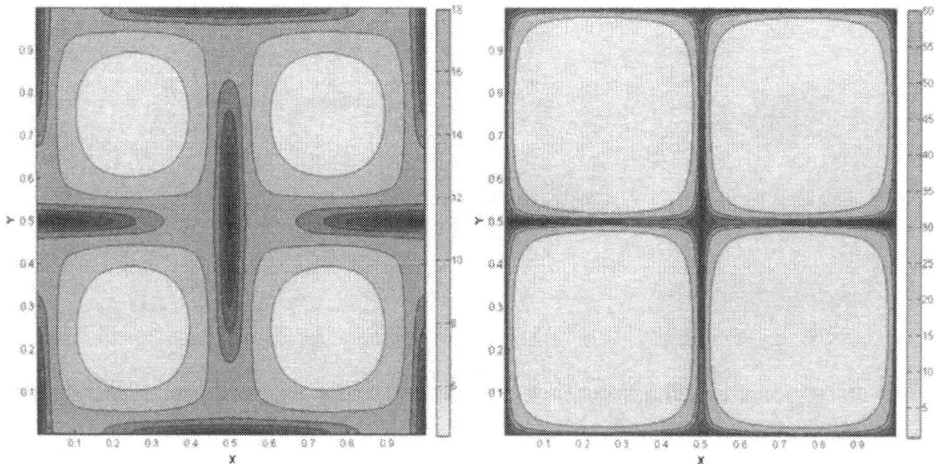

Figure 10. Snapshots of the droplet mass distribution at times $t = 0.33$ and $t = 1.0$ during the ejection from the center of the Taylor-Green vortices to the edges of the lattice for $\mathrm{St} = 0.9\mathrm{St}_c$.

numerical method in term of spatial transport in relationship to the possible singularities due to pressure-less gas dynamics.

Such dynamics are representative of what will occur in a turbulence flow with vortical structures such as the frozen one presented in Figure 11. Once again, we are able to observe with an Eulerian description the evolution of the spray mass density and momentum, the dynamics of segregation due to the vortices. Once again, the key issue is the robustness of the Eulerian numerical scheme that will be used in more complex configurations. The comparisons, in the two considered configurations with a Lagrangian solver developed by J. Reveillon, in order to characterize the quality of the approximation, are performed in de Chaisemartin et al. (2006) and show the ability of the Eulerian multi-fluid model to correctly reproduce the dynamics of sprays of various size with a limited computational cost.

Euler/Lagrange comparisons In fact we will present some comparisons with Lagrangian numerical simulations in the context of the non-uniform initial distribution in both the non-evaporating and the evaporating context. Again, precise quantitative comparisons are conducted in de Chaisemartin et al. (2006) and this is not the purpose of the present contribution to provide such detailed information but to show the behavior of the multi-fluid approach and provide references. In Figure 12, the non-evaporating case is considered at time $t = 0.4$ for $\mathrm{St} = 0.03$. On the left, we have represented the mass density spatial distribution during the ejection process, as well as a cut on the right allowing to compare the Lagrangian results projected on the Eulerian mesh at the considered time. For the evaporation case, in Figure 13 we take $t = 0.4$ for $\mathrm{St} = 0.015$ and $\mathrm{K} = 1/3$. Globally the Eulerian model is able to reproduce precisely the dynamics of the spray mass density and evaporation. However, two key points have to be underlined. The mesh size in order to be very precise (that is the order of 2% of the error compared to a reference very

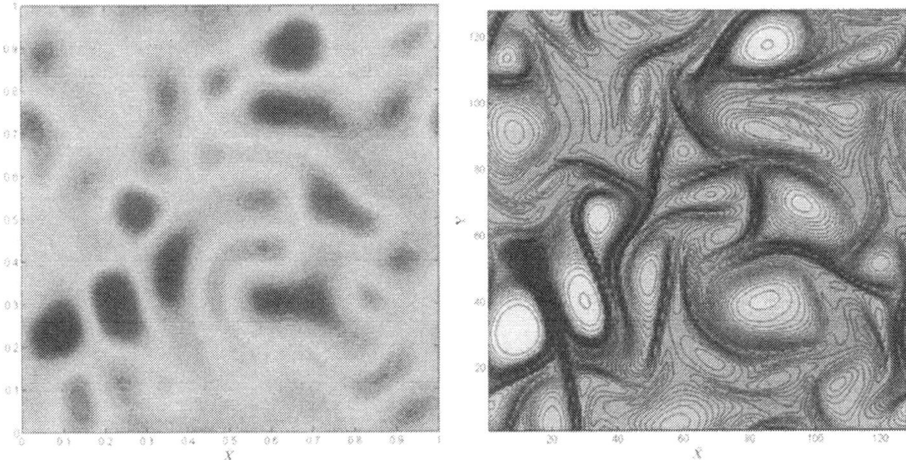

Figure 11. Left: Vorticity field for a frozen forced isotropic gaseous turbulent flow. Right: Droplet mass density at time $t = 1.5$ with a 256×256 spatial discretization and $\mathrm{St} = 0.9\mathrm{St}_c$ based on the maximal strain rate observed in the gas flow field.

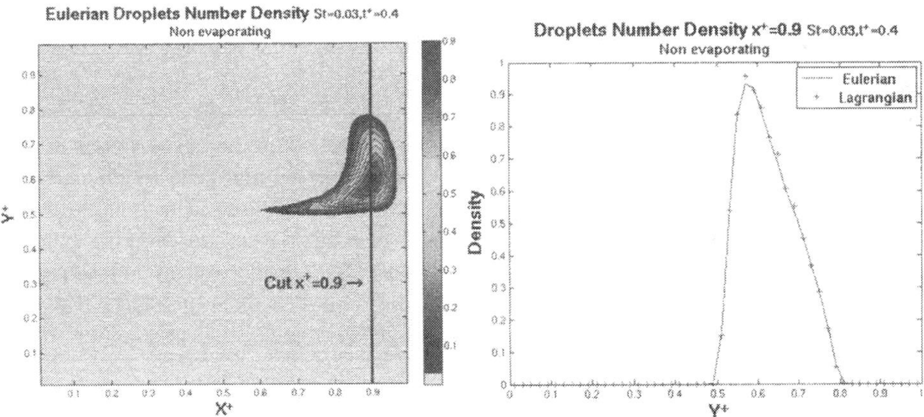

Figure 12. Non-evaporating spray dynamics at time $t = 0.4$ for $\mathrm{St} = 0.03$ on a grid of 100×100 cells and comparison with the Lagrangian solution projected on the Eulerian mesh.

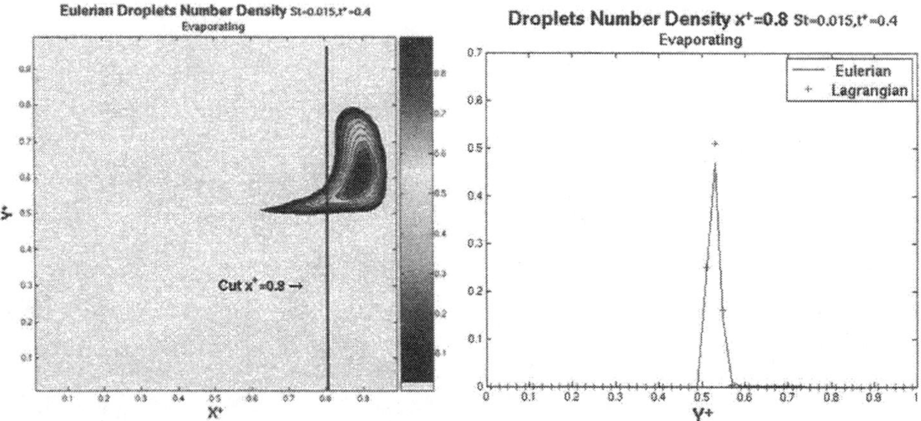

Figure 13. Evaporating spray dynamics at time $t = 0.4$ for $\mathrm{St} = 0.015$ on a grid of 100×100 cells and comparison with the Lagrangian solution projected on the Eulerian mesh.

finely discretized Lagrangian solution) has to be 400×400 cells in the non-evaporating case. Consequently, since we have to limit ourselves to a second order in space method for stability purposes, we have to compensate the numerical diffusion by a higher level of refinement. Besides, for the evaporating case, the order of the multi-fluid size discretization is one and has a strong impact on the global error. This is the reason why some higher-order methods have been designed, but still we need a further improvement at this level for which we will propose some research directions.

Conclusion We have designed and validated a Eulerian multi-fluid model as well as the associated numerical schemes, which have nice structure properties in order to be able to capture the high concentrations effects associated with spray segregation by vortices. The dynamics of such sprays are well captured as long as the critical Stokes number is not too far from the calculated critical value. For a very precise resolution in the spirit of DNS, the discretization level has to be high enough since the developed second-order methods inherit the usual numerical diffusion associated to TVD methods. However, the algorithm has been designed for its parallel optimization capabilities.

Let us also underline that the difficulty encountered in the multi-fluid model, that is the fact that we can only describe one mean velocity at each location is very common in most of the existing Eulerian models (two-fluid approaches for example); however, since we work with the mean velocity conditioned by size, our approach is much less restrictive since we can reproduce the fact that droplets of different sizes have various dynamics. This should not hide two facts: (1) the need for a Eulerian model in terms of velocity moments able to describe the presence of droplet of the same size with multiple velocity and, and (2) the need for another modeling associated with the Williams equation completed in the case of "crossing" of two clouds with opposite velocities. The global question is then: What model should really be used in region of very high droplet concentrations from both a kinetic and fluid point of view?

5.4 Spatially decaying turbulence

The purpose of this subsection is to validate a similar model for ensemble-averaged simulations (EAS) of sprays in turbulent flow fields through a comparison to DNS of a non-homogeneous turbulent flow. Such a scope has to rely on dedicated Euler/Lagrange DNS associated with a Eulerian statistical analysis for the spray characteristics. Much literature has been devoted to the development of analytical tools for the Eulerian analysis of the turbulent dispersion of both passive scalars and fixed size particles starting from the key paper of Taylor (1921). We refer to the paper by Mashayek and Pandya (2003) for a review in the field. The issue has also been tackled numerically in a number of papers (see Reveillon et al. (2004) and the references therein), however, few papers (Mashayek et al., 1997; Reveillon et al., 1998) have been devoted to the evaporation of polydisperse sprays and almost none to the configuration of a non-uniform turbulent flow field. However, it was not the purpose of those papers to analyze the resulting flow fields from a Eulerian point of view and to provide a comparison with a Eulerian numerical simulation.

This Eulerian perspective is the aim of the present subsection and we have selected, following the work in Reveillon et al. (2004), the configuration of a grid turbulence or "Spatially Decaying Turbulence" (SDT) for the gaseous carrier flow. Various monodisperse sprays are injected in "thermal equilibrium" with the surrounding turbulent flow after having been "thermalized" in a forced isotropic homogeneous turbulence within the framework of one-way coupling. As a result of the interaction between evaporation and turbulence, the spray becomes polydisperse along the streamwise direction. A comprehensive analysis of the Eulerian statistics of the spray provided by DNS simulations, for various initial Stokes numbers, has thus been carried out. The details of the configuration and of the Eulerian analysis are to be found in Reveillon et al. (2004). A special emphasis has been laid on the polydisperse character of the spray in the particular chosen configuration. It has been shown that the Eulerian dynamics of the droplets at given streamwise directions are conditioned by their size and the velocity distribution can be considered as Maxwellian thus justifying the use of a Eulerian multi-fluid model in order to reproduce such a behavior. This is an important result in the process of the development of Eulerian models allowing the description of the dispersion, the evaporation and thus the combustion of turbulent sprays in configurations where droplets of various sizes can have various dynamics.

Gas-spray Euler/Lagrange DNS – configuration We first briefly present the modeling and configuration of the two-phase flow we are going to study. The main results and conclusions from Reveillon et al. (2004) are then recalled. Through the Navier-Stokes equations, a classical Eulerian formulation is used to resolve the carrier phase, which is a compressible Newtonian fluid following the equation of state for a perfect gas. The instantaneous balance equations describe the evolution of the mass, the momentum, the total energy and can be found for example in Reveillon et al. (2004). The Eulerian/Lagrangian simulations used in this paper have been extensively described and analyzed in an Eulerian context in Reveillon et al. (2004) and even if it is questionable, these simulations are referred to as 'DNS'. For details in the numerical algorithms, the reader is referred to this reference, only the presentation of the configuration will be repeated here.

In order to implement a more realistic configuration than the usual fully homogeneous configurations, we decided to consider a Spatially Decaying Turbulence (SDT). This configuration was selected for three reasons. Firstly, it is non-homogeneous in the streamwise direction, offering a spatial evolution of the stationary statistics. Thus, a more realistic analogy with experimental or industrial flows may be carried out. Secondly, the evolution of a droplet inserted inside a turbulent flow may be divided into two sequences: first the droplet needs a delay while it adapts its characteristics to the flow properties (mean velocity, temperature, etc.) before a "wandering" stage where it undergoes local fluctuations of the flow. The choice has been made to consider droplets past their adaptation period in order to focus on the sole effects of the turbulence fluctuations on the spray dispersion. The third reason, as already mentioned, is to provide a good test case from which to extract the Eulerian properties of the spray in order to indicate what kind of Eulerian model should be used for predictability purposes. We have decided to focus on simplified evaporation laws and one-way coupling to isolate the key physical phenomena to be captured by a Eulerian description, which would be otherwise hidden by the intricacies of the many couplings occurring with more complex models. It is important to differentiate the direct impact of the turbulence on the dispersion from a side-effect due to droplet presence. Consequently, a one-way coupling has been implemented leading to a turbulent flow with constant main properties whatever the characteristics of the disperse spray. Moreover, non-reactive flows are considered in this work leading to an almost constant gas temperature in the whole computational domain. Thus, injected droplets maintain their saturation temperature at a constant level and a *d-square* law may be used to describe the time evolution of droplet surface which, by definition, decreases linearly with time. Thus polydispersity processes may be studied while using simple initial conditions for the spray description.

The permanent injection of an isotropic homogeneous turbulence into the domain seems to be an ideal way to simulate a grid turbulence with well-known statistics. The best way of approaching this problem consists in resolving, apart from the main computational domain, a forced isotropic homogeneous turbulence with prescribed properties (kinetic energy, dissipation, integral scale). Then this established and controlled turbulent flow is used as boundary condition for the main DNS problem. The fully compressible NS equations are then solved with periodic boundary conditions along the spanwise directions and NSCBC (Poinsot and Lele, 1992) for the inlet and the outlet along the streamwise direction in the physical-space domain where the SDT is simulated. A 2-D example of the procedure is shown in Figure 14 where both the spectral and physical domains are shown along with dispersing droplets. However, all the numerical simulations presented here will be carried out in 3-D configurations. In the following, $x = 0$ corresponds to the streamwise position where evaporation starts.

Because both turbulent and two-phase flows are involved, two sets of reference parameters may be used. A first set depends on the properties of the turbulence: the velocity root mean square U', the integral scale L_{turb} and, consequently, the eddy turn over time $\tau_{\mathrm{gas}} = L_{\mathrm{turb}}/U'$. These parameters are equal to the corresponding mean values in the statistically stationary spectral computation. A second set of normalization parameters is based on the properties of the spray. Droplet size is expressed in term of surface s which is normalized in the results by their unique (monodisperse) injection value s^0. The other parameters are the characteristic droplet evaporation time τ_{ev}, the carrier-phase mean velocity $A = \overline{U}$ and, consequently, the mean distance $L_{\mathrm{ev}} = \overline{U}\,\tau_{\mathrm{ev}}$ covered by a droplet before its total evaporation. A last parameter concerning the spray is the momentum or kinetic response time τ_{p} which has been defined above as a function varying linearly with the droplet surface $\tau_{\mathrm{p}} \propto s^0$.

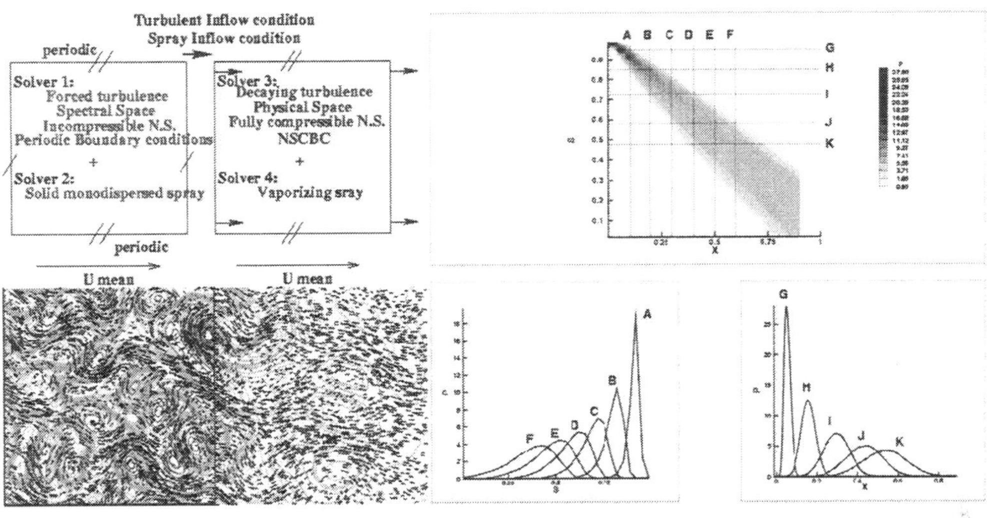

Figure 14. Left/Top: Schematic overview of the coupling between the four solvers. Left/Bottom: 2-D example of the injection procedure. Left: Spectral domain with forced turbulence. Right: Physical-space domain. Lines: vorticity levels. Vectors: droplet velocity. Right: Iso-contours (top) and cuts of the PDF $P(x, s)$. The profiles are made along both s (bottom right) and x (bottom left) directions for different positions shown in the top plot.

Both computational grids (spectral and physical) have the same streamwise (L_x) and spanwise (L_y, L_z) dimensions: $L_x = 0.7L_{\text{ev}} = 14L_{\text{turb}}$ and $L_y = L_z = 0.35L_{\text{ev}} = 7L_{\text{turb}}$. The following grid has been used for the 3-D spectral DNS: $64 \times 32 \times 32$ whereas $128 \times 64 \times 64$ nodes have been necessary for the 3D physical domain. Concerning the liquid phase, four configurations have been carried out. They are based on various initial Stokes numbers: St $= 0.03$ (T0), St $= 0.14$ (T1), St $= 0.39$ (T2) and St $= 1.95$ (T3). The mean evaporation delay is the same in the four cases and we will present results in the present contribution for the (T2) case. From a general point of view, statistics are considered temporally and spatially along the spanwise directions. The substance of the results obtained in Reveillon et al. (2004) reveals that the droplet mean velocity is conditioned by the droplet size; besides, at a given space location, and for a given size, the velocity distribution is Gaussian around the mean value and conditioned by size as well, which indicates that a Eulerian model should have the ability to capture the size-conditioned dynamics.

Phase-space diffusion coefficients The key point in order to perform relevant comparisons between numerical simulation involving the semi-kinetic model and the results from the Eulerian analysis of the DNS is the evaluation of the phase-space diffusion coefficients as stated in Massot et al. (2004). This is an important first step in order to validate such an approach and if it works, we will then be able to compare the values of these coefficients extracted from the DNS to model evaluation from local gaseous turbulent properties such as kinetic energy and Lagrangian time (see, for example, (Minier and Peirano, 2001)). In order to compute the coefficients from the

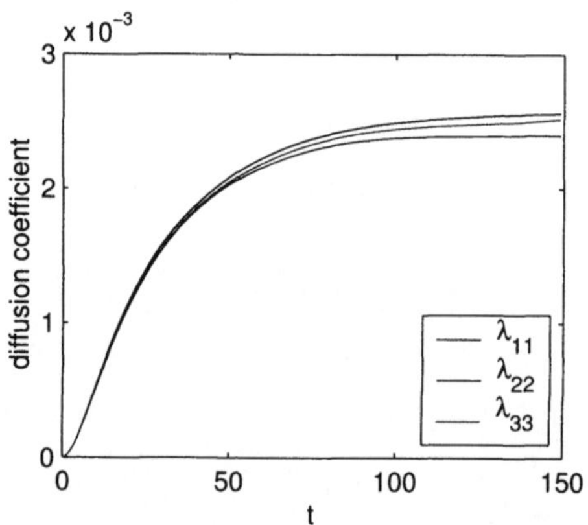

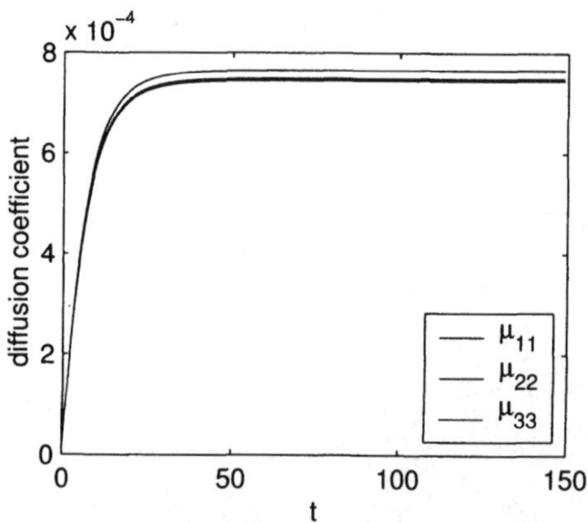

Figure 15. Phase-space diffusion coefficients in the spectral domain during a residence time of 150, which corresponds to a mean motion of one spectral box at velocity $\overline{U}$. Left: $\overline{\lambda}$. Right: $\overline{\mu}$.

DNS results, we need to sample a number of droplets and to keep track of the local gas velocity they have seen along their trajectories. We then create a so-called "life table" where we store, for ten thousand droplets, their complete history along their trajectories including the gas-phase velocity they are reacting to. Besides, for the preceeding relations providing the diffusion coefficients are valid, the initial condition has to satisfy that no initial correlation exists between the distribution in phase space and the stochastic field $\mathbf{U}'$. Consequently, we start with a uniformly spatially distributed cloud for the sampling droplets and let them "thermalize" in the spectral domain during $3\tau_{\text{ev}}$, that is a wandering motion across three spectral boxes without evaporation and the droplets then arrive in the physical domain and start evaporating at their entrance. We then have at our disposal the history of the gaseous fluctuations in both the spectral and physical domains.

The SDT case studied in this paper falls within the category of homogeneous mean flows which means that $\overline{\mathbf{U}}$ is a constant and the expressions given by equations (2.25-2.24) are valid. This leaves the velocity-integrated phase-space diffusion coefficients $\overline{\lambda}$ and $\overline{\mu}$ as the only parameters to be recovered from the "life table" issued from the DNS since $\overline{\gamma}$ can be shown to be negligible. From this table, we first evaluate the coefficients in the spectral domain. After wandering across one spectral box, we see that the coefficients reach an equilibrium value for which it can be checked that the actual droplet agitation internal energy corresponds to the Stokes number times the $\overline{\mu}$ diffusion coefficient divided by two (see Eq. 2.27). Besides, the isotropy in the three directions is confirmed by the results for the three spatial directions as observed in Figure 15. This isotropy is also found in the physical domain. However, in the physical domain, the effect of evaporation is coupled to the decrease of kinetic energy as well as to the increase of integral time scale and yields a rather constant $\overline{\lambda}$ coefficient in Figure 16, whereas the $\overline{\mu}$ coefficient first decreases and then increases further away from the injection point.

The effect of the $\overline{\mu}$ coefficient is rather classical since it is the source term in the droplet agitation internal energy and defines the dispersion of the velocity dispersion around the mean value. The $\overline{\lambda}$ coefficient however is related to the fact the turbulent motion is not a Markovian process

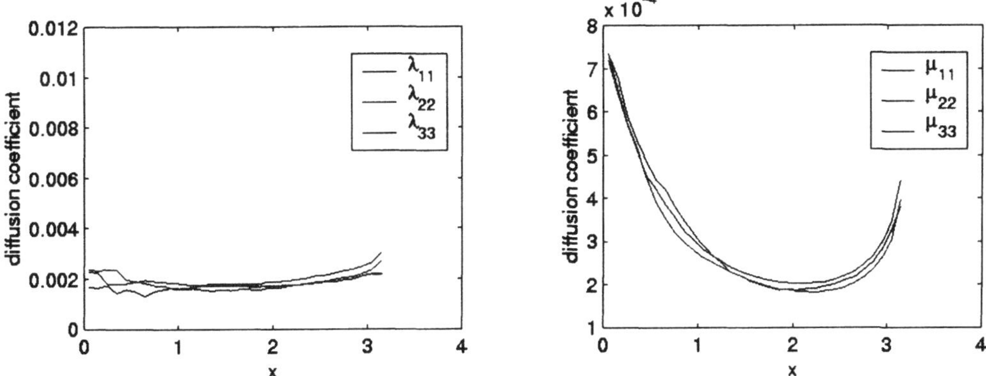

Figure 16. Phase-space diffusion coefficients in the physical domain versus position. Left: $\overline{\lambda}$. Right: $\overline{\mu}$.

but contains some memory associated to the turbulent scales. It creates an additional pressure term. Depending on the Stokes number, this term has various effects. For small Stokes numbers, on can show that system (2.27) becomes, using a singular perturbation process, a diffusion equation on the density :

$$\mathcal{D}_t \overline{n} + \partial_x \left(\overline{n}\,\overline{U} + \partial_x \left(\tilde{\lambda} n \right) \right), \tag{5.2}$$

where $\tilde{\lambda} = \mathrm{St}\, s \overline{\lambda}$ has a finite limit as St $\rightarrow 0$. This was already noticed in Reeks (1991). Thus $\overline{\lambda}$ play a major role in this context and is the leading pressure term in the state law. On the other end of the spectrum, for large Stokes number, the effect of $\overline{\lambda}$ become a minor one, since one can show that its contribution in the state law is very small compared to the agitation internal energy (see Massot et al., 2004).

Comparisons with Eulerian multi-fluid model We finally present the comparisons between the ensemble-averaged Eulerian fields of the disperse phase analyzed from the Euler-Lagrange DNS and the ones simulated from the Eulerian model and conclude on the ability of our approach to capture the physics involved. The SDT configuration yields a stationary set of conservation equations and introducing $\tau = (s_0 - s)/\mathrm{K}$, we end up with $\mathcal{D}_t Q = \partial_\tau Q$ in the system (2.27). The size of the droplets becomes equivalent to a Lagrangian time for the ensemble-averaged equations and this yields a set of conservation equations that is exactly the unstationary Euler equations for a real gas. Besides, we end up with a set of conservation in only the axial direction, with an equivalent $\gamma = 3$, as for an ideal gas in one dimension. We then integrate using standard WENO-5 schemes and the obtained results are presented in Figure 17. The symbols are extracted from the DNS and the lines come from the Eulerian simulation. In the first plot on the left, the density spatial distributions conditioned by various sizes are presented and the connection with Figure 14 is obvious. For the other plot representing the velocity and the droplet turbulent kinetic energy, we have "recentered" the profiles by switching to the referential associated with the mean velocity $\overline{U}$. It corresponds to describing the evolution of the velocity spatial distribution condi-

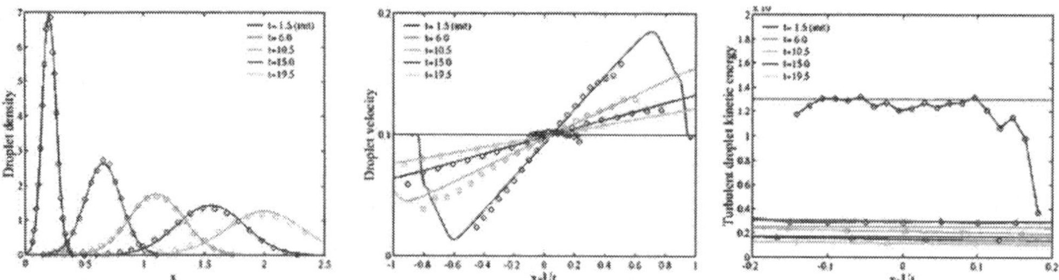

Figure 17. Time evolution of the droplet number density (left), velocity (middle) and agitation internal energy (right) for five times corresponding to the statistically stationary profiles of the NDF conditioned by droplet size. The symbols represent the data from the DNS and the solid lines the results from the Eulerian semi-kinetic model.

tioned by droplet size along the mean motion of the spray. The same is done for the turbulent kinetic energy, which we have also called previously the "droplet agitation internal energy" by analogy to kinetic theory. We then obtain very good agreement between the average data issued from the DNS and the calculated profiles for the three quantities.

6 Conclusions and Future Directions

In this contribution we have tried to provide a global picture of the status of Eulerian multi-fluid models for the description of polydisperse evaporating sprays, an important issue as far as many industrial applications are concerned. Its main advantage as compared to the usual two-fluid approaches is the ability to describe the dynamics, evaporation and heating conditioned by droplet size without having to cope with the difficult task of retrieving for example the size/velocity correlation from mixed global moments. After having described the derivation of the semi-kinetic system of conservation, we identified the underlying assumptions at the kinetic level, which are equivalent to a closure of the system of "fluid" or macroscopic equations. As with many moment methods, there is a loss of information; however, we are able to characterize it and to determine the mathematical singularities stemming from such assumptions. This is a major advantage in order to design efficient and robust numerical methods in order to solve the resulting systems of conservations equations. Even though this approach has recently been extended to higher order, the necessity to discretize the size phase space too finely because of numerical diffusion in the size discretization shows the necessity to create some new numerical schemes in the field.

Another way is to use the recently developed direct quadrature method of moments (DQ-MOM) (Marchisio and Fox, 2005) to treat Williams equation in a Eulerian framework. As its name implies, DQMOM is a moment method that closes the non-linear terms using weighted quadrature points (abscissas) in phase space. The method distinguishes itself from other quadrature methods (e.g., QMOM, McGraw (1997); Marchisio et al. (2003)) by solving transport equations for the weights and abscissas directly (instead of transport equations for the moments). We have recently compared the two approaches on several test cases and the advantages and drawbacks of both approaches are analyzed in Fox et al. (2006). Where the multi-fluid experiences some numerical diffusion in the size phase space, the principle of the DQMOM allows one to avoid such problems. However, the question of droplet evaporation and the associated flux of disappearing droplets is difficult to accurately tackle in the framework of moment methods, whereas it is an easy task for the multi-fluid model. We then believe that one way to really improve the capability of the actual methods would be to combine the flexibility of multi-fluid models to the efficiency of the DQMOM approach.

7 Acknowledgments

The present research was done thanks to a Young Investigator Award from the French Ministry of Research (New Interfaces of Mathematics - M. Massot, 2003-2006), the support of the French Ministry of Research (Direction of the Technology) in the program: "Recherche Aéronautique sur le Supersonique" (Project coordinator: M. Massot 2003-2006) and an ANR (National Research Agency - France) Young Investigator Award (M. Massot, 2006-2009). The author also acknowledges support from IDRIS-CNRS (Institut de Developpement et de Ressources en Informatique Scientifique, Centre National de la Recherche Scientifique) where some of the computations were performed. I would like to thank my present and former graduate students who work on this subject: S. de Chaisemartin, G. Dufour and a special mention to F. Laurent; I also acknowledge the precious help of my other collaborators: R. O. Fox, A. Gomez, R. Knikker, C. Mouhot, C. Péra, J. Reveillon, M. D. Smooke and P. Villedieu. Special thanks to D. L. Marchisio and R. O. Fox for the organization of the CISM course and their patience and professionalism in the editing process.

Bibliography

B. Abramzon and W. A. Sirignano (1989). Droplet vaporization model for spray combustion calculations. *International Journal of Heat and Mass Transfer*, 32:1605–1618.

P. Achim (1999). *Simulation de collisions, coalescence et rupture de gouttes par une approche lagrangienne: application aux moteurs à propergol solide*. PhD thesis, Faculté des Sciences de l'Université de Rouen.

A. A. Amsden, P. J. O'Rourke, and T. D. Butler (1989). Kiva II, a computer program for chemically reactive flows with sprays. Technical Report LA-11560-MS, Los Alamos National Laboratory.

N. Ashgriz and J. Y. Poo (1990). Coalescence and separation in binary collisions of liquid droplets. *Journal of Fluid Mechanics*, 221:183–204.

E. Aulisa, S. Manservisi, R. Scardovelli, and S. Zaleski (2003). A geometrical area-preserving volume-of-fluid method. *Journal of Computational Physics*, 192:355–364.

G. A. Bird (1994). Molecular Gas Dynamics and the Direct Simulation of Gas Flows. *Oxford Science Publications*, 42.

F. Bouchut (1994). On zero pressure gas dynamics. In *Advances in Kinetic Theory and Computing*, pages 171–190. World Scientific Publishing, River Edge, NJ.

F. Bouchut, S. Jin, and X. Li (2003). Numerical approximations of pressureless and isothermal gas dynamics. *SIAM Journal of Numerical Analysis*, 41:135–158.

A. Bracco, P. H. Chavanis, A. Provenzale, and A. Spiegel (1999). Particle aggregation in a turbulent Keplerian flow. *Physics of Fluids*, 11:2280–2286.

P. R. Brazier-Smith, S. G. Jennings, and J. Latham (1972). The interaction falling water drops: coalescence. *Proceedings of the Royal Society*, 326:393–408.

C. Cercignani, R. Illner, and M. Pulvirenti (1994). *The Mathematical Theory of Dilute Gases*. Springer-Verlag, New York.

G. Chanteperdrix, P. Villedieu, and J. P. Vila (2002). A compressible model for separated two-phase flows computations. In *ASME Fluids Engineering Division Summer Meeting*, number 31141, Montreal.

P. H. Chavanis (2000). Trapping of dust by coherent vortices in the solar nebula. *Astronomy and Astrophysics*, 356:1089–1111.

S. de Chaisemartin, F. Laurent, M. Massot, and J. Reveillon (2006). Evaluation of Eulerian Multi-Fluid versus Lagrangian methods for the ejection of polydisperse evaporating sprays by vortices. In *11th SIAM International Conference on Numerical Combustion, Granada, Spain, April*. submitted to Journal of Computational Physics.

S. Descombes and M. Massot (2004). Operator splitting for nonlinear reaction-diffusion systems with an entropic structure: singular perturbation and order reduction. *Numerische Mathematik*, 97(4):667–698.

K. Domelevo and L. Sainsaulieu (1997). A numerical method for the computation of the dispersion of a cloud of particles by a turbulent gas flow field. *Journal of Computational Physics*, 133(2):256–278.

D. A. Drew and S. L. Passman (1999). Theory of Multicomponent Fluids. *Applied Mathematical Sciences*, 135.

G. Dufour (2005). *Modélisation multi-fluide eulérienne pour les écoulements diphasiques à inclusions dispersées*. PhD thesis, Université Paul Sabatier Toulouse III.

G. Dufour, M. Massot, and C. Mouhot (2007). Eulerian semi-kinetic modeling of sprays in a weakly turbulent gaseous motion: singularity formation and relation to pressureless gas dynamics. Technical report, in preparation.

G. Dufour, M. Massot, and P. Villedieu (2003). Un modèle de fragmentation secondaire pour les brouillards de gouttelettes. *Comptes Rendus de l'Académie des Sciences de Paris, Série I (Mathématique)*, 336(5):447–452.

G. Dufour and P. Villedieu (2005). A second-order multi-fluid model for evaporating sprays. *M2AN Mathematical Modelling and Numerical Analysis*, 39(5):931–963.

J. K. Dukowicz (1980). A particle-fluid numerical model for liquid sprays. *Journal of Computational Physics*, 35(2):229–253.

R. O. Fox, F. Laurent, and M. Massot (2006). Numerical simulation of polydisperse, dense liquid sprays in an Eulerian framework: direct quadrature method of moments and multi-fluid method. *Journal of Computational Physics*. submitted.

R. O. Fox (2007). Introduction and fundamentals of modeling approaches for polydisperse multiphase flows. In *Computational Models for Turbulent Multiphase Reacting Flows,* Udine, July 2006. CISM Courses and Lectures. Springer Verlag.

V. Giovangigli (1999). *Multicomponent Flow Modeling*. Birkhäuser Boston Inc., Boston, MA.

J. B. Greenberg, I. Silverman, and Y. Tambour (1993). On the origin of spray sectional conservation equations. *Combustion and Flame*, 93:90–96.

R. J. Hall, M. D. Smooke, and M. B. Colket (1997). Predictions of soot dynamics in opposed jet diffusion flames, a tribute to Irvin Glassman. In *Physical and Chemical Aspects of Combustion*, chapter 8, pages 189–230. Gordon and Breach Science Publishers.

M. Herrmann (2005). A Eulerian level set/vortex sheet method for two-phase interface dynamics. *Journal of Computational Physics*, 203(2):539–571.

L.-P. Hsiang and G. M. Faeth (1993). Drop properties after secondary breakup. *International Journal of Multiphase Flow*, 19(5):721–735.

K. E. Hyland, S. McKee, and M. W. Reeks (1999). Derivation of a pdf kinetic equation for the transport of particles in turbulent flows. *Journal of Physics A*, 32(34):6169–6190.

J. Hylkema (1999). *Modélisation cinétique et simulation numérique d'un brouillard dense de gouttelettes. Application aux propulseurs à poudre*. PhD thesis, ENSAE.

J. Hylkema and P. Villedieu (1998). A random particle method to simulate coalescence phenomena in dense liquid sprays. In *Lecture Notes in Physics*, volume 515, pages 488–493, Arcachon, France. Proceedings of the 16th International Conference on Numerical Methods in Fluid Dynanics.

P.-E. Jabin (2002). Various levels of models for aerosols. *Mathematical Models and Methods in Applied Sciences*, 12(7):903–919.

C. Josserand, L. Lemoyne, R. Troeger, and S. Zaleski (2005). Droplet impact on a dry surface: triggering the splash with a small obstacle. *Journal of Fluid Mechanics*, 524:47–56.

A. Kaufmann (2004). *Vers la simulation des grandes échelles en formulation Euler-Euler des écoulements réactifs diphasiques*. PhD thesis, Institut National Polytechnique de Toulouse.

F. Laurent (2006). Numerical analysis of Eulerian multi fluid models in the context of kinetic formulations for dilute evaporating sprays. *M2AN Mathematical Modelling and Numerical Analysis*, 40(3):431–468.

F. Laurent and M. Massot (2001). Multi-fluid modeling of laminar poly-dispersed spray flames: origin, assumptions and comparison of the sectional and sampling methods. *Combustion Theory and Modelling*, 5:537–572.

F. Laurent, M. Massot, and P. Villedieu (2004a). Eulerian multi-fluid modeling for the numerical simulation of coalescence in polydisperse dense liquid sprays. *Journal of Computational Physics*, 194:505–543.

F. Laurent, V. Santoro, M. Noskov, A. Gomez, M. D. Smooke, and M. Massot (2004b). Accurate treatment of size distribution effects in polydispersed spray diffusion flames: multi-fluid modeling, computations and experiments. *Combustion Theory and Modelling*, 8:385–412.

R. J. LeVeque (2002). *Finite Volume Methods for Hyperbolic Problems*. Cambridge Texts in Applied Mathematics. Cambridge University Press, Cambridge.

D. L. Marchisio and R. O. Fox (2005). Solution of population balance equations using the direct quadrature method of moments. *Journal of Aerosol Science*, 36:43–73.

D. L. Marchisio, R. D. Vigil, and R. O. Fox (2003). Quadrature method of moments for aggregation-breakage processes. *Journal of Colloid and Interface Science*, 258(2):322–334.

F. Mashayek, F. A. Jaberi, R. S. Miller, and P. Givi (1997). Dispersion and polydispersity of droplets in stationary isotropic turbulence. *International Journal of Multiphase Flow*, 23(2): 337–355.

F. Mashayek and R. V. R. Pandya (2003). Analytical description of particle/droplet-laden turbulent flows. *Progess in Energy and Combustion Science*, 29:329–378.

M. Massot, R. Knikker, C. Péra, and J. Reveillon (2004). Lagrangian/Eulerian analysis of the dispersion of evaporating sprays in non-homogeneous turbulent flows. In *International Conference on Multiphase Flows, Japan*.

M. Massot, M. Kumar, A. Gomez, and M. D. Smooke (1998). Counterflow spray diffusion flames of heptane: computations and experiments. In *Proceedings of the 27th Symposium (International) on Combustion, The Combustion Institute* , pages 1975–1983.

M. Massot and P. Villedieu (2001). Modélisation multi-fluide eulérienne pour la simulation de brouillards denses polydispersés. *Comptes Rendus de l'Académie des Sciences de Paris, Série I (Mathématique)*, 332(9):869–874.

C. McEnally, A. Shaffer, M. B. Long, L. Pfefferle, M. D. Smooke, M. B. Colket, and R. J. Hall (1998). Computational and experimental study of soot formation in a coflow laminar ethylene diffusion flame. In *Proceedings of the 27th Symposium (International) on Combustion, The Combustion Institute*, pages 664–672.

R. McGraw (1997). Description of aerosol dynamics by the quadrature method of moments. *Aerosol Science and Technology*, 27:255–265.

J. P. Minier and E. Peirano (2001). The pdf approach to turbulent polydisperse two-phase flows. *Physics Reports*, 352:1–214.

A. Murrone and H. Guillard (2005). A five equation reduced model for compressible two-phase flow problems. *Journal of Computational Physics*, 202:664–698.

R. I. Nigmatulin (1991). *Dynamics of Multiphase Media*. Hemisphere, Washington D.C.

P. J. O'Rourke (1981). *Collective drop effects on vaporizing liquid sprays*. PhD thesis, Princeton University.

P. J. O'Rourke and A. Amsden (1987). The tab method for numerical calculation of spray droplet breakup. Technical Report 87545, Los Alamos National Laboratory, Los Alamos, New Mexico.

T. Poinsot and S. K. Lele (1992). Boundary conditions for direct simulations of compressible viscous flows. *Journal of Computational Physics*, 1(101):104–129.

S. B. Pope (2000). *Turbulent Flows*. Cambridge University Press.

D. Ramkrishna and A. G. Fredrickson (2000). *Population Balances: Theory and Applications to Particulate Systems in Engineering*. Academic Press.

P.-A. Raviart and L. Sainsaulieu (1995). A nonconservative hyperbolic system modeling spray dynamics. I. Solution of the Riemann problem. *Mathematical Models and Methods in Applied Science*, 5(3):297–333. ISSN 0218-2025.

M. W. Reeks (1991). On a kinetic equation for the transport of particles in turbulent flows. *Physics of Fluids A*, 3:446–456.

M. W. Reeks (1992). On the continuum equations for dispersed particles in nonuniform flows. *Physics of Fluids A*, 4(6):1290–1303.

M. W. Reeks (1993). On the constitutive relations for dispersed particles in nonuniform flows. I: Dispersion in simple shear flow. *Physics of Fluids A*, 5(3):750–761.

J. Reveillon, K. N. C. Bray, and L. Vervisch (1998). Dns study of spray vaporization and turbulent micro-mixing. In *AIAA 98-1028*, 36th Aerospace Sciences Meeting and Exhibit, January 12-15, Reno NV.

J. Reveillon, C. Péra, and M. Massot (2002). Lagrangian/Eulerian analysis of the dispersion of vaporizing polydispersed sprays in turbulent flows. CTR 2002 Summer Program, Center for Turbulence Research, Stanford University.

J. Reveillon, C. Péra, M. Massot, and R. Knikker (2004). Eulerian analysis of the dispersion of evaporating polydispersed sprays in a statistically stationary turbulent flow. *Journal of Turbulence*, 5(1):1–27.

M. Rüger, S. Hohmann, M. Sommerfeld, and G. Kohnen (2000). Euler/Lagrange calculations of turbulent sprays: the effect of droplet collisions and coalescence. *Atomization and Sprays*, 10 (1):47–81.

C.-W. Shu (1998). Essentially non-oscillatory and weighted essentially non-oscillatory schemes for hyperbolic conservation laws. In *Advanced Numerical Approximation of Nonlinear Hyperbolic Equations (Cetraro, 1997)*, volume 1697 of *Lecture Notes in Mathematics*, pages 325–432. Springer, Berlin.

M. D. Smooke, J. Crump, K. Seshadri, and V. Giovangigli (1990). Comparison between experimental measurements and numerical calculations of the structure of counterflow, diluted, methane-air, premixed flames. In *Proceedings of the 23rd Symposium (International) on Combustion, The Combustion Institute*, pages 503–511.

G. Strang (1968). On the construction and comparison of difference schemes. *SIAM Journal of Numerical Analysis*, 5:506–517.

S. Tanguy and A. Berlemont (2005). Development of a level set method for interface tracking: application to droplet collisions. *International Journal of Multiphase Flows*, 31(9):1015–1035.

G. I. Taylor (1921). Diffusion by continuous movements. *Proceedings of the Royal Society of London*, 20:421–478.

Y. Tsuji, T. Tanaka, and S. Yonemura (1998). Cluster patterns in circulating fluidized beds predicted by numerical simulation (discrete particle versus two-fluid models). *Powder Technology*, 95:254–264.

P. Villedieu and J. Hylkema (1997). Une méthode particulaire aléatoire reposant sur une équation cinétique pour la simulation numérique des sprays denses de gouttelettes liquides. *Comptes Rendus de l'Académie des Sciences de Paris, Série I (Mathématique)*, 325(3):323–328.

K. L. Wert (1995). A rationally-based correlation of mean fragment size for drop secondary breakup. *International Journal of Multiphase Flow*, 21(6):1063–1071.

F. A. Williams (1958). Spray combustion and atomization. *Physics of Fluids*, 1:541–545.

F. A. Williams (1985). *Combustion Theory (Combustion Science and Engineering Series)*. ed F. A. Williams (Reading, MA: Addison-Wesley).

D. L. Wright, R. McGraw, and D. E. Rosner (2001). Bivariate extension of the quadrature method of moments for modeling simultaneous coagulation and sintering of particle populations. *Journal of Colloid and Interface Science*, 236:242–251.

Ya. B Zel'dovich (1970). Gravitational instability: an approximate theory for large density perturbations. *Astronomy and Astrophysics*, 5:84–89.

A. Zucca, D. L. Marchisio, A. A. Barresi, and R. O. Fox (2006). Implementation of the population balance equation in CFD codes for modelling soot formation in turbulent flames. *Chemical Engineering Science*, 61.

Multi-fluid CFD Analysis of Chemical Reactors[1]

Bjørn H. Hjertager

Chemical Engineering Laboratory, Aalborg University Esbjerg, Esbjerg, Denmark

Abstract. An overview of modeling and simulation of flow processes in gas/particle and gas/liquid systems are presented. Particular emphasis is given to computational fluid dynamics (CFD) models that use the multi-dimensional multi-fluid techniques. Turbulence modeling strategies for gas/particle flows based on the kinetic theory for granular flows are given. Sub models for the interfacial transfer processes and chemical kinetics modeling are presented. An overview of a well established numerical solution method used is also given. Examples are shown for several gas/particle systems including flow and chemical reaction in risers as well as gas/liquid systems including bubble columns and stirred tanks.

1 Introduction

1.1 Basic considerations

Several different approaches are available for setting up the governing equations for multi-phase reactors. One method is the so-called PSIC (Particle-Source-In-Cell) procedure originally presented by Migdal and Agosta (1967).This method treats the continuous phase (liquid or gas) in a usual Eulerian description, whereas the dispersed phases (bubbles, droplets or particles) are described in a Lagrangian way. This means that the dispersed phase is tracked through the flow domain from inlet to outlet. In the original method it was assumed no interaction between the various dispersed phases and thus this method was only applicable to low volume fractions of the dispersed phase. In later years this limitation has been removed and the resulting method is called the discrete particle method (DPM).

The second method is the Volume of Fluid (VOF) for gas/liquid systems and was first proposed by Nichols et al. (1980). The method makes it possible to calculate the interphase between the gas

[1] This chapter is based on the work that the author and his colleagues have performed in the field of multiphase flow modeling over the last two decades at his two affiliations Telemark University College, Porsgrunn, Norway and his present affiliation Aalborg University Esbjerg, Denmark. The author will take this opportunity to thank and acknowledge his present and previous collaborators: Prof., Dr. T. Solberg, Dr. K. Morud, Dr. A. E. Samuelsberg, Dr. E. Manger, Dr. V. Mathiesen, Mr. T. Solbakken, Ms. T. Teppen, Dr. P. Chr. Friberg, Dr. C.H. Ibsen, Dr. N. G. Deen, Dr. K. Granly Hansen, Dr. S. Bove, Dr. J. Madsen, Mr. R. Hansen

and liquid and is thus able to resolve the details of the bubble shapes as they move through the liquid. The method assumes that the flow is fully segregated (resolved), which means that we do not have a mixture of gas and liquid anywhere in the computational cells. We either have pure gas or pure liquid present in the cells. This means that we solve only one set of balance equations: gas or liquid. The inter-phase is tracked by solving a transport equation for the VOF variable. This is a discontinuous variable that is either zero or one. The method is such that in order to resolve sharp interfaces it needs large grid resolution and hence large computer resources.

The third method is the Lattice Boltzman (LB) method that is able to model the flow around solid particles and is therefore able to fully resolve the solids fluid flow. Both the VOF and LB may be characterised by being so-called Direct Numerical Simulation (DNS) method for multi-phase flows.

The fourth method deduces the governing equations based on the Eulerian concept and is normally named the multi-fluid method (Spalding, 1985). This means that the phases are treated as interpenetrating fluids that share the space and interact through the source terms. All equations are such that the volume fraction may take a value between zero and one.

Figure 1-1 gives an overview of the four various methods for gas-liquid and gas-solids flow. We note that at the top row in the figure gives the most fundamental methods, i.e. methods that need no modelling since all aspects of the flow are calculated. These methods will however need large computational resources. We also see from the figure that DPM/PSIC approach is an intermediate method where some details are lost and needs to be modelled. The third row in the figure gives the multi-fluid method. This is the most demanding method with regard to modelling needs. The important aspect to note is that due to large dimensions and computing demands the multi-fluid method is the optimum choice for industrial scale reactors. However, the VOF, LB and DPM/PSIC methods are highly needed to feed better sub-models for the multi-fluid technique. We also see that in the last row it is envisaged a combination of multi-fluid and the DPM methods that may be optimum where large bubbles are flowing through a fluidised bed.

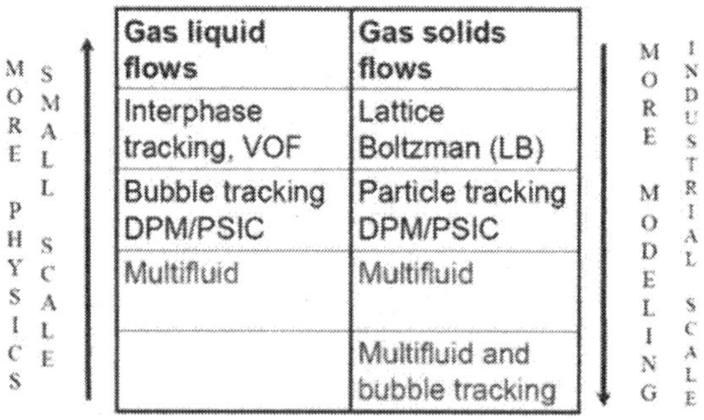

MORE PHYSICS / SMALL SCALE ↑

Gas liquid flows	Gas solids flows
Interphase tracking, VOF	Lattice Boltzman (LB)
Bubble tracking DPM/PSIC	Particle tracking DPM/PSIC
Multifluid	Multifluid
	Multifluid and bubble tracking

↓ MORE MODELING / INDUSTRIAL SCALE

Figure 1-1 Relations between description level for gas liquid and gas solids systems and 1) degree of modeling details needed and 2) scale of reactor (adapted from van der Hoef et al (2004)).

1.2 Objective and contents of Chapter

This Chapter will mainly deal with the multi-fluid technique. Details will be given on the derivation of the multi-fluid equations from the instantaneous single phase equations for the individual phases. Section 2 will deal with the basics of the multi-fluid approach including a brief mention of the mixture model. Section 3 will deal with the closures for bubble-liquid flows whereas Section 4 will handle the closure framework for solids fluid flows based on the kinetic theory of granular flow (KTGF). Additional sub models for bubble size, interfacial transfer processes, chemical reaction and stirred tank models are dealt with in Section 5. Numerical solution procedure is given in Section 6. Section 7 presents some applications of multi-fluid simulations for bubble columns, stirred tanks and fluidised beds. Section 8 gives a summary of the Chapter.

2 Basics of the Multi-fluid Approach

2.1 Introduction

A multiphase flow system consists of a number of single phase regions bounded by moving interfaces. In principle, a multiphase flow model could be formulated in terms of the local instantaneous variables pertaining to each phase and matching boundary conditions at the interfaces. The direct numerical solution of this kind of formulation is not feasible in practice. It would require an extremely fine mesh and a very small time step. The local instantaneous formulation can be used as starting point to derive macroscopic equations by an average procedure, whose numerical solution is feasible. The modelling of the unclosed terms arising from the average procedures is the price to pay for solving averaged multiphase flow systems. This Section starts with the presentation of the local instantaneous balance equations, based on the works of Drew (1983), Enwald et al. (1996) and Drew and Passman (1999) and generalized from two to N phases. Then, the time, volume and ensemble average approaches with their properties are briefly discussed. Employing the volume average technique the multi-fluid model is shown to be derived from the local instantaneous formulation.

2.2 Local instantaneous formulation

The balance of a scalar or vector variable ψ will be computed in a region in two- or three-dimensional space, V, fixed in space (Euler formulation), which can be shared by N phases as shown in Fig. 2.1.

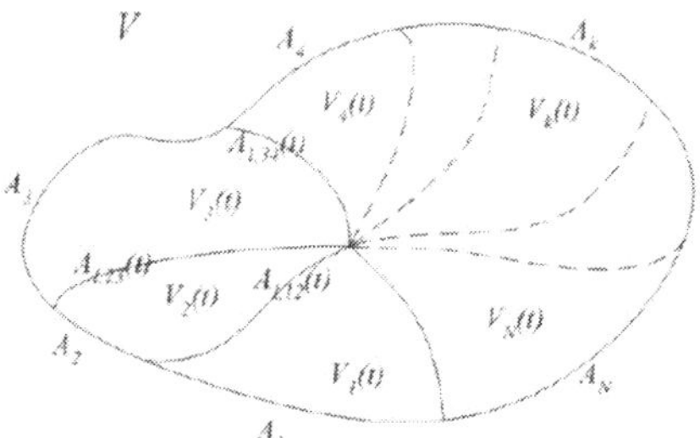

Figure 2-1 Eulerian control volume for balances of multi-fluid flow properties (Bove (2005))

A number of phases is present in the region V and each phase occupies a subregion $V_k \subseteq V$. V can then be written as the union of V_ks, where k = 1, . . . ,N.

$$V = \bigcup_{k=1}^{N} V_k(t) \tag{2.1}$$

The relation (2.1) holds also when some of the phases are not present in V. In fact, for such a case the measure of the sub regions for these phases is set equal to zero. The balance equation ([accumulation] = [net inflow] + [net sources]) of the variable ψ in V can be written as

$$\sum_{k=1}^{N}\left(\frac{\mathrm{d}}{\mathrm{d}t}\int_{V_k(t)} \rho_k\psi_k \mathrm{d}V\right) = -\sum_{k=1}^{N}\left(\int_{A_k(t)} \rho_k\psi_k(\vec{u}_k\cdot\vec{n}_k)\mathrm{d}A\right) - \sum_{k=1}^{N}\left(\int_{A_k(t)} (\vec{J}_k\cdot\vec{n}_k)\mathrm{d}A\right) + \\ +\frac{1}{2}\sum_{k=1}^{N}\sum_{j=1}^{N}(1-\delta_{jk})\int_{A_{I,jk}(t)} \Phi_{I,jk}\mathrm{d}A + \sum_{k=1}^{N}\int_{V_k(t)} \rho_k\Phi_k \mathrm{d}V \tag{2.2}$$

Here, $A_{I,jk}(t)$ is the interface between the jth and kth phases, $A_k(t)$ is the surface of the volume $V_k(t)$, ρ_k is the density, ψ_k is an extensive property of the flow, $\vec{n}_k$ is the outwardly directed normal unit vector to the interface of the volume occupied by phase k, $\vec{u}_k$ is the velocity of phase k, $\vec{J}_k$ is the molecular flux, Φ_k is the source term, $\Phi_{I,jk}$ is the interfacial source term and δ_{jk} is the Kronecker's delta (=1 when i=j and = 0 when $i \neq j$). In deriving Eq. (2.2), the interfaces were assumed to have no mass. In other words, the interfaces are treated as geometrical surfaces in which discontinuities in the variable may exist.
Defining the location of the surface $A_{I,jk}(t)$ as

$$\vec{r}_{I,jk} = \vec{r}_{I,jk}(x(\zeta,\eta,t), y(\zeta,\eta,t), z(\zeta,\eta,t)) \tag{2.3}$$

the velocity of the surface point, (ζ, η), is defined by:

$$\vec{u}_{I,jk} = \left(\frac{\partial \vec{r}_{I,jk}}{\partial t}\right)_{\zeta,\eta=const} \tag{2.4}$$

The LHS of Eq. (2.2) can be transformed into a volume and a surface integral using the Leibniz's theorem

$$\frac{\mathrm{d}}{\mathrm{d}t}\int_{V_k(t)} \rho_k\psi_k \mathrm{d}V = \\ \int_{V_k(t)} \frac{\partial}{\partial t}(\rho_k\psi_k)\mathrm{d}V + \sum_{j=1}^{N}(1-\delta_{jk})\int_{A_{I,jk}(t)} \rho_k\psi_k\vec{u}_{I,kj}\cdot\vec{n}_{kj}\mathrm{d}A + \int_{A_k(t)} \rho_k\psi_k\vec{u}_{I,k}\cdot\vec{n}_k\mathrm{d}A = \\ = \int_{V_k(t)} \frac{\partial}{\partial t}(\rho_k\psi_k)\mathrm{d}V + \sum_{j=1}^{N}(1-\delta_{jk})\int_{A_{I,jk}(t)} \rho_k\psi_k\vec{u}_{I,kj}\cdot\vec{n}_{kj}\mathrm{d}A \tag{2.5}$$

The surface integral over $A_k(t)$ vanishes because the velocity, $\vec{u}_{I,k}$ at a generic point of the surface $A_k(t)$ is zero as a consequence of control volume V being fixed in space. The convective and diffusion terms can be rewritten as sum of a volume and a surface integral using the Gauss' theorem:

$$\int_{V_k(t)} \nabla \cdot (\rho_k \psi_k \vec{u}_k) \mathrm{d}V = \int_{A_k(t)} \rho_k \psi_k \vec{u}_k \cdot \vec{n}_k \mathrm{d}A + \sum_{j=1}^{N} (1-\delta_{jk}) \int_{A_{I,kj}(t)} \rho_k \psi_k \vec{u}_k \cdot \vec{n}_{kj} \mathrm{d}A \tag{2.6}$$

$$\int_{V_k(t)} \nabla \cdot \vec{J}_k \mathrm{d}V = \int_{A_k(t)} \vec{J}_k \cdot \vec{n}_k \mathrm{d}A + \sum_{j=1}^{N} (1-\delta_{jk}) \int_{A_{I,kj}(t)} \vec{J}_k \cdot \vec{n}_{kj} \mathrm{d}A \tag{2.7}$$

Accordingly, Eq. (2.2) can be rewritten as:

$$\sum_{k=1}^{N} \int_{V_k(t)} \left[\frac{\partial}{\partial t}(\rho_k \psi_k) + \nabla \cdot (\rho_k \psi_k \vec{u}_k) + \nabla \cdot \vec{J}_k - \rho_k \Phi_k \right] \mathrm{d}V + \tag{2.8}$$

$$-\sum_{k=1}^{N} \sum_{j=1}^{N} (1-\delta_{jk}) \int_{A_{I,kj}(t)} \left[\left(\dot{m}_{I,kj} \psi_k + \vec{J}_k \cdot \vec{n}_{kj} \right) + \frac{1}{2} \Phi_{I,kj} \right] \mathrm{d}A = 0$$

Here, $\dot{m}_{I,kj}$ is the mass transfer per unit area of interface and unit time from the kth phase to the jth phase through the interface and is defined as:

$$\dot{m}_{I,kj} = \rho_k \left(\vec{u}_k - \vec{u}_{I,kj} \right) \cdot \vec{n}_{I,kj} \tag{2.9}$$

Eq. (2.8) must be satisfied for any $V_k(t)$ and $A_{I,kj}(t)$. Thus, the local instantaneous conservation equation is

$$\frac{\partial}{\partial t}(\rho_k \psi_k) + \nabla \cdot (\rho_k \psi_k \vec{u}_k) + \nabla \cdot \vec{J}_k - \rho_k \Phi_k = 0 \tag{2.10}$$

and the local instantaneous jump condition is

$$\psi_j \dot{m}_{I,jk} + \psi_k \dot{m}_{I,kj} + \vec{J}_j \cdot \vec{n}_{I,jk} + \vec{J}_k \cdot \vec{n}_{I,kj} = \Phi_{I,jk} \tag{2.11}$$

The general local instantaneous equations, Eqs. (2.10) and (2.11), can be specified for mass, momentum, energy, chemical species, etc. as indicated in Table 2-1.

BALANCE	Ψ_K	$\vec{J}_k$	Φ_K	$\Phi_{I,KJ}$
Total mass	1	0	0	0
Momentum	$\vec{u}_k$	$-\overline{\overline{T}}$	$\vec{b}_k$	$-\vec{m}^{\sigma}_{jk}$
Total energy	E_k	$\vec{q}_k - \overline{\overline{T}} \cdot \vec{u}_k$	$\vec{b}_k \cdot \vec{u}_k + Q_k$	$-\varepsilon^{\sigma}_{jk}$
Chemical specie	Y_k	$\vec{F}_{Y,k}$	Ψ_k	$-\Psi_{I,jk}$

Table 2-1 Value of Ψ_k, , $\vec{J}_k$, Φ_k and $\Phi_{I,kj}$ for balance equations and jump conditions (from Bove (2005))

The equations and jump conditions are expressed as:
Mass conservation

$$\frac{\partial \rho_k}{\partial t} + \nabla \cdot (\rho_k \vec{u}_k) = 0 \tag{2.12}$$

Mass jump condition

$$\dot{m}_{I,jk} + \dot{m}_{I,kj} = 0 \tag{2.13}$$

Momentum balance equation

$$\frac{\partial}{\partial t}(\rho_k \vec{u}_k) + \nabla \cdot (\rho_k \vec{u}_k \vec{u}_k) = \nabla \cdot \overline{\overline{T}}_k + \rho_k \cdot \vec{b}_k \tag{2.14}$$

Momentum jump condition

$$\dot{m}_{I,jk}\vec{u}_j + \dot{m}_{I,kj}\vec{u}_k + \overline{\overline{T}}_j \cdot \vec{n}_{I,jk} + \overline{\overline{T}}_k \cdot \vec{n}_{I,kj} = \vec{m}_{jk}^{\sigma} \tag{2.15}$$

Here, $\overline{\overline{T}}$ is the stress tensor, $\vec{b}_k$ represents the body forces and $\vec{m}_{jk}^{\sigma}$ is the surface traction, which has the dimension of stress and is defined as

$$\vec{m}_{jk}^{\sigma} = 2H_{jk}\sigma\vec{n}_{I,jk} - \nabla_{I,jk}\sigma$$

where σ is the surface tension, H_{jk} is the mean curvature of the interface and $\nabla_{I,jk}$ denotes the gradient on the interface between the jth and k^th^ phases in surface coordinates (Aris 1962).

Energy balance equation

$$\frac{\partial}{\partial t}(\rho_k E_k) + \nabla \cdot (\rho_k E_k \vec{u}_k) = -\nabla \cdot \left(\vec{q}_k - \overline{\overline{T}}_k \cdot \vec{u}_k\right) + \rho_k \left(\vec{b}_k \cdot \vec{u}_k + Q_k\right) \tag{2.16}$$

Energy jump condition

$$\dot{m}_{I,jk}E_j + \dot{m}_{I,kj}E_k + \left(q_j - \overline{\overline{T}}_j \cdot \vec{u}_j\right) \cdot \vec{n}_{I,jk} + \left(q_k - \overline{\overline{T}}_k \cdot \vec{u}_k\right) \cdot \vec{n}_{I,kj} = \varepsilon_{jk}^{\sigma} \tag{2.17}$$

Here, $\vec{q}_k$ is the heat flux, which is modelled by the Fourier's law, Q_k is the heating source per unit of mass and $\varepsilon_{jk}^{\sigma}$ is the surface energy associated with the interface jk and is formulated as

$$\varepsilon_{jk}^{\sigma} = \nabla_{I,jk} \cdot \vec{q}_{I,jk} - \nabla_{I,jk} \cdot \left(\sigma_{jk}\vec{u}_{I,jk}\right) \tag{2.18}$$

$\vec{u}_{I,jk}$ is the tangential component of the surface velocity.
Balance of a chemical specie

$$\frac{\partial}{\partial t}(\rho_k Y_k) + \nabla \cdot (\rho_k Y_k \vec{u}_k) = -\nabla \cdot \vec{F}_{Y,k} + \rho_k \Psi_k \tag{2.19}$$

Specie jump condition

$$\dot{m}_{I,jk}Y_j + \dot{m}_{I,kj}Y_k + \vec{F}_{Y,j} \cdot \vec{n}_{I,jk} + \vec{F}_{Y,k} \cdot \vec{n}_{I,kj} = \Psi_{I,jk} \tag{2.20}$$

Here, $\vec{F}_{Y,k}$ is the molecular or diffusive flux of the species Y in phase k, which is modelled by Fick's law; Ψ_k is the production or destruction rate of specie Y per unit of mass of phase k; $\Psi_{I,jk}$ is the production or destruction rate of specie Y per unit area of the interface jk.

The Eqs. (2.12)-(2.20) together with the appropriate constitutive equations describe completely the thermo fluid dynamics of multiphase systems. In cases where the details of the fluid dynamics of the phases is not of interest, as it is in several applications of process design, or when the solution of the detailed flow field is computationally unfeasible, averaging techniques are used.

2.3 Averaging techniques

In modelling multiphase flows, time, volume and ensemble averaging are the averaging techniques used. For that reason, the main lines of these techniques and their properties are given in the following section. For further details, the reader is referred to the literature: Nigmatulin (1979), Drew (1983), Joseph et al. (1990), Enwald et al. (1996) and Drew and Passman (1999). Defining a space domain $\mathcal{R}$, a time domain $\mathcal{T}$ and an event space $\mathcal{E}$ of a process $\mathcal{P}$, a microscopic formulation of a generic local instantaneous variable f defined in $\mathcal{P}$ can be written as $f(\vec{r},t;\mu)$. Here, $\vec{r} \in \mathcal{R}$ is the position vector, t $\in \mathcal{T}$ is the time and $\mu \in \mathcal{E}$ is a particular realization of the process $\mathcal{P}$

Phase indicator function. Before exposing the different averaging techniques, the phase indicator X_k for phase k needs to be introduced. It is a Heavyside function that picks the phase k while ignoring the other phases and the interfaces and is defined as

$$X_k = (\vec{r},\, t, \mu) = \begin{cases} 1 & \text{if } \vec{r} \in \mu \\ 0 & otherwise \end{cases} \tag{2.21}$$

An important relation for the phase indicator is

$$\frac{\partial X_k}{\partial t} + \vec{u}_{I,jk} \cdot \nabla X_k = 0 \tag{2.22}$$

which states that the material derivative of the phase indicator is zero. The gradient of the phase indicator can be expressed as

$$\nabla X_k = \left(\frac{\partial X_k}{\partial n}\right)\vec{n}_{I,kj} = \delta\left(\vec{r} - \vec{r}_{I,kj}\right)\vec{n}_{I,kj} \tag{2.23}$$

Here, $\delta\left(\vec{r} - \vec{r}_{I,kj}\right)$, is the Dirac's delta function associates with the interface between the kth and j^{th} phases. The phase indicator and its properties are useful in deriving the averaged equation or even only to sort out a particular phase microscopic description from the multiphase system.

Time averaging. Considering the variable $f(\vec{r},t;\mu)$ the time averaging is defined as

$$\langle f(\vec{r},t;\mu)\rangle^{T} = \frac{1}{T}\int_{t-\frac{T}{2}}^{t+\frac{T}{2}} f(\vec{r},t;\mu)\mathrm{d}\tau \tag{2.24}$$

In multiphase systems, f may represent a fluid property and be discontinuous between the phases. Hence, care must be taken in deriving f in time $\partial f/\partial t$ and space ∇f. Since the averaged balance equations are obtained multiplying the local balance equations by the phase indicator and then averaging, are useful the following relations.

Leibniz's theorem

$$\left\langle X_k \frac{\partial f}{\partial t}\right\rangle^{T} = \left\langle \frac{\partial (X_k f)}{\partial t}\right\rangle^{T} + \sum_{t_l \in \left(t-\frac{T}{2}, t+\frac{T}{2}\right)} \frac{1}{T\left|\vec{n}_{I,kj}\cdot\vec{u}_{I,kj}\right|}\vec{n}_{I,kj}\cdot\vec{u}_{I,kj} f\left(t_l^k\right) \tag{2.25}$$

Gauss' theorem

$$\left\langle X_k \nabla f\right\rangle^{T} = \left\langle \nabla (X_k f)\right\rangle^{T} - \sum_{t_l \in \left(t-\frac{T}{2}, t+\frac{T}{2}\right)} \frac{1}{T\left|\vec{n}_{I,kj}\cdot\vec{u}_{I,kj}\right|}\vec{n}_{I,kj} f\left(t_l^k\right) \tag{2.26}$$

Here, $f\left(t_l^k\right)$ denotes the value of f on the component k side of the interface. It should be also noted that $\vec{n}_{I,kj}\cdot\vec{u}_{I,kj}$ is positive when the interface is approaching the monitoring point and negative when is leaving it.

Volume averaging. The volume average of $f(\vec{r},t;\mu)$ is defined as

$$\langle f(\vec{r},t;\mu)\rangle^{V} = \frac{1}{V}\int_{V} f(\vec{r},t;\mu)\mathrm{d}V_{\eta} \tag{2.27}$$

Leibniz's theorem

$$\left\langle X_k \frac{\partial f}{\partial t}\right\rangle^{V} = \left\langle \frac{\partial (X_k f)}{\partial t}\right\rangle^{V} - \frac{1}{V}\int_{A_{I,kj}(\vec{r},t)} \vec{n}_{I,kj}\cdot\vec{u}_{I,kj} f(\vec{r}',t)\mathrm{d}A' \tag{2.28}$$

Gauss' theorem

$$\left\langle X_k \nabla f\right\rangle^{V} = \left\langle \nabla (X_k f)\right\rangle^{V} + \frac{1}{V}\int_{A_{I,kj}(\vec{r},t)} \vec{n}_{I,kj} f(\vec{r}',t)\mathrm{d}A' \tag{2.29}$$

Ensemble averaging. The ensemble average of f is

$$\langle f(\vec{r},t;\mu)\rangle^{\varepsilon} = \int_{\varepsilon} f(\vec{r},t;\mu)\mathrm{d}m(\mu) \tag{2.30}$$

where *dm(μ)* is the probability density on the set of all events ε.

Leibniz's rule

$$\left\langle X_k \frac{\partial f}{\partial t} \right\rangle^{\varepsilon} = \left\langle \frac{\partial (X_k f)}{\partial t} \right\rangle^{\varepsilon} - \left\langle f_{kj} \frac{\partial X_k}{\partial t} \right\rangle^{\varepsilon} \tag{2.31}$$

Gauss' rule

$$\left\langle X_k \nabla f \right\rangle^{\varepsilon} = \left\langle \nabla (X_k f) \right\rangle^{\varepsilon} - \left\langle f_{kj} \nabla X_k \right\rangle^{\varepsilon} \tag{2.32}$$

where f_{kj} is the value of the function f evaluated in the component k side of the interface kj.

Considerations on averaging. First of all, it should be noted that all averaging techniques exposed above satisfy the Reynold's averaging rules:

$$\langle f + g \rangle = \langle f \rangle + \langle g \rangle \tag{2.33}$$

$$\langle \langle f \rangle g \rangle = \langle f \rangle \langle g \rangle \tag{2.34}$$

$$\langle \text{constant} \rangle = \text{constant} \tag{2.35}$$

$$\left\langle \frac{\partial f}{\partial t} \right\rangle = \frac{\partial \langle f \rangle}{\partial t} \tag{2.36}$$

$$\langle \nabla f \rangle = \nabla \langle f \rangle \tag{2.37}$$

$$\langle \nabla \cdot f \rangle = \nabla \cdot \langle f \rangle \tag{2.38}$$

2.4 Averaged balance equations

To derive the averaged equations the volume average will be used. Mainly three reasons may be given for this: (i) the solution of the volume averaged equations can be validate against experimental volume averaged data; (ii) the volume average can be seen as a spatial filter technique closely related to the Large Eddy Simulation (LES) approach, and (iii) in numerical techniques such as finite volumes, the values of the variables in each finite volume are considered averaged on the finite volume. Hence, from here to the end of this Section, the average on a generic variable f has to be intended as volume average and will be denoted by $<f>$ without superscript to simplify the notation.

The averaging procedure will be applied to the local instantaneous equations to obtain the averaged balance equations. To obtain the general averaged balance equation, the instantaneous local balance equation, Eq. (2.10), is first multiplied by the phase indicator and then averaged. Recalling the relations (2.22), (2.23), (2.28), (2.29), (2.33)-(2.38), the general form of the averaged balance equation yields

$$\frac{\partial}{\partial t}\langle X_k \rho_k \psi_k \rangle + \nabla \cdot \langle X_k \rho_k \psi_k \vec{u}_k \rangle = -\langle X_k \vec{J}_k \rangle + \langle X_k \rho_k \Phi_k \rangle + \\ -\frac{1}{V}\sum_{j=1}^{N}\left(1-\delta_{jk}\right) \int_{A_{I,kj}} \left[\dot{m}_{I,kj}\psi_k + \vec{J}_k \cdot \vec{n}_{I,kj} \right] \mathrm{d}A \tag{2.39}$$

The averaged general jump condition is obtained multiplying the jump condition, Eq. (2.11), with the normal derivative of the phase indicator and averaging

$$\frac{1}{V}\int_V \left[\psi_j \dot{m}_{I,jk} + \psi_k \dot{m}_{I,kj} + \vec{J}_j \cdot \vec{n}_{I,jk} + \vec{J}_k \cdot \vec{n}_{I,kj} \right] \frac{\partial X_k}{\partial \vec{n}_{I,kj}} \mathrm{d}V = \frac{1}{V}\int_V \Phi_{I,jk} \frac{\partial X_k}{\partial \vec{n}_{I,kj}} \mathrm{d}V \tag{2.40}$$

which can be written as

$$\frac{1}{V}\int_V \left[\psi_j \dot{m}_{I,jk} + \psi_k \dot{m}_{I,kj} + \vec{J}_j \cdot \vec{n}_{I,jk} + \vec{J}_k \cdot \vec{n}_{I,kj} \right] \mathrm{d}A = \frac{1}{V}\int_V \Phi_{I,jk} \mathrm{d}A \tag{2.41}$$

Multi-fluid model. In cases where a detailed description of the interface dynamics is not relevant or is computationally too expensive, the averaging can be performed on a volume larger than the characteristic volume of the secondary phase as illustrated in Figure 2-2. Doing so, all the information lost on the scales smaller than the averaging scale need to be accounted for through sub-grid models, which may be derived empirically, analytically or numerically. The multi-fluid model is derived from the averaged local instantaneous balance equations with an averaging volume larger than the characteristic volume of the secondary phase, and is based on the assumption of inter-penetrating continuum media. That is, different phases can share the same spatial position.

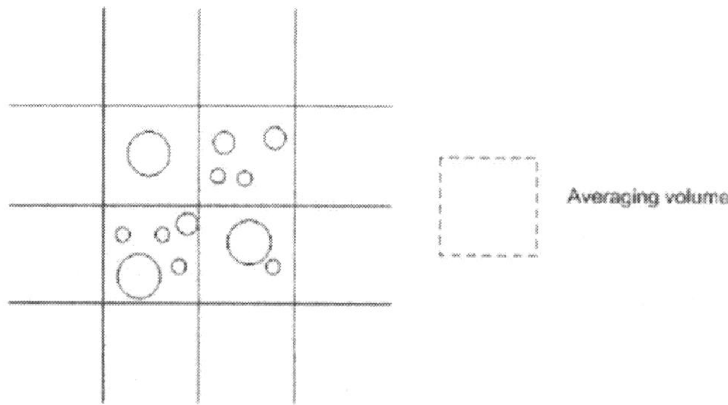

Figure 2-2: Averaging volume for a multi-fluid flow model framework (Bove (2005)).

The multi-fluid model balance equations can be written in a compact form if the following averages are used:

Phasic average

$$\overline{\psi}_k = \frac{\langle X_k \psi_k \rangle}{\alpha_k} \tag{2.42}$$

Here, α_k represents the volume fraction of phase k, and is defined as the volume average of the phase indicator X_k:

$$\alpha_k = \langle X_k \rangle \tag{2.43}$$

where

$$\sum_{k=1}^{N} \alpha_k = 1.0 \tag{2.44}$$

Another average of importance is the interfacial area per unit volume expressed as

$$A_k = -\langle \vec{n}_k \bullet \nabla X_k \rangle \tag{2.45}$$

Here $\vec{n}_k$ is the unit normal vector to phase k.

Mass weighted average (Favre averaging) is expressed as

$$\hat{\psi}_k = \frac{\langle X_k \rho_k \psi_k \rangle}{\alpha_k \overline{\rho}_k} \tag{2.46}$$

The averaged balance equation, Eq. (2.39), of a generic variable may then be written as

$$\begin{aligned}\frac{\partial}{\partial t}\left(\alpha_k \overline{\rho}_k \hat{\psi}_k\right) + \nabla \cdot \left(\alpha_k \overline{\rho}_k \hat{\vec{u}}_k \hat{\psi}_k\right) &= -\nabla \cdot \left[\alpha_k \left(\hat{\vec{J}}_k + \vec{J}_k^T\right)\right] + \left(\alpha_k \overline{\rho}_k \hat{\Phi}_k\right) + \\ &- \frac{1}{V}\sum_{j=1}^{N}\left(1-\delta_{jk}\right) \int_{A_{I,kj}} \left[\dot{m}_{I,kj}\psi_k + \vec{J}_k \cdot \vec{n}_{I,kj}\right] dA\end{aligned} \tag{2.47}$$

Here, $\vec{J}_k^T$ stems from the average of the convective term. It is an unclosed term and holds the information on the unresolved or residual turbulence scales. The unresolved vector (or tensor if ψ is a vector field) $\vec{J}_k^T$ can be expressed as

$$\vec{J}_k^T = \overline{\rho}_k \left(\widehat{\psi_k \vec{u}_k} - \hat{\psi}_k \hat{\vec{u}}_k\right) \tag{2.48}$$

Using Table 2-1, mass conservation and momentum balance equations can be derived from Eq. (2.47)

Mass conservation

$$\frac{\partial}{\partial t}\left(\alpha_k \overline{\rho}_k\right) + \nabla \cdot \left(\alpha_k \overline{\rho}_k \hat{\vec{u}}_k\right) = \Gamma_k \tag{2.49}$$

Momentum balance

$$\frac{\partial}{\partial t}\left(\alpha_k \overline{\rho}_k \hat{\vec{u}}_k\right) + \nabla \cdot \left(\alpha_k \overline{\rho}_k \hat{\vec{u}}_k \hat{\vec{u}}_k\right) = \nabla \cdot \left[\alpha_k \left(\hat{\overline{\overline{T}}}_k + \overline{\overline{T}}_k^T\right)\right] + \left(\alpha_k \overline{\rho}_k \vec{g}\right) + \Gamma_k \hat{\vec{u}}_k + \vec{M}_{I,k} \tag{2.50}$$

Here, Γ_k represents the exchange of mass between phases, $\overline{\overline{T}}_k^T$ is the residual stress tensor accounting for the unresolved turbulence scales and $\vec{M}_{I,k}$ accounts for the interfacial forces, i.e.

$$\Gamma_k = -\frac{1}{V}\sum_{j=1}^{N}\left(1-\delta_{jk}\right)\int_{A_{I,kj}} \dot{m}_{I,kj}\mathrm{d}A \tag{2.51}$$

and

$$\vec{M}_{I,k} = -\frac{1}{V}\sum_{j=1}^{N}\left(1-\delta_{jk}\right)\int_{A_{I,kj}}\left[\dot{m}_{I,kj}\vec{u}_k - \overline{\overline{T}}_k\cdot\vec{n}_{I,kj}\right]\mathrm{d}A \tag{2.52}$$

From the averaged jump condition for mass and momentum, i.e.

$$\frac{1}{V}\int_{A_{I,kj}}\left[\dot{m}_{I,jk}+\dot{m}_{I,kj}\right]\mathrm{d}A = 0 \tag{2.53}$$

and

$$\frac{1}{V}\int_{A_{I,kj}}\left[\vec{u}_j\dot{m}_{I,jk}+\vec{u}_k\dot{m}_{I,kj}-\overline{\overline{T}}_j\cdot\vec{n}_{I,jk}-\overline{\overline{T}}_k\cdot\vec{n}_{I,kj}\right]\mathrm{d}A = -\frac{1}{V}\int_{A_{I,kj}}\vec{m}_{jk}^{\sigma}\mathrm{d}A \tag{2.54}$$

The following constraints hold

$$\sum_{k=1}^{N}\Gamma_k = 0 \tag{2.55}$$

and

$$\sum_{k=1}^{N}\left[\Gamma_k\vec{u}_k+\vec{M}_{I,k}\right]=\frac{1}{2}\sum_{k=1}^{N}\sum_{j=1}^{N}\frac{1}{V}\left(1-\delta_{kj}\right)\int_{A_{I,kj}}\vec{m}_{jk}^{\sigma}\mathrm{d}A = \vec{M}_{\sigma} \tag{2.56}$$

The average energy equation expressed by enthalpy h as the dependent variable reads (kinetic energy is subtracted):

$$\begin{aligned}\frac{\partial}{\partial t}(\alpha_k\overline{\rho}_k\hat{h}_k)+\nabla\bullet(\alpha_k\overline{\rho}_k\hat{\vec{u}}_k\hat{h}_k) &= -\nabla\bullet\alpha_k(\hat{\vec{q}}_k+\hat{\vec{q}}_k^T)+\frac{\mathrm{D}_k}{\mathrm{D}t}(\alpha_k p_k)\\ &+\alpha_k(\hat{\overline{\overline{\tau}}}_k+\hat{\overline{\overline{\tau}}}_k^T):\nabla\hat{\vec{u}}_k+\hat{h}_{ki}\cdot\Gamma_k+\alpha_k\overline{\rho}_k S_k+E_k\end{aligned} \tag{2.57}$$

Here the terms on the RHS are heat transfer in phase k due to temperature gradients, reversible work in phase k, dissipation due to shear stress in phase k, energy transfer to phase k due to mass transfer between phases, energy generation in phase k due to e.g. chemical reaction and finally heat transfer to phase k from the other phases.

The average balance equation for a chemical specie is expressed as

$$\frac{\partial}{\partial t}(\alpha_k\overline{\rho}_k\hat{Y}_k)+\nabla\bullet(\alpha_k\overline{\rho}_k\hat{\vec{u}}_k\hat{Y}_k) = -\nabla\bullet\alpha_k(\hat{\vec{F}}_{Y,k}+\hat{\vec{F}}_{Y,k}^T)+\hat{Y}_{ki}\cdot\Gamma_k+\alpha_k\overline{\rho}_k\hat{\Psi}_k \tag{2.58}$$

Here the terms on the RHS are mass transfer in phase k due concentration gradients, mass transfer to phase k due to mass transfer from the other phases, mass generation of specie Y in phase k due to e.g. chemical reaction

Mixture model. Another approach that describes multi-fluid flows as a whole is the mixture model. The mixture model equations are still intended as averaged on volume (if volume averaging is considered) larger than the characteristic volume of the dispersed phases, as for the multi-fluid model, but the properties of the mixture are described rather than the properties of each phase. There are two approaches to model the mixture balance equations. One derives the equations from physical consideration on the mixture properties (Drew and Passman 1999), whereas the other derives the balance equations directly from the multi-fluid model.

Here, the latter approach is followed (Manninen and Taivassalo 1996). From Eq.(2.49) summing over all phases yields

$$\frac{\partial}{\partial t}\sum_{k=1}^{N}\left(\alpha_k\overline{\rho}_k\right)+\nabla\cdot\sum_{k=1}^{N}\left(\alpha_k\overline{\rho}_k\hat{\vec{u}}_k\right)=\sum_{k=1}^{N}\Gamma_k \tag{2.59}$$

Defining the mixture density as

$$\rho_m=\sum_{k=1}^{N}\alpha_k\overline{\rho}_k \tag{2.60}$$

and mixture velocity as

$$\vec{u}_m=\sum_{k=1}^{N}\frac{\alpha_k\overline{\rho}_k\hat{\vec{u}}_k}{\rho_m} \tag{2.61}$$

which can be regarded as the velocity of the mixture centre of mass, and considering the mass jump condition constraint, Eq. (2.55), Eq. (2.59) can be written as

$$\frac{\partial}{\partial t}\rho_m+\nabla\cdot\left(\rho_m\vec{u}_m\right)=0 \tag{2.62}$$

The mixture momentum balance equations can be obtained from the momentum equation (2.50) by summing over all phases.

$$\begin{aligned}&\frac{\partial}{\partial t}\sum_{k=1}^{N}\left(\alpha_k\overline{\rho}_k\hat{\vec{u}}_k\right)+\nabla\cdot\sum_{k=1}^{N}\left(\alpha_k\overline{\rho}_k\hat{\vec{u}}_k\hat{\vec{u}}_k\right)=\\&\nabla\cdot\sum_{k=1}^{N}\left[\alpha_k\left(\hat{\overline{T}}_k+\overline{\overline{T}}_k^{T}\right)\right]+\sum_{k=1}^{N}\left(\alpha_k\overline{\rho}_k\vec{g}\right)+\sum_{k=1}^{N}\left(\Gamma_k\hat{\vec{u}}_k+\vec{M}_{I,k}\right)\end{aligned} \tag{2.63}$$

The second term on the LHS can be written as

$$\sum_{k=1}^{N}\left(\alpha_k\overline{\rho}_k\hat{\vec{u}}_k\hat{\vec{u}}_k\right)=\nabla\cdot\left(\rho_m\vec{u}_m\vec{u}_m\right)+\nabla\cdot\sum_{k=1}^{N}\left(\alpha_k\overline{\rho}_k\hat{\vec{u}}_{Mk}\hat{\vec{u}}_{Mk}\right) \tag{2.64}$$

Where $\hat{\vec{u}}_{Mk}$ is the diffusion velocity or the velocity of phase k relative to the mixture centre of mass. The diffusion velocity $\hat{\vec{u}}_{Mk}$ can be expressed as

$$\hat{\vec{u}}_{Mk} = \hat{\vec{u}} - \vec{u}_m = \sum_{i=1}^{N} c_i \vec{u}_{s,ki} \tag{2.65}$$

where, c_i is the mass fraction of phase i and $\vec{u}_{s,ki}$ is the relative or slip velocity of phase k relative to phase i, $\vec{u}_{s,ki} = \hat{\vec{u}}_k - \hat{\vec{u}}_i$. The mixture stress tensor, $\overline{\overline{T}}_m$ and residual stress tensor, $\overline{\overline{T}}_m^T$ can be defined as

$$\overline{\overline{T}}_m = \sum_{k=1}^{N} \alpha_k \overline{\overline{T}}_k \tag{2.66}$$

and

$$\overline{\overline{T}}_m^T = \sum_{k=1}^{N} \alpha_k \overline{\overline{T}}_k^T \tag{2.67}$$

Considering the relations in Eqs. (2.60), (2.61), (2.64), (2.66), (2.67) and the constraint on the momentum jump condition in Eq. (2.56), Eq. (2.63) yields:

$$\frac{\partial}{\partial t}(\rho_m \vec{u}_m) + \nabla \cdot (\rho_m \vec{u}_m \vec{u}_m) = \nabla \cdot \left(\overline{\overline{T}}_m + \overline{\overline{T}}_m^T \right) + \rho_m \vec{g} - \nabla \cdot \left(\sum_{k=1}^{N} \alpha_k \overline{\rho}_k \vec{u}_{Mk} \vec{u}_{Mk} \right) + \vec{M}_\sigma \tag{2.68}$$

With the last two terms in the Eq. (2.68) considered as source terms, mixture continuity equation, Eq. (2.62), and the mixture momentum balance equation, Eq. (2.80), are formally identical to the respective single fluid phase equations of fluid dynamics. It should be noted that in deriving the mixture equations no assumption or approximation has been made. Therefore, at this stage the mixture equations have the same potential to describe multiphase flows as the multi-fluid model equations from which they have been derived. For completeness, it should be mentioned that the mixture model is sometimes formulated in terms of drift velocity (Ishii 1975, Manninen and Taivassalo 1996, Hibiki and Ishii 2002, Nigam 2003). In this case, it is referred to as the drift-flux model.

3 Closure Framework

3.1 General

So far, all the multiphase models described are not yet in a solvable form. They still include unclosed terms that are not explicitly depending on the averaged fluid properties. Unclosed terms, for multiphase flow models, can be grouped in three different classes:

(i) phase interaction terms $\left(\Gamma_k, \vec{M}_{I,k}, \vec{M}_\sigma, \hat{\vec{u}}_{Mk}\right)$,

(ii) self-interaction terms $\left(\overline{\overline{T}}_k, \overline{\overline{T}}_m\right)$ and

(iii) turbulence terms $\left(\overline{\overline{T_k^T}}, \overline{\overline{T_m^T}}\right)$.

Furthermore the thermodynamic state of the system needs to be specified through state equations, which link the thermodynamic variables. The closures of the terms in (i)-(iii) together with state equations go under the name of closure laws or constitutive laws (Though, some authors refer to constitutive laws as a subclass of closure laws). In search for constitutive laws, one must be guided by the following principles:

1. equipresence
2. well-posedness
3. frame indifference
4. determinism
5. fulfilment of the entropy inequality (or second law of thermodynamics).

Equipresence means that any variable described by a closure law should be a function of all the other variables. The principle of well-posedness states that a solution of the model equations should exist and be unique. Frame indifference means that the closure laws should not depend on the reference frame. Determinism states the predictability of a present state from a past history. The fulfilment of the second law of thermodynamics makes sure that the solution has a physical meaning. To solve the multiphase model equations, initial conditions to specify how the flow starts and boundary conditions to specify how the flow interacts with its environment are needed. Once closure laws, initial conditions and boundary conditions are set, the multiphase model equations are closed and ready to be solved. In the next Section, the closure laws are derived for multi-fluid and mixture models in the specific case of bubbly flows.

3.2 Bubble liquid

Self-interaction The viscous stress tensor of phase k, $\hat{\overline{\overline{T}}}_k$ can be split up into a pressure term and a shear stress term as

$$\hat{\overline{\overline{T}}}_k = -p_k \overline{\overline{I}} + \overline{\overline{\tau}}_k \tag{3.1}$$

The shear stress tensor is often modelled using the Newtonian strain-stress relation

$$\overline{\overline{\tau}}_k = \xi_k \left(\nabla \bullet \hat{\vec{u}}_k\right) \overline{\overline{I}} + 2\mu_k \left[\overline{\overline{S}}_k - \frac{1}{3} \nabla \bullet \hat{\vec{u}}_k \overline{\overline{I}}\right] \tag{3.2}$$

where the strain-rate tensor is defined by

$$\overline{\overline{S}}_k = \frac{1}{2}\left(\nabla \bullet \hat{\vec{u}}_k + \left(\nabla \bullet \hat{\vec{u}}_k\right)^T\right) \tag{3.3}$$

The bulk viscosity, ξ_k, is usually set to zero for all phases, whereas the dynamic viscosity, μ_k, is usually set to a constant value (laminar viscosity) for the primary phase (liquid) and to zero for the secondary phases (classes of bubble). The assumption of constant viscosity for the primary phase may be questionable, since there should be dependence from the concentration of the secondary phases and temperature when non isothermal process is studied. In the present analysis, a constant laminar viscosity is assumed.

The mixture stress tensor can also be decomposed into a mixture pressure term and a mixture viscous stress term as

$$\hat{\hat{T}}_m = -p_m \bar{\bar{I}} + \bar{\bar{\tau}}_m \tag{3.4}$$

where

$$p_m = \sum_{k=1}^{N} \alpha_k p_k \tag{3.5}$$

The mixture viscous stress can either be derived from the viscous stress tensor of the phases,

$$\bar{\bar{\tau}}_m = \sum_{k=1}^{N} \alpha_k \tau_k \tag{3.6}$$

or be formulated in terms of the mixture strain-rate tensor as

$$\bar{\bar{\tau}}_m = \mu_m \left[\nabla \bullet \hat{\vec{u}}_m + \left(\nabla \bullet \hat{\vec{u}}_m \right)^T - \frac{2}{3} \nabla \bullet \hat{\vec{u}}_m \bar{\bar{I}} \right] \tag{3.7}$$

Phase interaction

The interactions between the phases are accounted for by the terms $\overrightarrow{M}_{I,k}$, Γ_k in the multi-fluid model and $\nabla \bullet \left(\alpha_k \rho_k \vec{u}_{Mk} \vec{u}_{Mk} \right)$, Γ_k in the mixture model. The formulation of the exchange mass rate between phases, Γ_k, depends on the particular process under consideration (evaporation, solidification, condensation, etc.) and will not be discussed in this Section.

The interfacial force for phase k is usually decomposed in terms of a generalized drag force and averaged interfacial pressure and shear stress.

$$\overrightarrow{M}_{I,k} = \overrightarrow{M}_{I,k}^{gd} + p_{I,k} \nabla \alpha_k - \bar{\bar{\tau}}_{I,k} \nabla \alpha_k \tag{3.8}$$

By this decomposition the RHS of Eq. (2.50) it may be rewritten as follows

$$\begin{aligned} \nabla \bullet \left(\alpha_k \hat{\hat{T}}_k \right) + \left(\alpha_k \bar{\rho}_k \vec{g} \right) + \Gamma_k \hat{\vec{u}}_k + \overrightarrow{M}_{I,k} = -\alpha_k \nabla p_k + \nabla \bullet \left(\alpha_k \bar{\bar{\tau}}_k \right) + \\ + \left(p_{I,k} - p_k \right) \nabla \alpha_k - \bar{\bar{\tau}}_{I,k} \nabla \bullet \alpha_k + \left(\alpha_k \bar{\rho}_k \vec{g} \right) + \Gamma_k \hat{\vec{u}}_k + \overrightarrow{M}_{I,k}^{gd} \end{aligned} \tag{3.9}$$

The generalized drag force accounts for several interaction forces. Here, only the main contributions for bubbly flows, i.e. drag, virtual mass and lift forces, will be considered. In case of bubbly flows, the multiphase system can be modelled with a primary phase (k=1), consisting of the liquid phase, and (N-1) secondary phases each one representing a class of bubbles with volumes ranging within an interval. Assuming that the interfacial forces are only acting between the primary phase and the secondary phases and not among the secondary phases, which is reasonable for dispersed systems, the generalized drag form can be formulated for the primary phase as

$$\overrightarrow{M}_{I,1}^{gd} = \sum_{k=2}^{N} \left(\overrightarrow{M}_{I,k}^{D} + \overrightarrow{M}_{I,k}^{L} + \overrightarrow{M}_{I,k}^{VM} \right) \tag{3.10}$$

and for the secondary phases as

$$\overrightarrow{M}_{I,k}^{gd} = -\overrightarrow{M}_{I,k}^{D} - \overrightarrow{M}_{I,k}^{L} - \overrightarrow{M}_{I,k}^{VM} \qquad k = 2,.....N \tag{3.11}$$

This formulation satisfies the averaged jump constraint for the momentum, Eq. (2.56). The drag, lift and virtual mass are modelled as follows

Drag force

$$\overrightarrow{M}_{I,k}^{D} = \frac{3}{4}\overline{\rho}_1\alpha_k \frac{C_{D,k}}{\overline{d}_{b,k}}\left|\hat{u}_k - \hat{u}_1\right|\left(\hat{u}_k - \hat{u}_1\right) \tag{3.12}$$

Lift force

$$\overrightarrow{M}_{I,k}^{L} = -C_{L,k}\,\overline{\rho}_1\alpha_k\left(\nabla\times\hat{u}_1\right)\times\left(\hat{u}_k - \hat{u}_1\right) \tag{3.13}$$

Virtual mass force

$$\overrightarrow{M}_{I,k}^{L} = C_{VM,k}\,\overline{\rho}_1\alpha_k\left(\frac{\mathrm{D}}{\mathrm{D}t}\hat{u}_k - \frac{\mathrm{D}}{\mathrm{D}t}\hat{u}_1\right), \tag{3.14}$$

Considering the mixture model, for the same type of flow and assumptions introduced previously, the only term to be written in closed form is $\nabla\bullet\left(\alpha_k\rho_k\vec{u}_{Mk}\vec{u}_{Mk}\right)$. The volume fractions, α_k, are derived from the phase mass conservation equation (2.49)

$$\frac{\partial}{\partial t}\left(\alpha_k\overline{\rho}_k\right) + \nabla\cdot\left(\alpha_k\overline{\rho}_k\hat{\tilde{u}}_k\right) = \Gamma_k \tag{3.15}$$

The phase velocity, $\overline{\overline{u}}_k$ is not yet available but can be derived by the relation (2.65),

$$\vec{u}_{Mk} = \overline{\overline{\vec{u}}}_k - \overrightarrow{u_m} = \sum_{i=1}^{N} c_i\vec{u}_{s,ki} \tag{3.16}$$

once the slip velocities are known. There are several models available in literature for describing the slip velocity (Ishii 1975, Manninen and Taivassalo 1996, Sokolichin and Eigenberger 1999, Lapin et al. 2001, Sokolichin et al. 2004). When the slip velocity is given by an algebraic expression, the mixture model is usually referred to as the algebraic slip model (ASM). In deriving a model for the slip velocity several assumptions are made. The most restrictive is the local equilibrium approximation, which requires that the particles are rapidly accelerated to the terminal velocity or the particle relaxation time should be smaller than the characteristic time scale of the flow field (Clift et al. 1978, Manninen and Taivassalo 1996). However, the slip velocity can be rigorously derived from the multi-fluid model equations without making any severe assumption. The derivation of the slip velocity between phase k and phase i, $u_{s,ki}$ is accomplished by combining first the momentum and the continuity equations for each phase and then subtracting the momentum equations. The result is (from Bove, 2005):

$$\begin{aligned}\frac{\partial}{\partial}\vec{u}_{s,ki} + \hat{u}_k\bullet\nabla\hat{u}_k - \hat{u}_i\bullet\nabla\hat{u}_i &= \frac{1}{\overline{\rho}_i}\nabla p_i - \frac{1}{\overline{\rho}_k}\nabla p_k + \frac{1}{\rho_k\alpha_k}\nabla\bullet\left(\alpha_k\hat{\hat{\tau}}_k\right) - \frac{1}{\rho_i\alpha_i}\nabla\bullet\left(\alpha_i\hat{\hat{\tau}}_i\right) + \\ &+ \frac{1}{\overline{\rho}_k\alpha_k}\overrightarrow{M}_{I,k}^{gd} - \frac{1}{\overline{\rho}_i\alpha_i}\overrightarrow{M}_{I,i}^{gd} + \frac{1}{\overline{\rho}_k\alpha_k}\left(p_{I,k} - p_k\right)\nabla\bullet\alpha_k - \frac{1}{\overline{\rho}_i\alpha_i}\left(p_{I,i} - p_i\right)\nabla\bullet\alpha_i + \\ &- \frac{1}{\overline{\rho}_k\alpha_k}\hat{\hat{\tau}}_{I,k}\nabla\bullet\alpha_k + \frac{1}{\overline{\rho}_i\alpha_i}\hat{\hat{\tau}}_{I,i}\nabla\bullet\alpha_i\end{aligned} \tag{3.17}$$

Turbulence closure The averaging process on the convective terms of phase k generates the so called residual stress tensor (or Reynolds stress tensor when a time or ensemble average is performed), $\overline{\overline{T}}_k^T$, which accounts for the residual or unresolved turbulence filtered out by the averaging process. Due to the nature of turbulence, which is still a non well understood problem of physics, the modelling of the residual tensor is not trivial. Kataoka and Serizawa (1989) and successively Lopez de Bertodano et al. (1994) have derived and extended from single phase to multiphase the k−ε turbulence model. However, in deriving the equations for the turbulent kinetic energy, k, and the turbulent dissipation rate, ε, several cross-correlation terms have been neglected mainly because the lack and the difficulties of understanding them. Hence care must be taken when using this model. The work of Chahed et al. (2003) is one of the few if not the only one in which the Reynolds stress model is applied for multiphase flows. In the last years there has been some attempt to use very large eddy simulations (VLES) for computing multiphase flows, in particular bubbly flows, (Milelli 2002, Deen et al. 2001, Bove et al. 2004). This technique has shown to be able to capture the unsteadiness and the main characteristic of the flow quite successfully. LES and VLES are based on spatial filtered Navier-Stokes equations. Since the volume average of the multiphase model equations is equivalent to filter in space by a particular filter, they can be regarded as LES/VLES. Here VLES will be considered and the residual stress tensor will be modelled consequently. For the continuous phase, k=1, the residual stress tensor can be written as

$$\overline{\overline{T}}_1^T = \overline{\rho}_1 \left(\hat{u}_1 \hat{u}_1 - \widehat{\vec{u}_1 \vec{u}_1} \right) = -\frac{2}{3} k_1 \overline{\overline{I}} + \overline{\overline{\tau}}_1 \tag{3.18}$$

Here, k_1 is the residual turbulent kinetic energy for the continuous phase and is defined as

$$k_1 = \tfrac{1}{2} \overline{\rho}_1 \left(\widehat{\vec{u}_1 \vec{u}_1} - \hat{u}_1 \hat{u}_1 \right) \tag{3.19}$$

τ_1 is the anisotropic residual stress tensor and accounts for two different turbulence contributions, one induced by the shear stress in the liquid phase and the other deriving from the bubble liquid interaction or bubble induced turbulence (Sato and Sekoguchi 1975). The linear eddy-viscosity model is used to relate the anisotropic residual stress to the rate of strain and the superposition of the two turbulence effects is assumed. The anisotropic residual stress tensor can then be modelled as

$$\overline{\overline{\tau}}_1 = 2\left(\mu_T + \mu_{T,BIT}\right) \overline{\overline{S}}_1 \tag{3.20}$$

By analogy with the mixing-length hypothesis, the turbulent viscosity, μ_T, is modelled as (Smagorinsky 1963)

$$\mu_T = \overline{\rho}_1 \left(C_s \Delta \right)^2 S_1 \tag{3.21}$$

where, C_s is the Smagorinsky coefficient, Δ is the filter length or the volume average characteristic length and S_1 is the characteristic filtered rate of strain. It is defined as

$$S_1 = \left(2 \overline{\overline{S}}_1 : \overline{\overline{S}}_1 \right)^{1/2} \tag{3.22}$$

The bubble induced viscosity was introduced and modelled by Sato and Sekoguchi (1975) as

$$\mu_{T,BIT} = \overline{\rho}_1 C_{\mu,BIT} \sum_{k=2}^{N} \alpha_k d_k \left| \vec{u}_k - \vec{u}_1 \right| \tag{3.23}$$

where the model constant, $C_{\mu,BIT}$, is set equal to 0.6. For the secondary phases, the residual stress tensor should account for the dispersion of the bubbles (or particles in general) due to the unresolved eddies. It has been shown by Bove (2005) that the modelling of this term by the linear eddy-viscosity assumption is not able to describe the desired physics. Several works account for the turbulent dispersion in the generalized drag force, among others Lopez de Bertodano (1998) and Drew and Passman (1999). This is questionable. In fact, if the turbulent dispersion effect is considered in the generalized drag force, what is the meaning of the residual stress tensor for the secondary phases? When VLES is applied to bubbly flows, eddies down to a length scales comparable with the bubble size (two times the bubble diameter if the grid spacing is set equal to the bubble diameter) are resolved. These eddies are responsible of the turbulent dispersion of the gas phase. With an appropriate choice of the grid spacing the most of the turbulent dispersion effect is then resolved and only a small part needs to be modelled. Therefore the residual stress tensor for the secondary phases will be neglected. The reader can refer to the work of Moraga et al. (2003) for a review of various turbulent dispersion models.

Neglecting the residual stress tensor for the secondary phases, the mixture anisotropic residual stress tensor is equal to the anisotropic residual stress tensor of the primary phase,

$$\overline{\overline{\tau}}_m = \overline{\overline{\tau}}_1 \tag{3.24}$$

Another popular model for bubble liquid flows is to model the continuous phase using the so-called k-ε turbulence model. The dispersed phase influence is taken account by introducing additional source terms. The turbulence viscosity is modelled as

$$\mu_{T,k} = C_\mu \bar{\rho}_k \frac{k_1^2}{\varepsilon_1} \tag{3.25}$$

The equation for the turbulent kinetic energy for the continuous phase (k=1) reads

$$\frac{\partial}{\partial t}(\alpha \bar{\rho} k)_1 + \nabla \bullet (\alpha \bar{\rho} \hat{\vec{u}} k)_1 = \nabla \bullet \left(\alpha \frac{\mu_T}{\sigma_k} \nabla k \right)_1 + \alpha_1 (G + P_{b,k} - \bar{\rho} \varepsilon)_1 \tag{3.26}$$

The equation for dissipation of turbulent kinetic energy for the continuous phase (k=1) reads

$$\frac{\partial}{\partial t}(\alpha \bar{\rho} \varepsilon)_1 + \nabla \bullet (\alpha \bar{\rho} \hat{\vec{u}} \varepsilon)_1 = \nabla \bullet \left(\alpha \frac{\mu_T}{\sigma_\varepsilon} \nabla \varepsilon \right)_1 + \alpha_1 \left[\left(\frac{\varepsilon}{k} \left(C_1 G - C_2 \bar{\rho} \varepsilon \right) + P_{b,\varepsilon} \right) \right]_1 \tag{3.27}$$

Production of turbulence due to mean strain gradients is given by:

$$G_1 = \overline{\overline{\tau}}_1 : \nabla \hat{\vec{u}}_1 \tag{3.28}$$

Additional turbulent kinetic energy is produced or dissipated due to the work induced by the bubbles when they move through the liquid phase

$$P_{b,k} = C_b \overrightarrow{M}_1^D \cdot (\hat{\vec{u}}_b - \hat{\vec{u}}_1) \qquad P_{b,\varepsilon} = C_{bd}\, \alpha_g \, \bar{\rho}_1 \frac{k_1^{3/2}}{d_b} \tag{3.29}$$

C_b range from 0.02 to 0.75. This means that from 2 to 75 percent of the bubble-induced turbulence goes into the large eddy structure of the continuous phase. C_{bd} range from 0.02 to 0.2. These large variations in the two constants indicate the problems with the k-ε turbulence model for bubble-liquid flows. The values of the other constants applied in the k-ε model are given in Table 3-1.

C_μ	C_1	C_2	σ_k	σ_ε
0.09	1.44	1.92	1.0	1.3

Table 3-1 Values of constants in the k-ε model.

Lahey et al. (1993) propose an extension to the k-ε model given above. Their extension introduces two time scales, namely one which is the normal turbulent turnover time of the liquid eddies $(k/\varepsilon)_l$ and one which is related to the relative velocity and the bubble diameter $(d_b/|u_{rel}|)$. The kinetic energy of turbulence is derived from transport equations similar to Eq. 3.26, one for shear induced turbulence k_{SI}, and one for bubble induced turbulence k_{BI}. These two contributions are added to get the total turbulent kinetic energy in the liquid phase, k_l.

Closed multi-fluid model Assuming that the surface tension effect is not important, as it is for non-separated flows, the pressure can be considered to have locally the same value for all phases, $p_k = p_{I,k} = p$. The mass conservation and momentum balance yields

Mass conservation

$$\frac{\partial}{\partial t}\left(\alpha_k \bar{\rho}_k\right) + \nabla\cdot\left(\alpha_k \bar{\rho}_k \hat{u}_k\right) = \Gamma_k, \quad k = 1,......,N \tag{3.30}$$

$$\sum_{k=1}^{N} \alpha_k = 1; \quad \sum_{k=1}^{N} \Gamma_k = 0 \tag{3.31}$$

Momentum balance for the primary phase

$$\begin{aligned}
&\frac{\partial}{\partial t}\left(\alpha_1 \bar{\rho}_1 \hat{u}_1\right) + \nabla\cdot\left(\alpha_1 \bar{\rho}_1 \hat{u}_1 \hat{u}_1\right) = -\alpha_1 \nabla p + \alpha_1 \bar{\rho}_1 \vec{g} + \\
&+\nabla\cdot\left(\alpha_1 \mu_{eff,1}\left(\nabla \hat{u}_1 + \left(\nabla \hat{u}_1\right)^T\right)\right) - \nabla\cdot\left(\alpha_1 \mu_1\left(\frac{1}{3}\nabla \hat{u}_1 \bar{\bar{I}}\right)\right) + \\
&+\Gamma_1 \hat{u}_1 + \sum_{k=2}^{N} \frac{3}{4} \bar{\rho}_1 \alpha_k \frac{C_{D,k}}{d_{b,k}}\left|\hat{u}_k - \hat{u}_1\right|\left(\hat{u}_k - \hat{u}_1\right) + \\
&-\sum_{k=2}^{N} C_{L,k} \bar{\rho}_1 \alpha_k \left(\nabla \times \hat{u}_1\right) \times \left(\hat{u}_k - \hat{u}_1\right) + \\
&+\sum_{k=2}^{N} C_{VM,k} \bar{\rho}_1 \alpha_k \left(\frac{D}{Dt}\hat{u}_k - \frac{D}{Dt}\hat{u}_1\right)
\end{aligned} \tag{3.32}$$

Here, $\mu_{eff,1}$ represents the sum of the dynamic, μ_1, residual, μ_T, and residual bubble induced viscosity, $\mu_{T,BIT}$.

Momentum balance for the secondary phases

$$
\begin{aligned}
&\frac{\partial}{\partial t}\left(\alpha_k \bar{\rho}_k \hat{u}_k\right)+\nabla\bullet\left(\alpha_k \bar{\rho}_k \hat{u}_k \hat{u}_k\right)=-\alpha_k \nabla p+\alpha_k \bar{\rho}_k \vec{g}+\Gamma_1 \hat{u}_1 + \\
&\quad -\frac{3}{4}\bar{\rho}_1 \alpha_k \frac{C_{D,k}}{\bar{d}_{b,k}}\left|\hat{u}_k-\hat{u}_1\right|\left(\hat{u}_k-\hat{u}_1\right)+ \\
&\quad +C_{L,k}\bar{\rho}_1 \alpha_k\left(\nabla\times\hat{u}_1\right)\times\left(\hat{u}_k-\hat{u}_1\right) \\
&-C_{VM,k}\bar{\rho}_1 \alpha_k\left(\frac{D}{Dt}\hat{u}_k-\frac{D}{Dt}\hat{u}_1\right) \qquad k=2,\ldots\ldots,N
\end{aligned}
\tag{3.33}
$$

Normally, the virtual mass coefficient, $C_{VM,k}$, is set equal to 0.5. The lift coefficient, $C_{L,k}$, is either set to a constant value or modelled as suggested by Tomiyama (2004),

$$
C_{L,k}=\begin{cases}\min\left[0.288\tanh\left(0.121\cdot \mathrm{Re}\right), f\left(Eo\right)\right] & \text{if} \quad \mathrm{Eo}<4 \\ f(Eo) & \text{if} \quad 4< Eo\end{cases} \tag{3.34}
$$

Here,

$$
f\left(Eo\right)=0.00105\cdot Eo^3-0.00159\cdot Eo^2-0.204\cdot Eo+0.474 \tag{3.35}
$$

The drag coefficient, $C_{D,k}$, is modelled either by the distorted model (Ishii and Zuber 2002)

$$
C_{D,k}=\frac{2}{3}Eo_k^{1/2} \tag{3.36}
$$

or by different correlations depending on the mean diameter and the initial deformation of the bubbles as well as the water contamination level, which affects the surface tension (Tomiyama et al. 2002, Tomiyama 2004).

For non-contaminated water or high initial shape deformation, the drag coefficient is well correlated by

$$
C_{D,k}=\begin{cases}\frac{16}{\mathrm{Re}_k}\left(1.0+0.15\right)\cdot \mathrm{Re}_k^{0.687} & \text{if} \quad 0\,mm<\bar{d}_{b,k}<0.5\,mm \\ \frac{48}{\mathrm{Re}_k} & \text{if} \quad 0.5\,mm<\bar{d}_{b,k}<1.3\,mm \\ \frac{8}{3}\frac{\mathrm{Eo_k}}{\mathrm{Eo_k}+4} & \text{if} \quad 1.30\,mm<\bar{d}_{b,k}\end{cases} \tag{3.37}
$$

For contaminated water or low initial shape deformation, the following correlation holds

$$
C_{D,k}=\begin{cases}\frac{16}{\mathrm{Re}_k}\left(1.0+0.15\right)\cdot \mathrm{Re}_k^{0.687} & \text{if} \quad 0\,mm<\bar{d}_{b,k}<0.8\,mm \\ \text{Correlation} & \text{if} \quad 0.8\,mm<\bar{d}_{b,k}\end{cases} \tag{3.38}
$$

Tomiyama (2004) expresses the drag coefficient as

$$C_{D,k}=\begin{cases}\frac{8}{3}\frac{\mathrm{Eo_k}}{ER_k^{2/3}\left(1-ER_k^2\right)^{-1}\mathrm{Eo_k}+16ER_k^{4/3}}\mathrm{F}\left(ER_k\right)^{-2} & \text{if} \quad \mathrm{ER_k}<1.0\\ 6 & \text{if} \quad \mathrm{ER_k}=1.0\\ \frac{8}{3}\frac{\mathrm{Eo_k}}{ER_k^{2/3}\left(ER_k^2-1\right)^{-1}\mathrm{Eo_k}-16ER_k^{4/3}}G\left(ER_k\right)^{-2} & \text{if} \quad \mathrm{ER_k}>1.0\end{cases} \tag{3.39}$$

Where,

$$Eo_k=\frac{\Delta\rho g d_{b,k}^2}{\sigma} \quad (\text{Eotvos number}) \tag{3.40}$$

$$ER_k=\frac{1}{1+0.163Eo_k^{0.757}} \quad (\text{Mean aspect ratio}) \tag{3.41}$$

$$\mathrm{Re}_k=\frac{\rho\left|\vec{u}_{s,kl}\right|\bar{d}_{db,k}}{\mu_1} \quad (\text{Bubble Reynolds number}) \tag{3.42}$$

$$F\left(ER_k\right)=\frac{\cos^{-1}ER_k-ER_k\sqrt{1-ER_k^2}}{1-ER_k^2} \tag{3.43}$$

$$G\left(ER_k\right)=\frac{ER_k\sqrt{ER_k^2-1}-\tanh^{-1}\left(ER_k^{-1}\sqrt{ER_k^2-1}\right)}{ER_k^2-1} \tag{3.44}$$

The system of Eqs (3.25)-(3.28) is determined. In fact it has the same number of equations and unknowns. As stated previously, in order to be solved, initial and boundary conditions need to be specified. In general, they depend on the particular type of process to be solved. Therefore, they will be given when the type of flow and its operative environment is known.

Closed mixture model

Using the assumptions made in the previous section, the mixture model can be summarized as follows:

Mixture continuity

$$\frac{\partial}{\partial t}(\rho_m)+\nabla\cdot\left(\rho_m\vec{u}_m\right)=0 \tag{3.45}$$

Mixture momentum balance

$$\begin{aligned}&\frac{\partial}{\partial t}\left(\rho_m\vec{u}_m\right)+\nabla\cdot\left(\rho_m\vec{u}_m\vec{u}_m\right)=-\nabla p+\rho_m\vec{g}+\\&+\nabla\cdot\left[\alpha_1\left(\mu_{eff}+\mu_{T,BIT}\right)\left(\nabla\hat{\vec{u}}_1+\left(\nabla\hat{\vec{u}}_1\right)^T-\frac{2}{3}\nabla\hat{\vec{u}}_1\bar{\bar{I}}\right)\right]+\\&+\nabla\cdot\left(\sum_{k=1}^{N}\alpha_k\overline{\rho\vec{u}_{Mk}}\vec{u}_{Mk}\right)\end{aligned} \tag{3.46}$$

Diffusion velocity

$$\vec{u}_{Mk} = \hat{u}_k - \vec{u}_m = \sum_{i=1}^{N} c_i \vec{u}_{s,ki} \tag{3.47}$$

Phase continuity

$$\frac{\partial}{\partial t}\left(\alpha_k \bar{\rho}_k\right) + \nabla \cdot \left(\alpha_k \bar{\rho}_k \hat{u}_k\right) = \Gamma_k, \quad k = 1,......, N \tag{3.48}$$

$$\sum_{k=1}^{N} \alpha_k = 1; \quad \sum_{k=1}^{N} \Gamma_k = 0 \tag{3.49}$$

Slip velocity

$$\begin{aligned} \frac{\partial}{\partial t}\vec{u}_{s,ki} + \hat{u}_k \nabla \cdot \hat{u}_k - \hat{u}_i \nabla \cdot \hat{u}_i = \left(\frac{1}{\bar{\rho}_i} + \frac{1}{\bar{\rho}_k}\right)\nabla p + \frac{1}{\alpha_k \bar{\rho}_k}\nabla \cdot \left(\alpha_k \bar{\bar{\tau}}_k\right) - \frac{1}{\alpha_i \bar{\rho}_i}\nabla \cdot \left(\alpha_i \bar{\bar{\tau}}_i\right) \\ \frac{1}{\alpha_k \bar{\rho}_k}\vec{M}_{I,k}^{gd} - \frac{1}{\alpha_i \bar{\rho}_i}\vec{M}_{I,k}^{gd} \end{aligned} \tag{3.50}$$

Here, the constitutive relations, Eqs. (3.10)-(3.14), hold for the generalized drag force in Eq. (3.50).

4 Closure Framework for Particle (Solids) Gas Flows

4.1 Kinetic theory of granular flow (KTGF) for mono-sized particles

Introduction. Statistical approach used to formulate the constitutive equations of the solid phase equations come from the interactions of the fluctuating motion of the particles with the mean motion of the particles. Such interactions generate stresses in the solids phase and give rise to an effective viscosity of the solids phase. To be able to calculate these stresses and an effective solids viscosity, a model is proposed by Ding and Gidaspow (1990) and Gidaspow (1994). Model based on the kinetic theory of dense gases, as presented by Chapman and Cowling (1970) and the work of Jenkins and Savage (1983) and Lun et al. (1984).The thermal temperature in the kinetic theory of dense gases is here replaced with a granular temperature for which a transient differential equation is derived. The solid viscosity and solid stresses are a function of this granular temperature that varies with time and space in a fluidized bed. Derivation given is mainly based on the work of Gidaspow (1994). The present derivation is based on mono-sized particles. The generalised model for multi-sized particles given by Manger (1996) will be summarized in the Section 4.2 below.

The starting point is the Boltzman integral-differential equation for a velocity distribution of particles. A single particle velocity distribution function $f^{(1)}(\vec{r},\vec{c},t)$ for a large collection of granular particles may be defined to satisfy the Boltzman integral differential equation:

$$\frac{\partial f}{\partial t} + \vec{c} \bullet \frac{\partial f}{\partial \vec{r}} + \vec{F} \bullet \frac{\partial f}{\partial \vec{c}} = \left(\frac{\partial f}{\partial t}\right)_{coll} \tag{4.1}$$

where function $f^{(1)}$ is defined such that at t, $\vec{r}$ and $fd\vec{c}$ is the differential number of particles per unit volume with instantaneous velocities within the range $\vec{c}$ and $\vec{c} + d\vec{c}$, $\vec{F}$ is the external force per unit mass acting on each particle, and the term on the right side is the rate of change for the distribution function due to the particle collisions.

Collisional theory and the Jenkins-Savage transport theorem A transport equation for a quantity ψ can be obtained starting with the Boltzman equation, multiply it with ψ and integrate over $\vec{c}$, as done in Chapter 3 of Chapman and Cowling

$$\int \psi \left(\frac{\partial f}{\partial t} + \vec{c} \bullet \frac{\partial f}{\partial \vec{r}} + \vec{F} \bullet \frac{\partial f}{\partial \vec{c}} \right) \mathrm{d}\vec{c} = \int \psi \left(\frac{\partial f}{\partial t} \right)_{coll} \mathrm{d}\vec{c} \tag{4.2}$$

All these integrals can be transformed, as in Chapter 3 Chapman and Cowling. The net results of these transformations are then

$$\frac{\partial n < \psi >}{\partial t} + \frac{\partial}{\partial \vec{r}} n < \psi \vec{c} > - n \vec{F} \bullet < \frac{\partial \psi}{\partial \vec{c}} > = \int \psi \left(\frac{\partial f}{\partial t} \right)_{coll} \bullet \mathrm{d}\vec{c} = \phi_c(\psi) \tag{4.3}$$

Where n is the particle number density

$$n = \int f \, \mathrm{d}\vec{c} \tag{4.4}$$

and $<\psi>$ is the averaged particle quantity defined as

$$< \psi > = \frac{1}{n} \int \psi f \, \mathrm{d}\vec{c} \tag{4.5}$$

In order to evaluate the right-hand side of Equation 4.3, $\Phi_c(\psi)$, the collisional rate of change of ψ, the details of a collision must be studied. We assume that binary collisions between hard, smooth but inelastic particles are dominant. Then, the probable number of collisions between particles labelled 1 and 2 per unit time with velocities of particle 1 and 2 are in the range of $\mathrm{d}\vec{c}_1$ and $\mathrm{d}\vec{c}_2$ and the centre of particle 1 is in the volume $\mathrm{d}\vec{r}$ is (Ding, 1990; Jenkins and Savage, 1983):

$$f^{(2)}\left(\vec{c}_1, \vec{r}_1, \vec{c}_2, \vec{r}_2 + d_p \vec{k}\right) d_p^2 \left(\vec{c}_{12} \bullet \vec{k}\right) \mathrm{d}\vec{k} \mathrm{d}\vec{c}_1 \mathrm{d}\vec{c}_2 \tag{4.6}$$

Where $f^{(2)}\left(\vec{c}_1, \vec{r}_1, \vec{c}_2, \vec{r}_2 + d_p \vec{k}\right) d_p^2 \left(\vec{c}_{12} \bullet \vec{k}\right) \mathrm{d}\vec{k} \mathrm{d}\vec{c}_1 \mathrm{d}\vec{c}_2$ is the pair distribution function, which is a function of two particles velocities, positions and time. It is defined such that $f^{(2)} \mathrm{d}\vec{c}_1 \mathrm{d}\vec{c}_2 \mathrm{d}\vec{r}_1 \mathrm{d}\vec{r}_2$ at time t, is the probable number of pairs of particles located in the volume element $\mathrm{d}\vec{r}_2 \mathrm{d}\vec{r}_2$ centred at the points $\vec{r}_1$ $\vec{r}_2$ with velocities between $\vec{c}_1$ to $\vec{c}_1 + \mathrm{d}\vec{c}_1$ and $\vec{c}_2$ to $\vec{c}_2 + \mathrm{d}\vec{c}_2$. $\vec{c}_{12}$ is the relative velocity of particle 1 and 2 and $\vec{k}$ is the unit vector along the line from the centre of particle 1 to the centre of particle 2 at collision. If a particle with location at r_2 with velocity $\vec{c}_2$ is to collide with a second particle with velocity $\vec{c}_1$ in a time interval dt, so that, at collision, the line of centres $\vec{r}_1 - \vec{r}_2 = d_p \vec{k}$ is within the angle $\mathrm{d}\vec{k}$ centred at $\vec{k}$, it is necessary that the centre of the second particle lie in a collision cylinder of volume $d_p^2 \left(\vec{c}_{12} \bullet \vec{k}\right) \mathrm{d}\vec{k} \mathrm{d}t$.

As a result of the collision, the property ψ_1 of particle 1 changes to ψ_1'. This gives the collisional rate of change of ψ per unit volume (Jenkins and Savage, 1983) :

$$\phi_c(\psi) = \iiint_{\vec{c}_{12} \bullet \vec{k} > 0} (\psi'_1 - \psi_1) f^{(2)}\left(\vec{c}_1, \vec{r}_1, \vec{c}_2, \vec{r}_2 + d_p \vec{k}\right) d_p^2 \left(\vec{c}_{12} \bullet \vec{k}\right) \mathrm{d}\vec{k} \mathrm{d}\vec{c}_1 \mathrm{d}\vec{c}_2 \tag{4.7}$$

$\vec{c}_{12} \bullet \vec{k}$ means that the integrations to be taken over all values for which a collision is imbedding. Since the collision is symmetric, the collisional rate of change for ψ_2 is:

$$\phi_c(\psi) = \iiint_{\vec{c}_{12} \bullet \vec{k} > 0} (\psi'_2 - \psi_2) f^{(2)}\left(\vec{c}_1, \vec{r}_1, \vec{c}_2, \vec{r}_2 - d_p \vec{k}\right) d_p^2 \left(\vec{c}_{12} \bullet \vec{k}\right) \mathrm{d}\vec{k} \mathrm{d}\vec{c}_1 \mathrm{d}\vec{c}_2 \tag{4.8}$$

The values of $f^{(2)}$ in equation 4.7 and 4.8 are separated by a distance $d_p \vec{k}$ and they are related to each other by means of a Taylor series expansion. Expanding the pair distribution function $f^{(2)}\left(\vec{c}_1, \vec{r}_1, \vec{c}_2, \vec{r}_2 + d_p \vec{k}\right)$ using only spatial derivatives of first order and combining equation 4.7 and 4.8, the result is:

$$\phi(\psi) = -\nabla \bullet \overline{\overline{P}}_c + \chi(\psi) \tag{4.9}$$

where the collisional stress contribution is

$$\overline{\overline{P}}_c = -\frac{d_p}{2} \iiint_{\underline{c}_{12} \bullet \underline{k} > 0} (\psi'_1 - \psi_1) f^{(2)}\left(\vec{c}_1, \vec{r}_2, \vec{c}_2, \vec{r}_2 + d_p \vec{k}\right) d_p^2 \left(\vec{c}_{12} \bullet \vec{k}\right) \mathrm{d}\vec{k}\mathrm{d}\vec{c}_1 \mathrm{d}\vec{c}_2 \tag{4.10}$$

and

$$\chi = \frac{1}{2} \iiint_{\underline{c}_{12} \bullet \underline{k} > 0} (\psi'_2 + \psi'_1 - \psi_2 - \psi_1) f^{(2)}\left(\vec{c}_1, \vec{r}_1, \vec{c}_2, \vec{r}_2\right) d_p^2 \left(\vec{c}_{12} \bullet \vec{k}\right) \mathrm{d}\vec{k}\mathrm{d}\vec{c}_1 \mathrm{d}\vec{c}_2 \tag{4.11}$$

Substituting the expression for the collisional rate of change, $\varphi(\psi)$, into equation 4.3, the result is

$$\frac{\partial n < \psi >}{\partial t} + \frac{\partial}{\partial \vec{r}} \left(n < \vec{c}\, \psi > + \overline{\overline{P}}_c(\psi) \right) = \chi(\psi) + < n \vec{F} \bullet \frac{\partial \psi}{\partial \vec{c}} > \tag{4.12}$$

$\vec{F}$ is here the force acting on the system. Gidaspow (1994) called this equation for the Jenkins - Savage transport theorem.

Continuity equation for solids flow. Take ψ to be mass, m. Since mass is conserved in the collision and using the standard relation $\rho_p \alpha_p = mn$, the continuity equation for solids flow is written

$$\frac{\partial(\alpha\,\rho)_p}{\partial t} + \frac{\partial(\alpha\,\rho\,u_i)_p}{\partial x_i} = 0 \tag{4.13}$$

Here, the mean velocity is defined by equation 4.5

$$u_i = \langle c_i \rangle = \frac{1}{n} \int c_i f \,\mathrm{d}\vec{c} \tag{4.14}$$

Momentum equation for solid flow. The solids momentum equation is derived using a more convenient form of the Jenkins -Savage transport theorem. Defining a relative velocity $\vec{C}$, which Chapman and Cowling called the peculiar velocity, $\vec{C} = \vec{c} - \vec{u}(x_i, t)$ and using equation 4.12, the transport theorem can take a more useful form

$$\frac{D n < \psi >}{D t} + n < \psi > \frac{\partial u_j}{\partial x_i} + \frac{\partial}{\partial x_i}(n < C_i\, \psi > + n\, P_{c,i}(\psi)) - n\left(\vec{F} - \frac{\mathrm{D}\vec{u}}{\mathrm{D}t} \right) < \frac{\partial \psi}{\partial \vec{C}} >$$
$$- n < \frac{\partial \psi}{\partial \vec{C}} \vec{C} : \frac{\partial u_j}{\partial x_i} > = \chi(\psi) \tag{4.15}$$

Let $\psi = m\vec{C}$, where $< \vec{C} > = < \vec{c} > - \vec{u} = 0$, and where the hydrodynamic velocity u_i is defined by Equation 4.14. Then, the two first terms in the Jenkins - Savage transport theorems are both zero because $<C>=0$. The third term can be written

$$\frac{\partial}{\partial x_i}(n < C_i \psi >) = \frac{\partial}{\partial x_i} < \rho\, C_i\, C_i > = \overline{\overline{P}}_k \tag{4.16}$$

and from the definition $\overline{\overline{P}}_k$ is the kinetic part of the stress. The last term on the left-hand side of the equation gives

$$n < \frac{\partial \psi}{\partial \vec{C}} \vec{C} : \frac{\partial u}{\partial x_i} > = < \frac{\partial m \vec{C}}{\partial \vec{C}} \vec{C} > = m < \vec{C} > = 0 \tag{4.17}$$

$\chi(\psi)$ equals zero because momentum is conserved during a collision.

In a fluidized bed the forces $\vec{F}$ that are of interest are the gravity force, the buoyancy force, which appears through the gradient of fluid pressure, and the drag between the phases, then $\rho_p\ \vec{F} = \alpha_p\rho_p\ \vec{g}\ +\ \alpha_p \nabla P\ +\alpha_p \rho\ D\ (\ \vec{u}_f - \vec{u}_p\ )/m$. Where D is the drag force coefficient, the two phase drag coefficient can then be written as $\beta = \alpha_p\rho\ D/m$.

The solids momentum equation can then be written as

$$\frac{\partial(\alpha\,\rho\,u_j\,)_p}{\partial t}+\frac{\partial(\alpha\,\rho\,u_i u_j\,)_p}{\partial x_i}+\frac{\partial}{\partial x_i}(\overline{\overline{P}}_k+\overline{\overline{P}}_c(\psi\,))= \\ =-\alpha_p\frac{\partial P}{\partial x_j}+\alpha_p\,\rho_p\,g_j+\beta_j(u_{j,g}-u_{j,p}) \tag{4.18}$$

The total stress tensor is the sum of a kinetic part $\overline{\overline{P}}_k$ and a collisional part $\overline{\overline{P}}_c$.

Conservation of fluctuating energy. Take ψ to be $\frac{1}{2}m\vec{c}^{\,2}$ and substitute into the Jenkins - Savage transport theorem. The first term, together with the relations $\rho=mn$ $\Theta =<\vec{C}^2>$ and $<\vec{C}>=<\vec{c}> - \vec{u}$, gives

$$n<\psi> = \frac{1}{2}\rho<\vec{c}^{\,2}> = \frac{1}{2}\rho\left(<\vec{C}\bullet\vec{C}>+u_i^2\right) = \rho\left(\frac{3}{2}\Theta+\frac{1}{2}u_i^2\right) \tag{4.19}$$

The second term is

$$n<\psi\,\vec{c}> = \frac{mn}{2}<\vec{c}^{\,2}\,\vec{c}> = \frac{\rho}{2}<(\vec{C}+\vec{u}\ )^2(\vec{C}+u_i)> \\ =\frac{\rho}{2}<(\ \vec{C}^2\vec{C}+2\,\vec{C}u_iu_i+u_i^2\,)> = \frac{\rho}{2}\int\!\left(\vec{C}^2\vec{C}+(\,2\,\vec{C}\,\vec{C}+\vec{C}^2+u_i^2\,)u_i\right)f\ d\vec{c} \tag{4.20}$$

The heat flux $\vec{q}_k$ is defined as $\vec{q}_k = \frac{1}{2}n\,m<\vec{C}^2\,\vec{C}> = \frac{3}{2}n<\Theta\,\vec{C}>$, then

$$n<\psi\,\vec{c}> = \vec{q}_k + (\frac{1}{2}\rho\vec{u}^2+\frac{3}{2}\rho\,\Theta\,)\vec{u}+\overline{\overline{P}}_k\bullet\vec{u} \tag{4.21}$$

where $\overline{\overline{P}}_k$ is defined by Equation 4.16. The collisional part, the $\overline{\overline{P}}_c$ of the Jenkins - Savage theorem can be written by using the standard substitution $\vec{c}=\vec{C}+\vec{u}$ and decomposed into two collisional heat fluxes,

$\overline{\overline{P}}_c(\frac{1}{2}m\vec{c}^{\,2}) = \vec{q}_c + \vec{u}\bullet\overline{\overline{P}}_c$, where $q_c = P_c(\frac{1}{2}\,m\vec{C}^2\,)$

The result is then

$$\frac{\partial\rho\left(\frac{3}{2}\Theta+\frac{1}{2}\vec{u}^2\right)}{\partial t}+\frac{\partial}{\partial x_i}\left[\ \vec{q}_k+\vec{q}_c\ +\rho\left(\frac{3}{2}\Theta+\frac{1}{2}\vec{u}^2\right)+\vec{u}\bullet(\,\overline{\overline{P}}_k+\overline{\overline{P}}_c\,)\ \right]= \\ \rho<\vec{F}\bullet\vec{u}> + \chi(\frac{1}{2}m\,\vec{c}^{\,2}\,) \tag{4.22}$$

A mechanical energy equation is obtained by forming the scalar product of the mean velocity $\vec{u}$ with the momentum equation and rearranging gives:

$$\frac{\partial\left(\frac{1}{2}\rho\vec{u}^2\right)}{\partial t}+\frac{\partial}{\partial x_i}\left[\frac{1}{2}\rho\vec{u}^2\,\vec{u}+\vec{u}(\overline{\overline{P}}_k+\overline{\overline{P}}_c)\right]=\vec{u}\bullet\rho\vec{F}\;-(\overline{\overline{P}}_k+\overline{\overline{P}}_c):\frac{\partial u_j}{\partial x_i} \quad (4.23)$$

Subtracting equation 4.23 from equation 4.22 will give the fluctuating energy equation,

$$\frac{3}{2}\left[\frac{\partial}{\partial t}(\alpha\,\rho\,\Theta)_p+\frac{\partial}{\partial x_i}(\alpha\,\rho\,u_i\Theta)_p\right]=-\frac{\partial}{\partial x_i}(q_{k,i}+q_{c,i})+(\overline{\overline{P}}_k+\overline{\overline{P}}_c):\frac{\partial u_j}{\partial x_i}+$$
$$+\chi(\frac{1}{2}m\vec{c}^2)+\rho<\vec{F}\bullet\vec{c}>-\rho\vec{F}\bullet\vec{u} \quad (4.24)$$

The terms including the drag can be expressed as

$$\rho<\rho\vec{F}\bullet\vec{c}>-\rho\vec{F}\bullet\vec{u}=\beta\left[<\vec{C}_p\bullet(\vec{C}_g-\vec{C}_p)>-\vec{u}_p(\vec{u}_g-\vec{u}_p)\right]=\beta<\vec{C}_g\bullet\vec{C}_p>-3\beta\Theta \quad (4.25)$$

The term $\left\langle\vec{C}_p\bullet\vec{C}_g\right\rangle$ is the correlation between the fluctuations of the gas and the particles, the gas oscillations will produce particle oscillations or vice versa. Ding and Gidaspow (1990) neglect this term. The fluctuating energy equation can then be written as:

$$\frac{3}{2}\left[\frac{\partial}{\partial t}(\alpha\,\rho\,\Theta)_p+\frac{\partial}{\partial x_i}(\alpha\,\rho\,u_i\,\Theta)_p\right]=-\frac{\partial}{\partial x_i}(q_{k,i}+q_{c,i})+\left(\overline{\overline{P}}_k+\overline{\overline{P}}_c\right):\frac{\partial u_j}{\partial x_i}-3\beta\,\Theta-\gamma \quad (4.26)$$

The terms on the RHS are: the diffusive terms due to kinetic and collosional transport; the production of fluctuations due to the shear; dissipation due to interaction with fluid and dissipation due to inelastic collisions given by the term $\gamma=\chi\,(\,½\,m\vec{c}^2)$.

Constitutive relations. To close the momentum equation and the fluctuating energy equation, explicit expressions for the kinetic and collisional stress contribution, kinetic and collisional heat flow and the collisional energy dissipation are needed. Ding and Gidaspow (1990) developed a dense phase model and Gidaspow (1994) developed this model further and made it valid for both dense and dilute particle flow. Following Gidaspow`s (1994) deduction, these expressions will be deduced, skipping some of the underlying theory. The collisional part of the stress tensor can be split into two parts, $\overline{\overline{P}}_{c,tot}=\overline{\overline{P}}_{c,1}+\overline{\overline{P}}_{c,2}$. This is done by expanding $f^{(1)}$ and $f^{(2)}$ in the usual way in a Taylor series using only terms of order one and introduce the radial distributions function. This function, called χ, was introduced by Chapman and Cowling (1970) gives a quantity which is one for ideal gases and becomes infinite when the molecules are so closely packed that motion is impossible. A form of the radial distribution function used by Ding and Gidaspow (1990) is

$$\chi(\vec{r}-\frac{1}{2}d_{12}\,\vec{k})=g_0(\alpha_p)=\frac{3}{5}\left[1.0-\left(\frac{\alpha_p}{\alpha_{p,\max}}\right)^{\frac{1}{3}}\right]^{-1} \quad (4.27)$$

Here, an empirical input, the maximum solid volume fraction is needed. Then the expansion of $f^{(2)}$ is, written in a compact form

$$f^{(2)}=g_o[f_1\bullet f_2+\frac{1}{2}\vec{k}\,d_p\,f_1\,f_2\nabla\ln\frac{f_2}{f_1}] \quad (4.28)$$

A more convenient form of Equation 4.11, the collisional part of the vector flow of ψ, can be written as, using Equation 4.27 and 4.28

$$\overline{\overline{P}}_c = -\frac{1}{2} d_p^3 g_0 \iiint_{\vec{c}_{12}\bullet\vec{k}>0} (\psi_1 - \psi_1') f_1 f_2 \vec{k} \left(\vec{c}_{12}\bullet\vec{k}\right) \mathrm{d}\vec{k}\mathrm{d}\vec{c}_1\mathrm{d}\vec{c}_2 \; + \\ -\frac{1}{4} d_p^4 g_0 \iiint_{\vec{c}_{12}\bullet\vec{k}>0} (\psi_1 - \psi_1') f_1 f_2 \vec{k}\bullet\nabla \ln\frac{f_2}{f_1} \vec{k}\left(\vec{c}_{12}\bullet\vec{k}\right) \mathrm{d}\vec{k}\mathrm{d}\vec{c}_1\mathrm{d}\vec{c}_2 \tag{4.29}$$

When using $\psi = m(\vec{c} - \vec{u})$ instead of $\psi = m(\vec{C})$ and the relation $\vec{c}_1{}' - \vec{c}_1 = ½(1+e)(\vec{c}_{12}\bullet\vec{k})\,\vec{k}$, where e is the restitution coefficient, the collisional stress tensor can be written as

$$\overline{\overline{P}}_{c\ tot} = \frac{1}{4} d_p^3 g_0 m(1+e) \iiint \left(\vec{c}_{12}\bullet\vec{k}\right)^2 \vec{k}\,\vec{k}\, f_1 f_2 \mathrm{d}\vec{k}\mathrm{d}\vec{c}_1\mathrm{d}\vec{c}_2 \\ +\frac{1}{8} d_p^4 g_0 m(1+e) \iiint \left(\vec{c}_{12}\bullet\vec{k}\right)^2 \vec{k}\,\vec{k}\, f_1^{(0)} f_2^{(0)} \vec{k}\bullet\nabla \ln\frac{f_2^{(0)}}{f_1^{(0)}} \mathrm{d}\vec{k}\mathrm{d}\vec{c}_1\mathrm{d}\vec{c}_2 = \overline{\overline{P}}_{c1} + \overline{\overline{P}}_{c2} \tag{4.30}$$

In the second integral, $f^{(0)}$ is the Maxwellian distribution or the normal distribution for velocities used. In terms of granular temperature, the Maxwellian distribution $f^{(0)}$ is written

$$\ln f^{(0)} = const + \ln\frac{n}{\Theta^{\frac{1}{2}}} - \frac{\vec{C}^2}{2\Theta} \tag{4.31}$$

One can then deduce that $\overline{\overline{P}}_{c1}$, which is the collisional contribution to the solid phase pressure equals:

$$P_p = 2(1+e)\rho_p \alpha_p^2 g_0\, \Theta \tag{4.32}$$

When adding the kinetic part of the pressure $P_k = \rho\Theta$ to the collisional part, the solid phase pressure, as derived by Ding and Gidaspow (1990) is:

$$P_p = \alpha_p \rho_p \left[1 + 2(1+e)\alpha_p g_0\right] \Theta \tag{4.33}$$

This pressure equals the van der Waals equation of state, as derived by Chapman and Cowling (1970).

Then the final expression for $\overline{\overline{P}}_{c2}$ can be written

$$\overline{\overline{P}}_{c2} = \frac{4\alpha_p^2 \rho_p D_p g_0 (1+e)}{3\sqrt{\pi}} \Theta^{\frac{1}{2}} \left[\frac{6}{5} S + \frac{2}{3} \nabla\bullet\vec{u}\,\delta_{ij}\right] \\ \text{where } S = \frac{1}{2}\left[\frac{\partial u_j}{\partial x_i} + \frac{\partial u_i}{\partial x_j}\right] - \frac{1}{3}\frac{\partial u_i}{\partial x_i}\delta_{ij} \tag{4.34}$$

These expressions for $\overline{\overline{P}}_{c2}$ gives rise to the solid dense phase shear and bulk viscosity, and can be written in a more convenient form as

$$\overline{\overline{P}}_{c2} = 2\alpha_p \mu_p S + \alpha_p \zeta \nabla\bullet\vec{u}\,\delta_{ij} \\ \text{where } \mu_p = \frac{4}{5}\alpha_p \rho_p d_p g_0 (1+e)\left(\frac{\Theta}{\pi}\right)^{\frac{1}{2}} \text{ and } \zeta_p = \frac{4}{3}\alpha_p \rho_p d_p g_0 (1+e)\left(\frac{\Theta}{\pi}\right)^{\frac{1}{2}} \tag{4.35}$$

The kinetic part of the stress is defined to be $P_k = \rho < \vec{C}\bullet\vec{C} >$. Using the second approximation to the frequency distribution, as done by Chapman and Cowling (1970) and Gidaspow (1994), the net result is

$$\rho<\vec{C}\bullet\vec{C}>=-\frac{2}{(1+e)g_0}\left[1+\frac{4}{5}\alpha_p g_0(1+e)\right]2\mu_{p,dil}S \tag{4.36}$$

The total stress tensor, valid for both dense and dilute phase, can then be written as

$$\overline{\overline{P}}=\alpha_p\rho_p[1+2(1+e)\alpha_p g_0]\Theta\delta_{ij}-\frac{4\alpha_s^2\rho_p d_p g_0(1+e)}{3\sqrt{\pi}}\Theta^{\frac{1}{2}}\nabla\bullet\vec{u}\,\delta_{ij}+$$

$$-\left[\frac{2\mu_{p,dil}}{(1+e)g_0}\left[1+\frac{4}{5}(1+e)g_0\alpha_p\right]^2+\frac{4\alpha_p^2\rho d_p g_0(1+e)}{5\sqrt{\pi}}\Theta^{\frac{1}{2}}\right]2S \tag{4.37}$$

$$\text{where}\quad \mu_{\mathrm{p,dil}}=\frac{5}{96}\rho_{\mathrm{p}}\mathrm{d_p}\sqrt{\pi\Theta}$$

The viscosity is a product of the mean free path times an oscillation velocity times a density, as in kinetic theories of gases. The kinetic and collisional heat flow can be derived in much the same way as the stress tensor was deduced. It consists of a kinetic part and two collisional parts, $\vec{q}_{tot}=\vec{q}_k+\vec{q}_{c1}+\vec{q}_{c2}$. The kinetic part can be showed to be valid for dilute flow only, and the result is

$$\vec{q}_k=-\frac{75}{384}\rho_p d_p\sqrt{\pi\Theta}\nabla\Theta \tag{4.38}$$

The collsional part of the granular heat flux can be found, using $\psi=\frac{1}{2}m\vec{c}^2$ and an expression for the collisional part of the vector flow of ψ, Equation 4.29, the result is:

$$\begin{aligned}\vec{q}_c&=\frac{1}{4}d_p^3 g_0 m\iiint(C_1'^2-C_1^2)f_1 f_2\;\vec{k}\left(\vec{c}_{12}\bullet\vec{k}\right)d\vec{k}d\vec{c}_1 d\vec{c}_2\\&+\frac{1}{8}\iiint(C_1'^2-C_1^2)f_1 f_2\vec{k}\nabla\ln\frac{f_2^{(0)}}{f_1^{(0)}}\;\vec{k}\left(\vec{c}_{12}\bullet\vec{k}\right)d\vec{k}d\vec{c}_1 d\vec{c}_2\\&=\vec{q}_{c1}+\vec{q}_{c2}\end{aligned} \tag{4.39}$$

q_{c1} can be written as, Gidaspow (1994):

$$\vec{q}_{c1}=\frac{3}{5}\alpha_p g_0(1+e)\rho_p\alpha_p<C^2\underline{C}>=-\frac{2}{g_0(1+e)}\left[1+\frac{6}{5}\alpha_p g_0(1+e)\right]\kappa_{dil}\nabla\Theta$$

$$\text{where}\quad \kappa_{dil}=\frac{75}{384}\rho_{\mathrm{p}}\mathrm{d_p}\sqrt{\pi\Theta} \tag{4. 40}$$

Introducing the gradient of granular temperature, the second integral can be written as:

$$\vec{q}_{c2}=-2\rho_p\alpha_p^2 d_p(1+e)g_0\left(\frac{\Theta}{\pi}\right)^{\frac{1}{2}}\nabla\Theta \tag{4.41}$$

The expression for total heat flux $\vec{q}$ can then be expressed as

$$\vec{q}=-\left[\frac{2}{g_0(1+e)}\left[1+\frac{6}{5}\alpha_p g_0(1+e)\right]^2\kappa_{dil}+2\alpha_p^2\rho_p d_p g_0(1+e)\left(\frac{\Theta}{\pi}\right)^{\frac{1}{2}}\right]\nabla\Theta \tag{4.42}$$

In the analogue of Fourier's law of conduction, the total flux of fluctuating energy can be represented as:

$$\vec{q}=-\kappa\nabla\Theta \tag{4.43}$$

To find the collisional energy dissipation, an expression for the change in energy due to the collision, is needed. For particles of equal mass, it can be written in terms of the relative velocity $\vec{c}_{12}$ and the coefficient of restitution e as:

$$\Delta E = \frac{1}{4} m (e^2 - 1) \left(\vec{c}_{12} \bullet \vec{k} \right)^2 \tag{4.44}$$

Using equation 4.11 and equation 4.44 together with Equation 4.42, the result is

$$\begin{aligned} \gamma = \chi (\frac{1}{2} m \vec{c}^2) = \frac{1}{2} d_p^2 \iiint \frac{1}{4} m (e^2 - 1) \left(\vec{c}_{12} \bullet \vec{k} \right)^2 \cdot \\ \cdot g_0 \left[f_1^{(0)} f_2^{(0)} + \frac{1}{2} d_p f_1^{(0)} f_2^{(0)} \nabla \ln \frac{f_2^{(0)}}{f_1^{(0)}} \right] \mathrm{d}\vec{k} \mathrm{d}\vec{c}_1 \mathrm{d}\vec{c}_2 \end{aligned} \tag{4.45}$$

Jenkins and Savage (1983) and Ding and Gidaspow (1992) evaluated this integral and the final result for the collisional energy dissipation is

$$\gamma = \chi (\frac{1}{2} m \vec{c}^2) = 3 (1 - e^2) \alpha_p^2 \rho_p g_0 \Theta \left[\frac{4}{d_p} \sqrt{\frac{\Theta}{\pi}} - \frac{\partial U_{k,p}}{\partial x_k} \right] \tag{4.46}$$

For restitution coefficient of one, no energy is lost in the collision and γ equals zero.

Multi-sized particles

In gas/solid systems, particle segregation due to different size and/or density will play a significant role on the flow behaviour. To describe such phenomena, an extension to multiple particle phases is essential. Jenkins and Mancini (1987) extended the kinetic theory for granular flow to binary mixtures. The basic assumption was equal turbulent kinetic energy with a small correction for the individual phase temperatures. Mathiesen et al. (1996) developed a model based on this work and performed a simulation with one gas and three solid phases. The model predicted segregation effects fairly well, and good agreement with experimental data was obtained. Gidaspow et al. (1996) and Manger (1996) extended the kinetic theory to binary mixtures of solid with unequal granular temperatures between the phases. Based on their research, a generalized multiphase gas/solid model will be given in the next section.

4.2 Summary of Governing Equations for Fluid/Multi-sized particle flows

The three-dimensional, finite-volume, multiphase Eulerian/Eulerian CFD code FLOTRACS-MP-3D (see also Mathiesen et al., 2000 a and b, Ibsen et al., 2001 and Hansen et al. 2003) uses the generalized fluid/multi-sized particle mentioned above. The turbulent motion of the particulate phase is modeled using the kinetic theory of granular flow described by Manger (1996) and the gas phase turbulence is modeled using a LES/Sub-Grid-Scale model.

Mass balances and momentum balances. The governing equations may be written as:
Continuity equation for phase k without mass transfer:

$$\frac{\partial}{\partial t} \left(\alpha_k \rho_k \right) + \frac{\partial}{\partial x_i} \left(\alpha_k \rho_k U_{i,k} \right) = 0 \tag{4.47}$$

Here α, ρ and U are the volume fraction, density and velocity of phase k, respectively. The momentum equation for phase k is written as:

$$\frac{\partial}{\partial t}\left(\alpha_k \rho_k U_{j,k}\right)+\frac{\partial}{\partial x_i}\left(\alpha_k \rho_k U_{i,k} U_{j,k}\right)=-\alpha_k \frac{\partial p}{\partial x_j}+\frac{\partial}{\partial x_i}\tau_{ij,k}+\alpha_k \rho_k g_j+\sum_{m=1,m\neq k}^{M}\beta_{km}\left(U_{j,m}-U_{j,k}\right) \quad (4.48)$$

Here p, τ_{ij}, g and β are pressure, stress tensor, gravity and the inter-phase drag coefficient, respectively. There are N solids phases (s) and one gas phase (g) and the total number of phases are therefore M=N+1.

Auxiliary relations

The gas phase stress tensor is given by:

$$\tau_{ij,g}=\mu_{eff,g}\left[\left(\frac{\partial U_j}{\partial x_j}+\frac{\partial U_i}{\partial x_j}\right)-\frac{2}{3}\delta_{ij}\frac{\partial U_k}{\partial x_k}\right]_g \quad (4.49)$$

where δ_{ij} is the Kroenecker delta. The effective viscosity, $\mu_{eff,g}$, is derived from a sub-grid-scale (SGS) model based on Smagorinsky (1963), where the effective viscosity is a sum of a laminar and a turbulent part.

$$\mu_{eff,g}=\alpha_g\left(\mu_{lam,g}+\mu_{turb,g}\right)=\alpha_g\mu_{lam,g}+\alpha_g\rho_g\left(c_t\Delta\right)^2\sqrt{S_{ij,g}:S_{ij,g}} \quad (4.50)$$

The SGS eddy coefficient, c_t, is set to 0.079 based on Deardoff (1971). The length scale, Δ, and the strain rate tensor of the resolved field, $S_{ij,g}$, are given by:

$\Delta=\left(\Delta x\Delta y\Delta z\right)^{1/3}$ and

$$S_{ij,g}=\frac{1}{2}\left[\frac{\partial U_j}{\partial x_i}+\frac{\partial U_i}{\partial x_j}\right]_g \quad (4.51)$$

The solid phase stress tensor is given by:

$$\tau_{ij,s}=-P_s\delta_{ij}+\xi_s\delta_{ij}\frac{\partial U_{k,s}}{\partial x_k}+\mu_s\left[\left(\frac{\partial U_j}{\partial x_i}+\frac{\partial U_i}{\partial x_j}\right)-\frac{2}{3}\delta_{ij}\frac{\partial U_k}{\partial x_k}\right]_s \quad (4.52)$$

The solids phase pressure, P_s, the bulk viscosity, ξ_s, and the shear viscosity, μ_s, are derived from the kinetic theory of granular flow. The solids pressure is found from, Ding and Gidaspow (1990):

$$P_s=\sum_{n=1}^{N}P_{C,sn}+\alpha_s\rho_s\Theta_s \quad (4.53)$$

where $P_{C,n}$ is the pressure caused by collisions between the solids phases s and n and has the expression:

$$P_{C,sn} = \frac{\pi}{3}(1+e_{sn})d_{sn}^3 g_{sn} n_s n_n \left[\frac{m_0\theta_s\theta_n}{(m_s/m_n)\theta_s + (m_n/m_s)\theta_n}\right]\left[\frac{(m_0/m_s)^2\theta_s\theta_n}{\left(\theta_s + (m_s/m_n)^2\theta_n\right)(\theta_s+\theta_n)}\right]^{3/2} \quad (4.54)$$

$$e_{sn} = \frac{1}{2}(e_n + e_s),\quad d_{sn} = \frac{1}{2}(d_n + d_s) \quad \text{and} \quad m_0 = m_s + m_n$$

e, d, n and m are coefficient of restitution, diameter of the particle, number of particles and mass of a particle, respectively. The coefficient of restitution is unity for fully elastic, and zero for inelastic collisions. By using the assumption of spherical particles, number of particles and mass of a particle are, respectively:

$$n_s = \frac{6\alpha_s}{\pi d_s^3} \quad \text{and} \quad m_s = \frac{\pi d_s^3 \rho_s}{6} \quad (4.55)$$

g_{sn} is the radial distribution function, which is close to one for dilute flow and approach infinity for dense flow making motion impossible. Based on the single solid phase model given implicitly by Bagnold (1954), a new binary radial distribution function is proposed here:

$$g_0 = \left[1 - \left(\frac{\alpha_s}{\alpha_{s,\max}}\right)^{1/3}\right]^{-1} \qquad g_{sn} = \frac{N}{2}\frac{g_0}{1-\alpha_g}(\alpha_s + \alpha_n) \quad (4.56)$$

The maximum solids packing, $\alpha_{s,max}$, is often put to 0.65. The solids bulk viscosity is calculated as (Ding and Gidaspow, 1990):

$$\xi_s = \sum_{n=1}^{N} P_{C,sn}\frac{d_{sn}}{3}\left(\theta_s + (m_n/m_s)\theta_n\right)\sqrt{\frac{2}{\pi\theta_s\theta_n\left(\theta_s + (m_n/m_s)^2\theta_n\right)}} \quad (4.57)$$

The solids phase shear viscosity consists of a kinetic term:

$$\mu_{col,s} = \sum_{n=1}^{N} P_{C,sn}\frac{d_{sn}}{3}\left(\theta_s + (m_n/m_s)\theta_n\right)\sqrt{\frac{2}{\pi\theta_s\theta_n\left(\theta_s + (m_n/m_s)^2\theta_n\right)}} \quad (4.58)$$

and a collisional part:

$$\mu_{kin,s} = \frac{2\mu_{dil,s}}{\frac{1}{N}\sum_{n=1}^{N}(1+e_{sn})g_{sn}}\left[1 + \frac{4}{5}\sum_{n=1}^{N} g_{sn}\alpha_n(1+e_{sn})\right]^2 \quad (4.59)$$

where the dilute solids viscosity, $\mu_{dil,s}$, is found from:

$$\mu_{dil,s} = \frac{15}{8d_s^3}\alpha_s l_s\sqrt{\frac{2m_s\theta_{s,av}}{\pi}} \quad \text{and} \quad l_s = \frac{1}{6\sqrt{2}}\frac{d_s}{\alpha_s} \quad (4.60)$$

To ensure that the dilute viscosity is finite as the volume fraction of solid approaches zero, the mean free path l_s is limited by a characteristic dimension. The average granular temperature $\theta_{s,av}$ is obtained from:

$$\theta_{s,av} = \frac{2m_s\theta_s}{\left[\sum_{n=1}^{N}\frac{n_n}{n_s}\left(\frac{d_{sn}}{d_s}\right)^2\sqrt{\frac{(m_0/m_s)^2\theta_n}{\left(\theta_s+(m_0/m_s)^2\theta_n\right)}}S^{3/2}\right]^2} \tag{4.61}$$

$$S = \frac{(m_0/m_s)^2\theta_s\theta_n}{\left(\theta_s+(m_0/m_s)^2\theta_n\right)(\theta_s+\theta_n)}$$

Three solids gas inter-phase drag models will be presented.

Gidaspow (1994):

For $\alpha_g \le 0.8$ the inter-phase drag coefficient is found from the Ergun equation (Ergun, 1952):

$$\beta_{sg} = 150\frac{\alpha_s^2\mu_{lam,g}}{\alpha_g(\Phi_s d_s)^2} + 1.75\frac{\alpha_s\rho_g|\vec{u}_g-\vec{u}_s|}{\Phi_s d_s} \tag{4.62}$$

where Φ_s is the sphericity of the particles (i.e. 1.0 for spheres; 0.81 for cubes; 0.6 - 0.7 for crushed materials; 0.3 for Raschig rings).

For $\alpha_g > 0.8$ the drag formulation of Wen and Yu (1966) is used

$$\beta_{sg} = \frac{3}{4}C_D\frac{\alpha_s\alpha_g\rho_g|\vec{u}_g-\vec{u}_s|}{\phi_s d_s}\alpha_g^{-2.65} \tag{4.63}$$

with a drag coefficient, C_D, from Rowe (1961):

$$C_D = \begin{cases} \frac{24}{\mathrm{Re}_s}\left(1+0.15\,\mathrm{Re}^{0.687}\right) & \text{for} \quad \mathrm{Re}_s \le 1000 \\ 0.44 & \text{for} \quad \mathrm{Re}_s > 1000 \end{cases} \tag{4.64}$$

where the particle Reynolds number is defined as:

$$\mathrm{Re}_s = \frac{d_s\rho_g\alpha_g|\vec{u}_g-\vec{u}_s|}{\mu_{lam,g}} \tag{4.65}$$

Gibilaro et al. (1985):

$$\beta_{sg} = \left(\frac{17.3}{\mathrm{Re}_p}+0.336\right)\frac{\rho_g}{d_p}|\vec{u}_g-\vec{u}_s|\alpha_s\alpha_g^{-1.8} \tag{4.66}$$

Syamlal and O'Brien (1988):

$$\beta_{sg} = \frac{3}{4d_p}C_D\rho_g\frac{1}{R_t^2}\alpha_g(1-\alpha_g)|\vec{u}_g-\vec{u}_s| \tag{4.67}$$

R_t is the ratio between the falling velocity of a suspension and the terminal velocity of a single sphere. R_t, C_D, A and B are given by the following expressions:

$$2R_t = A - 0.06\,\mathrm{Re}_p + \sqrt{0.0036\,\mathrm{Re}_p^2 + 0.12\,\mathrm{Re}_p\left(2B - A\right) + A^2}$$

$$A = \alpha_g^{4.14} \qquad B = \begin{cases} 0.8\alpha_g^{-1.28} & \text{for } \alpha_g < 0.85 \\ \alpha_g^{2.65} & \text{for } \alpha_g > 0.85 \end{cases} \tag{4.68}$$

$$C_D = \left(0.63 + 4.8\sqrt{\frac{R_t}{\mathrm{Re}_p}}\right) \qquad \mathrm{Re}_p = \frac{d_s \rho_g \left|\vec{u}_g - \vec{u}_s\right|}{\mu_{lam,g}}$$

The particle/particle drag coefficient may be expressed as (Manger (1996)):

$$\beta_{s,g} = P_{C,sn}\left[\frac{3}{d_{sn}}\sqrt{\frac{2\left(m_s^2\theta_s + m_n^2\theta_n\right)}{\pi m_0^2 \theta_s \theta_n}} + \frac{1}{\left|\vec{u}_n - \vec{u}_s\right|}\left(\nabla\left|\ln\frac{\alpha_s}{\alpha_n}\right| + \frac{\theta_s\theta_n}{\theta_s + \theta_n}\left|\frac{\nabla\theta_n}{\theta_n^2} - \frac{\nabla\theta_s}{\theta_s^2}\right| + 3\nabla\left|\frac{\ln\left(m_n\theta_n\right)}{\ln\left(m_s\theta_s\right)}\right|\right)\right] \tag{4.69}$$

Granular temperature

A transport equation for the solids phases' turbulent kinetic energy or granular temperature is solved:

$$\frac{3}{2}\left[\frac{\partial}{\partial t}\left(\alpha_s\rho_s\theta_s\right) + \frac{\partial}{\partial x_i}\left(\alpha_s\rho_s U_{i,s}\theta_s\right)\right] = \frac{\partial}{\partial x_i}\left(\kappa_s\frac{\partial\theta_s}{\partial x_i}\right) + \tau_{ij,s} : \frac{\partial U_{j,s}}{\partial x_i} - \gamma_s - 3\beta_{sg}\theta_s \tag{4.70}$$

Here, the terms on the right side of the equation represent diffusive transport, production due to shear, dissipation due to inelastic collisions and dissipation due to fluid friction. The production/dissipation term due to fluctuations in drag has been assumed as negligible. This is a reasonable assumption for the relatively large and heavy particles. Hence, the particle response time is assumed to be much longer than the characteristic time scale for the turbulent fluid motion.

The conductivity of granular temperature κ_s, and the dissipation due to inelastic collisions γ_s are determined from the kinetic theory for granular flow (Manger (1996)). The conductivity is given by a dilute and a dense part as:

$$\kappa_s = \frac{2\kappa_{dil,s}}{\frac{1}{N}\sum_{n=1}^{N}\left(1 + e_{sn}\right)g_{sn}}\left(1 + \frac{6}{5}\frac{1}{N}\sum_{n=1}^{N}\left(1 + e_{sn}\right)g_{sn}\alpha_n\right)^2 + 2\alpha_s^2\rho_s d_s\sqrt{\frac{\theta_s}{\pi}}\sum_{n=1}^{N}\left(1 + e_{sn}\right)g_{sn}\alpha_n \tag{4.71}$$

where

$$\kappa_{dil,s} = \frac{225}{32}\alpha_s l_s\sqrt{\frac{2m_s\theta_{s,av}}{\pi}}$$

The dissipation due to inelastic collisions is given by (Manger (1996)):

$$\gamma_s = \sum_{n=1}^{N} \frac{3}{4} P_{C,sn} \frac{1-e_{sn}}{d_{sn}} \cdot$$
$$\cdot \left[4\sqrt{\frac{2\theta_s\theta_n}{\pi\left((m_s/m_0)^2\theta_s + (m_n/m_0)^2\theta_n\right)}} - d_{sn}\left(\frac{(m_s/m_0)\theta_s + (m_n/m_0)^2\theta_n}{(m_s/m_0)^2\theta_s + (m_n/m_0)^2\theta_n}\right)\frac{\partial U_{k,s}}{\partial x_k}\right] \tag{4.72}$$

The divergence term is often neglected. However, if the term is retained it may cause production instead of dissipation, and must be handled with special care.

5 Additional Sub Models

5.1 Bubble size models

Five different bubble size models will be examined. These are denoted models A to E.

Model A. The simplest model assumes that the bubble size is constant and given by overall flow conditions. Johansen and Boyson (1988) have used an estimation of the bubble diameter which is dependant on the inlet flow rate:

$$d_b = 0.35\left(\frac{\dot{V}_g}{g}\right)^{0.2} \tag{5.1}$$

Model B. Jakobsen (1993) has used a bubble size model which sets the diameter proportional to the length scale of the turbulent liquid eddies according to:

$$d_b = C_{SMD}\frac{k_l^{\frac{3}{2}}}{\varepsilon_l} \tag{5.2}$$

The constant C_{SMD} is determined to be 0.04 for flow in bubble columns. The two models given above do not include properties of the interface. This is included in the next model.

Model C. Calderbank (1958) has proposed an expression based on dissipation length scale and a critical Weber number as follows:

$$We_{crit} = \frac{\tau\, d_b}{\sigma};\ \tau = \rho_l u_{t,l}^2\ ;\ \varepsilon_l = \frac{u_{t,l}^3}{d_b} \tag{5.3}$$

Here τ is the shear stress, $u_{t,l}$ is the turbulence velocity, σ is the surface tension of the gas liquid interface and ε is the dissipation rate of turbulence. The expression for d_b from the above relations gives:

$$d_b = C_{We}\left(\frac{\sigma}{\rho_l}\right)^{0.6}\frac{1}{\varepsilon_l^{0.4}} \tag{5.4}$$

Experimental verification against data from stirred tanks has given the following correlation based on the above expression:

$$d_b = C_{We}\left(\frac{\sigma}{\rho_l}\right)^{0.6}\frac{1}{\varepsilon_l^{0.4}}\alpha_g^{0.5}\left(\frac{\mu_g}{\mu}\right)^{0.25} + 0.0009 \tag{5.5}$$

When C_{We} is taken to be 4.15, d_b is given in metres.

Model D. Both models B and C give bubble size as function of local properties. However, in many cases the bubble size is governed by history effects. A model that takes account of this is given by Cook and Harlow (1986). This model calculates the transport of the number concentration of bubbles N, according to:

$$\frac{\partial N}{\partial t} + \nabla\bullet(\vec{u}_g N) = \omega_b(N_e - N) \tag{5.6}$$

Here the equilibrium number concentration of bubbles N_e, is calculated based on a critical Weber number as:

$$N_e = \frac{6\,\alpha_g}{\pi\, d_{be}^3} \quad \text{with} \quad We_{crit} = \frac{(\rho_l - \rho_g)|\vec{u}_{rel}|^2 d_{be}}{\sigma} \tag{5.7}$$

The bubble diameter is related to the void fraction and the bubble number concentration as:

$$d_b = \left(\frac{6\,\alpha_g}{\pi N}\right)^{\frac{1}{3}} \tag{5.8}$$

The two constants in the model, the critical Weber number We_{crit}, and the relaxation parameter ω_b, are given the following values: 3.6 and between 4 and 20, respectively. These constants are determined for vertical gas-liquid flow past an obstacle.

Model E. The most general and comprehensive method is to calculate bubble size distribution by solving the Population Balance Equation. Here there are several methods available including: Classes/Sectional Methods (CM/SM) and methods of moments (QMOM, DQMOM, PPDC). Details of these methods are given in other Chapters of this book and will not be discussed here.

5.2 Interfacial heat transfer

Gas/liquid Heat transfer between the two phases without mass transfer may be treated as follows: T_1 and T_2 are the temperatures in the bulk of phase 1 and 2, respectively. T_s is the interface temperature. Energy balance over the phase separation surface: heat in = heat out.

$$Q_{1s} = (\lambda_1 a)(T_1 - T_s); \quad Q_{s2} = (\lambda_2 a)(T_s - T_2)$$

Solving for T_s , $Q_{1s} = Q_{2s}$

$$Q_{1s} = Q_{12} = a\frac{\lambda_1\lambda_2(T_1 - T_2)}{\lambda_1 + \lambda_2}$$

Or source term in the phase 2 and phase 1 enthalpy equations reads:

$$S_{h_2} = -S_{h_1} = aU(T_1 - T_2) \text{ where } 1/U = 1/\lambda_1 + 1/\lambda_2 \tag{5.9}$$

Here U is the total heat transfer coefficient between phase 1 and 2; λ_1 and λ_2 are the individual heat transfer coefficients between phase 1 and the interface and phase 2 and the interface, respectively; a is the specific surface area of the interface, i.e. area per unit volume.

Gas/particle Heat may be generated by the catalytic exothermic solid phase chemical reaction (Samuelsberg and Hjertager, 1996). This heat is transported between the phases (particles, p and gas, g) and may in addition be cooled by e.g. generation of steam in a submerged heat exchanger (wall, w).

Source term for the gas phase

$$S_{h_g} = h_v(T_p - T_g) + h_w(T_w - T_g) \tag{5.10}$$

Source term for the solids phase

$$S_{h_p} = h_v(T_g - T_p) + h_w(T_w - T_p) + \sum_{k=1}^{k=nreac} \Delta H_{rx,k} \bullet r_k \tag{5.11}$$

The effective transport coefficients are related to the turbulent viscosities as:

$$\Gamma_{h,g} = \frac{\kappa_{lam,g}}{c_{p,g}} + \frac{\mu_g}{0.7} \quad \text{and} \quad \Gamma_{h,p} = \frac{\mu_p}{0.7} \tag{5.12}$$

Where $\kappa_{lam,g}$ is the laminar conductivity and $c_{p,g}$ is the specific heat capacity of the gas.

The volumetric heat transfer coefficient h_v may be calculated by using several different correlations for the Nusselt number N_p in different flow regimes, characterized by the Reynolds number Re and the gas phase volume fraction α_g.

For $\alpha_g \leq 0.8$

$$N_p = (2 + 0.106\, Re)\, S_p; \quad Re \leq 200 \quad N_p = 0.123 \left(\frac{4\, Re}{d_p}\right)^{0.83} S_p^{0.17}; \quad 200 < Re \leq 2000$$
$$N_p = 0.61\, Re^{0.67}\, S_p; \quad Re > 2000 \tag{5.13}$$

For $\alpha_g > 0.8$

$$N_p = (2 + 0.16\, Re^{0.67})\, S_p; \quad Re \leq 200 \quad N_p = 8.2\, Re^{0.60}\, S_p; \quad 200 < Re < 1000$$
$$N_p = 1.06\, Re^{0.457}\, S_p; \quad Re \geq 1000 \tag{5.14}$$

where

$$Re = \frac{\rho_g d_p |\vec{u}_g - \vec{u}_p|}{\mu_{lam,g}}, \quad S_p = \frac{\alpha_p 6}{d_p}, \quad N_p = \frac{h_v d_p}{\kappa_{lam,g}} \tag{5.15}$$

The wall to bed heat transfer coefficient, h_w, may for many cases be taken to be a constant, only proportional to the respective volume fractions of the phases. Typical values for the bed to wall heat transfer coefficient may be found in Geldart (1986).

5.3 Mass transfer in bubble liquid flows

The interface mass transfer driven by concentration gradients is taken account of through the source terms for the species equation. For the liquid phase, $Y_{j,l}$, the source term reads:

$$S_{Y_{j,l}} = (k_l a)_l \, (Y_{j,g} - Y^*_{j,l}) \tag{5.16}$$

The $(k_l\, a)$ is related to the dissipation rate, ε_l gas fraction, α_g and bubble diameter, d_b (Trägårdh, 1988):

$$(k_l a) = C_m \varepsilon_l^{0.2} \frac{6\, \alpha_g}{d_b} \tag{5.17}$$

Modelling of mass transfer in single component two-phase flow where evaporation/condensation occurs, may be expressed as:

$$\Gamma = \frac{(\lambda_1 a)(T_1 - T_{1s}) + (\lambda_2 a)(T_2 - T_{2s})}{h_{12}} \tag{5.18}$$

Here T_{1s} and T_{2s} are saturation temperatures, dependant only on pressure; λ_1 is the heat transfer coefficient between phase 1 and the interface surface. λ_2 is the heat transfer coefficient between phase 2 and the interface surface, h_{12} enthalpy of evaporation and a specific surface area of interface.

5.4 Chemical reaction

Bioreaction

A microbial kinetic model describing the reaction rates of the fed-batch fermentation of *Saccharomyces cerevisiae* is proposed by Enfors et al (1992). The biomass growth rate is assumed to be of the Monod type and include ethanol formation. Mass balances are made over each component in the liquid phase using the reaction rates given in Table 5-1. The microbial kinetic model is based on the assumption that oxygen limitations are prevented for bakers' yeast processes operated as aerobic fed-batch fermenters. Thus, the dissolved oxygen does not influence the reaction rate of any species. Consequently, the oxygen transport is not calculated and the conservation equation of oxygen is not listed below.

Conservation of	$\Phi_{i,l}$	S_C
Biomass	C_X	μC_X
Sugar	C_S	$-q_S C_X$
Ethanol	C_E	$(q_{EP}-q_{EC})C_X$
$q_S=q_{SMAX}\cdot C_S/(C_S+K_S)$		Saturation constants
$q_{OS}=q_S\cdot Y_{OS}$		$K_S=0.2$ g/l
if $q_{OS}>q_{OMAX}$ then $q_{OS}=q_{OMAX}$		$K_{SE}=0.5$ g/l
$q_{SXA}=q_{OS}/Y_{OS}$		
$q_{SXF}=q_S-q_{SXA}$		
$q_{EC}=(q_{OMAX}-q_{OS})/Y_{OE}\cdot C_E/(C_E+K_{SE})$		Specific consumption rates
$q_{EP}=q_{SXF}\cdot Y_{ES}$		$q_{SMAX}=1.4$ gS/gX,h
$\mu=q_{SXA}\cdot Y_{XSA}+q_{SXF}\cdot Y_{XSF}+q_{EC}\cdot Y_{XE}$		$q_{OMAX}=0.3$ gO/gX,h
Yield coefficients		
$Y_{OS}=0.5$ gO/gS $Y_{ES}=0.35$ gE/gS $Y_{OE}=0.8$ gO/gE		
$Y_{XE}=0.8$ gX/gE $Y_{XSA}=0.5$ gX/gS(aerobe) $Y_{XSF}=0.15$ gX/gS(anaerobe)		

Table 5-1 Source terms for chemical species in the liquid phase.

Catalytic reaction

The ozone decomposition reaction. The ozone decomposition reaction is a simple irreversible first order catalytic reaction. The reaction can be written as:

$$2\,O_3 \rightarrow 3\,O_2$$

Due to the low concentration of ozone in a riser, the heat produced by the chemical reaction is negligible and the reaction is looked upon as being isothermal.

Species conservation equations A transport equation for the mass fraction of ozone in the gas phase $Y_{k,g}$ is solved

$$\frac{\partial}{\partial t}(\alpha\rho Y_k)_g + \frac{\partial}{\partial x_i}(\alpha\rho U_i Y_k)_g = \frac{\partial}{\partial x_i}\left[\Gamma_{Y_{k,g}}\frac{\partial Y_k}{\partial x_i}\right] + \alpha_g \cdot r \tag{5.19}$$

where the transport coefficient Γ_Y is expressed in terms of the diffusion coefficient, D_Y, and the turbulent viscosity of the gas phase:

$$\Gamma_{Y_{k,g}} = D_{Y_{k,g}} + \frac{\mu_{turb,g}}{0.7} \tag{5.20}$$

The reaction rate constant for the catalytic reaction was measured in a fixed bed reactor. The rate constant is expressed pr. unit volume of catalyst and was measured to be 3.96 s^{-1} (Ouyang et al., 1993). The reaction rate is expressed as:

$$r = -3.96 \cdot \alpha_s \cdot C_{Ozone} \quad [\text{kg Ozone/m}^3\ \text{s}] \tag{5.21}$$

Oxy-chlorination reaction. Energy equations for the two phases are solved together with equations for the chemical species (Samuelsberg and Hjertager, 1996). The catalytic reactions take place on the surface of the catalyst. The simple one step homogeneous catalytic oxy-chlorination reaction can be represented by the following equation:

$$2\,C_2H_4 + 4\,HCl + O_2 \rightarrow 2\,\underbrace{C_2H_4Cl_2}_{EDC} + 2\,H_2O, \quad -\Delta H_{rx} = -240\left[kJ/mol\right] \tag{5.22}$$

The rate of reaction is given as a function of the partial pressure of ethene as:

$$r = k\,p_{C_2H_4} \quad \left[mol\ EDC/h\ g_{cat}\right] \tag{5.23}$$

where the specific reaction rate is given by the usual Arrhenius equation:

$$k = k_1\,e^{\left(\frac{E}{RT}\right)} \tag{5.24}$$

To convert the component's partial pressures into mass fractions (Y_j), the ideal gas law is used. The rate of reaction may then be written as:

$$S_{Y_{EDC}} = k\,\alpha_p\,\rho_p\,M_{EDC}\,Y_{C_2H_4}\frac{R_i}{R}\,p_{tot}\left[kg\ EDC/sm^3\right] \tag{5.25}$$

5.5 Stirred tank modeling

There are many different methods available to simulate the flow in stirred vessels. A detailed review of these models was made by Brucato et al. (1998). Here the different methods will only be briefly introduced. The different methods can be divided into two distinct classes: one in which the whole tank is described with a fixed coordinate system and one in which a part of the co-ordinate system co-rotates with the impeller.

Fixed coordinate system One of the simplest ways to model the flow in a stirred tank is to use a fixed coordinate system, along with imposed boundary conditions (IBC) in the impeller swept region. This approach was used by Ranade & Joshi (1990b); Kresta & Wood (1991) and Jenne & Reuss (1999). The main disadvantage of this method is that the imposed boundary conditions are based on experimental data or on empirical models. Often, this data is not available, like for example in the case of multiphase flows. A way to overcome this problem is to model the presence of the impeller by adding source terms to the momentum equations. An implementation of this approach, called the snapshot-method was used by Ranade (1997) and Ranade et al. (2002) to simulate the flow in a stirred tank, equipped with a Rushton turbine. Using a standard k-ε model, they found good agreement with experimental data. The advantage of this method is that a few snapshots of the flow suffice to get a good description of the flow. That is, steady-state simulations are carried out for only a few impeller positions. The overall statistics are obtained by calculating the ensemble average of the simulated impeller positions. Revstedt et al. (1998) and Derksen and Van den Akker (1999) also used source terms to describe the impeller. Both authors used large eddy simulations to describe the turbulence. Due to the transient nature of the simulations a number of full rotations were needed to be simulated in order to obtain constant statistics.

Rotating coordinate system In the other class of methods, the inner part of the tank is described with a coordinate system that co-rotates with the impeller. In that case, normal wall boundary conditions can be used to set the velocities at the impeller. The advantage of this method is, that no empirical information is needed to set the boundary conditions. The outer part of the tank (i.e. the tank walls and the baffles) is described in a fixed coordinate frame. Using a rotating coordinate system, there are three different methods. The first two are very similar and are called the multiple reference frames method (MRF) and the inner-outer method (IO). The former was introduced by Luo et al. (1994) and the latter by Brucato et al. (1994). Both methods are similar to the snapshot method, in that steady-state simulations are carried out for only a few impeller positions. In the simulations, the equations for the fixed and the rotating part of the geometry are solved separately. There is only one distinct difference between MRF and IO; in IO, the calculation domains of the two parts have a small overlap. This is not the case for MRF. In the IO method, a number of outer iterations are required to ensure continuity across the interface between the two parts. This implies extra calculation time compared to the snapshot and the MRF method. The third method using a rotating coordinate system is called the sliding grid method (SG). This method was first applied to the flow in a stirred tank by Luo et al. (1993). In this method both the coordinate system and the grid of the inner part are rotating. This implies that the simulations need to be transient, making them computationally more expensive than the steady state methods. The difference between SG and the MRF and IO methods is that the entire domain can be solved as a whole, so no extra iterations are necessary. Continuity across the interface between the two parts is ensured by interpolating the variables on either side of the interface.

6 Numerical Solution Procedure

6.1 General

The governing partial differential equations are discretized by the finite volume method (Patankar, 1980 and Hjertager, 1986). Accordingly, the computational domain is divided by non-overlapping control volumes. A staggered grid arrangement is applied as depicted in Figure 6-1 where control volumes for the continuity and the momentum equations are shown.

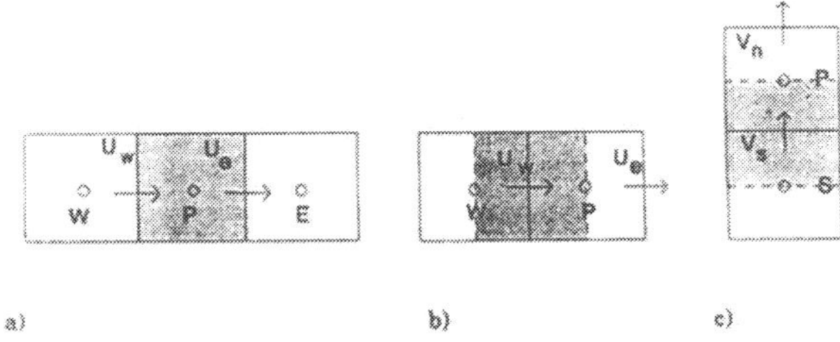

Figure 6-1 Control volumes for (a) mass and scalar variables, (b) x-direction momentum and (c) y-direction

The conservation equations for each phase are integrated over the control volumes with respect to time and space. The integrals are reduced to algebraic form by fully implicit temporal and hybrid spatial discretizations.

6.2 Phase mass conservation

For a one dimensional case, the continuity equation for one of the phases may be discretized using the control volume in Figure 6-1a:

$$\Delta V \frac{(\alpha\rho-(\alpha\rho)^{o})_{P}}{\Delta t}+A_{e}(\alpha\rho U)_{e}-A_{w}(\alpha\rho U)_{w}=\Gamma_{P}\Delta V \quad (6.1)$$

Upwind differencing:

$$\begin{aligned}(\alpha\rho U)_{e}&=\alpha_{P}\rho_{P}U_{e}\text{..........}when\text{........}U_{e}>0.\\(\alpha\rho U)_{e}&=\alpha_{E}\rho_{E}U_{e}\text{..........}when\text{........}U_{e}<0\\(\alpha\rho U)_{w}&=\alpha_{W}\rho_{W}U_{w}\text{.........}when\text{.........}U_{w}>0.\\(\alpha\rho U)_{w}&=\alpha_{P}\rho_{P}U_{w}\text{.........}when\text{.........}U_{w}<0\end{aligned} \quad (6.2)$$

We obtain by use of < x > = max(x, 0) => always > 0:

$$\alpha_P \rho_P\left(\frac{\Delta V}{\Delta t} + A_e < U_e > + A_w < -U_w > \right) =$$

$$A_e \alpha_E \rho_E < -U_e > + A_w \alpha_W \rho_W < U_w > + \Gamma_P \Delta V + \frac{(\rho^o \alpha^o)_P \Delta V}{\Delta t}$$

$$\downarrow \tag{6.3}$$

$$a_P^\alpha \cdot \alpha_P = \sum_{all\ neighbours} a_n^\alpha \cdot \alpha_n + S_P^\alpha$$

$$NB!\ a_P^\alpha \neq \sum_{all\ neighbours} a_n^\alpha$$

Simplified notation:

$$a_P \alpha_P = b_P$$

This states that: Outflow x volume fraction = inflow. Solution of such an equation together with the relation:

$$\alpha_1 + \alpha_2 = 1.0$$

gives both volume fractions.

Questions to be addressed: Is it the equation for phase no. 1 or phase no. 2 that should be solved? If the volume fraction for one of the phases is close to zero, the numerical properties for the equation are bad. Round-off errors may destroy the solution.

One possibility is to solve the equation for the phase that has the highest volume fraction. Another possibility is to combine the two equations according to:

$$a_1 \alpha_1 = b_1 \ldots\ldots | \ a_2$$

$$a_2 \alpha_2 = b_2 \ldots\ldots | \ a_1$$

Multiplication and addition give:

$$a_1 a_2 \overbrace{(\alpha_1 + \alpha_2)}^{=1.0} = b_1 a_2 + b_2 a_1 \quad \Rightarrow \quad a_1 = \frac{b_1 a_2 + b_2 a_1}{a_2}$$

For fluid 1, the volume fraction equation then reads:

$$\alpha_1 = \frac{b_1 a_2}{b_1 a_2 + b_2 a_1} \tag{6.4}$$

The terms in the denominator are of the same order of magnitude which is a good property.

Generalisation

For more phases than two:

$$a_1 \alpha_1 = b_1 \ldots\ldots | \ a_2 a_3 \cdots a_n$$

$$a_2 \alpha_2 = b_2 \ldots\ldots | \ a_1 a_3 \cdots a_n$$

$$\ldots.. = \ldots\ldots.. | \ \ldots\ldots\ldots..$$

$$a_n \alpha_n = b_n \ldots\ldots.. | \ a_1 a_2 \cdots a_{n-1}$$

Multiplication and addition:

$$a_1 a_2 \cdots a_n \overbrace{(\alpha_1+\alpha_2+\ldots+\alpha_n)}^{=1.0} = \sum_{i=1}^{n} b_i \prod_{j=1;j\neq i}^{n} a_j \quad \Rightarrow \quad a_i = \frac{\sum_{i=1}^{n} b_i \prod_{j=1;j\neq i}^{n} a_j}{\prod_{j=1;j\neq i}^{n} a_j}$$

Finally the volume fraction may be computed as:

$$\alpha_i = \frac{b_i \prod_{j=1;j\neq i}^{n} a_j}{\sum_{i=1}^{n} b_i \prod_{j=1;j\neq i}^{n} a_j} \tag{6.5}$$

6.3 Momentum equations

The x-direction momentum equations for two phases may be discretized as

$$(a_1+c_{12})U_1 = \sum_1 a_{nb}U_{nb} + d_1(p_W - p_P) + c_{12}U_2 + s_1 \tag{6.6}$$

$$(a_2+c_{12})U_2 = \sum_2 a_{nb}U_{nb} + d_2(p_W - p_P) + c_{12}U_1 + s_2 \tag{6.7}$$

where the coefficients *a, c, d* and *s* appear from the discretization of convection/diffusion, drag, pressure and source terms, respectively. Neighbouring points are denoted *nb*. The drag terms introduce a strong coupling between the two phases. In a sequential solution algorithm, such a strong coupling may cause slow numerical convergence if the drag terms are dominant. This may be avoided by implementing the Partial Elimination Algorithm (PEA) (Spalding, 1977 and 1980; Manger, 1996). For the x-direction momentum equations, the PEA-algorithm gives

$$\underbrace{\left(\frac{a_1 a_2+(a_1+a_2)c_{12}}{a_2+c_{12}}\right)}_{a_{P,1}} U_1 = \sum_1 a_{nb}U_{nb} + d_1(p_W - p_P) + s_1 + \\ + \frac{c_{12}}{a_2+c_{12}}\left(\sum_2 a_{nb}U_{nb} + d_2(p_W - p_P) + s_2\right) \tag{6.8}$$

and

$$\underbrace{\left(\frac{a_1 a_2+(a_1+a_2)c_{12}}{a_1+c_{12}}\right)}_{a_{P,2}} U_2 = \sum_2 a_{nb}U_{nb} + d_2(p_W - p_P) + s_2 + \\ + \frac{c_{12}}{a_1+c_{12}}\left(\sum_1 a_{nb}U_{nb} + d_1(p_W - p_P) + s_1\right) \tag{6.9}$$

Generalisation

Generalized to n phases (after Manger (1996)):

$$(a_j + \sum_{i=1;i\neq j}^{n} c_{ij})U_j = H_j + d_j\Delta p + s_j + \sum_{i=1;i\neq j}^{n} c_{ij}U_i \tag{6.10}$$

Here $\Delta p = p_W - p_P$ and $H_j = \Sigma a_{nb}U_{nb}$ for phase j.

$$(a_i + \sum_{k=1;k\neq i}^{n} c_{ki})U_i = H_i + d_i\Delta p + s_i + \sum_{k=1;k\neq i}^{n} c_{ki}U_k \tag{6.11}$$

Solves for U_i from equation 6.11:

$$U_i = \frac{1}{a_i + \sum_{k=1;k\neq i}^{n} c_{ki}}\left(H_i + d_i\Delta p + s_i + \sum_{k=1;k\neq i}^{n} c_{ki}U_k\right) \tag{6.12}$$

Insert equation 6.12 into 6.10 and obtain:

$$(a_j + \sum_{i=1;\neq j}^{n} c_{ij})U_j = H_j + d_j\Delta p + s_j + \sum_{i=1;i\neq j}^{n}\left[\frac{c_{ij}}{a_i + \sum_{k=1;k\neq i}^{n} c_{ki}}\left(H_i + d_i\Delta p + s_i + \sum_{k=1;k\neq i}^{n} c_{ki}U_k\right)\right] \tag{6.13}$$

Rearrange last term on the right side of equation 6.13:

$$\begin{aligned}(a_j + \sum_{i=1;\neq j}^{n} c_{ij})U_j &= H_j + d_j\Delta p + s_j + \\ &+ \sum_{i=1;i\neq j}^{n}\left[\frac{c_{ij}}{a_i + \sum_{k=1;k\neq i}^{n} c_{ki}}\left(H_i + d_i\Delta p + s_i + \sum_{k=1;k\neq i;k\neq j}^{n} c_{ki}U_k\right)\right] + \sum_{i=1;i\neq j}\left(\frac{c_{ij}}{a_i + \sum_{k=1;k\neq i}^{n} c_{ki}} c_{ij}U_j\right)\end{aligned} \tag{6.14}$$

Move last term on the right side in the above equation to the left side:

$$\begin{aligned}\left[a_j + \sum_{i=1;\neq j}^{n}\left(c_{ij} - \frac{c_{ij}^2}{a_i + \sum_{k=1;k\neq i}^{n} c_{ki}}\right)\right]U_j &= H_j + d_j\Delta p + s_j + \\ &+ \sum_{i=1;i\neq j}^{n}\left[\frac{c_{ij}}{a_i + \sum_{k=1;k\neq i}^{n} c_{ki}}\left(H_i + d_i\Delta p + s_i + \sum_{k=1;k\neq i;k\neq j}^{n} c_{ki}U_k\right)\right]\end{aligned} \tag{6.15}$$

Rearranging the term on the left side of equation 6.15 to obtain:

$$\left[a_j+\sum_{i=1;i\neq j}^{n}\frac{c_{ij}(a_i+\sum_{k=1;k\neq i}^{n}c_{ki})-c_{ij}^2}{a_i+\sum_{k=1;k\neq i}^{n}c_{ki}}\right]U_j=H_j+d_j\Delta p+s_j+ \\ +\sum_{i=1;i\neq j}^{n}\left[\frac{c_{ij}}{a_i+\sum_{k=1;k\neq i}^{n}c_{ki}}\left(H_i+d_i\Delta p+s_i+\sum_{k=1;k\neq i;k\neq j}^{n}c_{ki}U_k\right)\right] \tag{6.16}$$

Final equation for U_j:

$$\underbrace{\left[a_j+\sum_{i=1;\neq j}^{n}\frac{c_{ij}(a_i+\sum_{k=1;k\neq i;k\neq j}^{n}c_{ki})}{a_i+\sum_{k=1;k\neq i}^{n}c_{ki}}\right]}_{a_{P,j}}U_j=H_j+d_j\Delta p+s_j+ \\ +\sum_{i=1;i\neq j}^{n}\left[\frac{c_{ij}}{a_i+\sum_{k=1;k\neq i}^{n}c_{ki}}\left(H_i+d_i\Delta p+s_i+\sum_{k=1;k\neq i;k\neq j}^{n}c_{ki}U_k\right)\right] \tag{6.17}$$

6.4 Equation for the pressure

We can as for single phase flows (Patankar, 1980), construct an equation for the pressure from the continuity equations and the linearized momentum equations. Product terms are linearized as:

$$U_k=U_k^*+U_k'$$
$$\rho_k=\rho_k^*+\rho_k'$$
$$\alpha_k=\alpha_k^*+\alpha_k'$$
$$\Downarrow$$
$$\alpha_k\rho_kU_k\approx\alpha_k^*\rho_k^*U_k^*+\alpha_k^*\rho_k^*U_k'+\alpha_k^*U_k^*\rho_k'+\rho_k^*U_k^*\alpha_k'$$

From the momentum equations we obtain a relation between pressure variation and velocity variation:

$$U'_k=d_k^{p_k}\Delta p'_k$$

From the equation of state we get the relation between density and pressure:

$$\rho'_k=\frac{\partial\rho_k}{\partial p}\Big|_{s=const}\ p'_k=\frac{1}{c_s^2}\ p'_k;\quad \frac{p_k}{\rho_k}=R\ T_k$$

Here c_s is the sound speed of phase k. From the relation between pressure and volume fraction we get:

$$\alpha'_k = \frac{\partial \alpha_k}{\partial p_k} p'_k; \quad p_k = p + G(\alpha_k) \cdot \alpha_k$$

With the same arguments as for the volume fractions, we should use a combination of both continuity equations. It is most easy to add them. Because the heaviest phase has much larger density than the light one, the heaviest phase will in most cases dominate the coefficients in the equation for the pressure correction. The pressure will therefore be less dominated by the lighter phase. One possibility is to divide the continuity equations by a relevant reference density before they are added:

$$\frac{1}{\rho_{1,ref}}\left(\frac{\partial}{\partial t}\alpha_1\rho_1 + \nabla\cdot(\alpha_1\rho_1\vec{V}_1)\right) + \frac{1}{\rho_{2,ref}}\left(\frac{\partial}{\partial t}\alpha_2\rho_2 + \nabla\cdot(\alpha_2\rho_2\vec{V}_2)\right) = \frac{\Gamma_1}{\rho_{1,ref}} + \frac{\Gamma_2}{\rho_{2,ref}} \quad (6.18)$$

Extension to more phases than 2 is as follows:

$$\frac{\partial}{\partial t}\left(\sum_{i=1}^{n}\frac{\alpha_i\rho_i}{\rho_{i,ref}}\right) + \nabla\cdot\left(\sum_{i=1}^{n}\frac{\alpha_i\rho_i\vec{V}_i}{\rho_{i,ref}}\right) = \sum_{i=1}^{n}\frac{\Gamma_i}{\rho_{i,ref}} \quad (6.19)$$

Insertion of all relations that are dependent on p_k' into the individual phase mass balances gives the following equation for the phase pressure correction:

$$a_P^{p'_k} p'_{k,P} = \sum_{neighbours} a_n^{p'_k} p'_{k,n} + S^{p'_k} \quad (6.20)$$

6.5 Summary of Solution method

General solution method

The equations are solved sequentially. Because of the nonlinearities and coupling between the equations, there is a need for an iterative solution procedure. The single phase solution algorithm SIMPLE (semi implicit method for pressure linked equations) by Patankar and Spalding (1972), which is extended to two-phase flow by Spalding (1985), is used. The solution algorithm is summarized as follows

1. Initialize.
2. Update auxiliary variables such as density, temperature, etc. Apply boundary conditions.
3. Solve for α_k using modified expression for the continuity equation (Eq. 6.5). Solve eq. using Jacobi-iteration and find:

$$\alpha_N = 1.0 - \sum_{k=1}^{N-1}\alpha_k \quad (6.21)$$

4. Solve momentum equations using the PEA-algorithm and the latest available solution for the pressure (Eq. 6.17). Use Jacobi-iteration or TDMA line iteration. Calculate coefficients for the relation between pressure-correction and velocity, density and volume-fraction variation.
5. Solve the equations for p_k' (Eq. 6.20) using line-iteration (TDMA - algorithm)
6. Correct variables according to:

$$
\begin{aligned}
p_{k,P}^{n+1} &= p_{k,P}^{n} + p'_{k,P} \\
U_{k,e}^{n+1} &= U_{k,e}^{n} + d_{u_k}^{p_k}(p'_{k,P} - p'_{k,E}) \\
\rho_{k,P}^{n+1} &= \rho_{k,P}^{n} + d_{\rho_k} p'_{k,P} \\
\alpha_{k,P}^{n+1} &= \alpha_{k,P}^{n} + d_{\alpha_k} p'_{k,P}
\end{aligned}
\tag{6.22}
$$

7. Repeat from 5 until a specified convergence criterion is satisfied.
8. If steps 3, 4, 5 or 6 are not converged in a given number of steps repeat from 3, a prescribed maximum times.
9. Solve for the other equations, e.g. enthalpy, turbulence parameters, chemical species, etc.
10. If the simulation is finished, stop the programme. If more steps are needed go to 2.

Special arrangements for bio reaction simulation

After a converged solution of the gas/liquid flow is obtained, the flow variables are fixed, and only the equations governing the biochemical species are solved. This is to enable different time steps for the simulation of flow and biochemical reactions. The time steps for the flow simulations are calculated from a specified Courant number. The bioconversions to be simulated last for typically several hours, and the time steps were chosen based on a sensitivity test of the actual process.

7 Applications

The next subsections will give some examples of use of some of the models described above.

7.1 Bubble columns

In the work of Deen et al. (2001) the use of large eddy simulations (LES) in numerical simulations of the gas–liquid flow in bubble columns was studied. The Euler–Euler approach was used to describe the equations of motion of the two-phase flow.

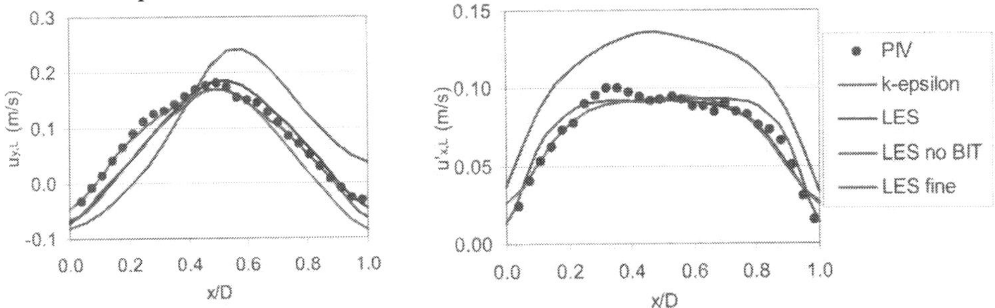

Figure 7-1 Comparison of simulated and experimental profiles of (left) the axial average liquid velocity and (right) the transverse liquid velocity fluctuations for different conditions.

It was found that, when the drag, lift and virtual mass forces are used, the transient behaviour that was observed in experiments can be captured. Good quantitative agreement with experimental data is obtained both for the mean velocities and the fluctuating velocities (Figure 7-1). The LES showed better agreement with the experimental data than simulations using the k–ε model (Figure 7-1). Bove (2005) extended the above modelling by including solution of the Bubble population equation using three nodes in the size coordinate. This enabled calculation of the bubble size distribution in bubble columns. An example of Bove's results is given in Figure 7-2.

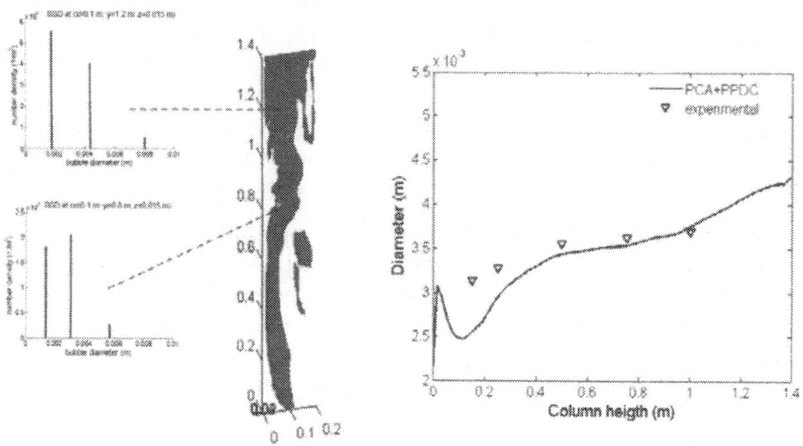

Figure 7-2 (left) A snap shot of the predicted size distribution at two position in the bubble column of van den Hengel et al (2005) and (right) the predicted and measured bubble diameter along the column centreline.

7.2 Stirred tanks

Deen et al (2003) did a study of flow in a stirred tank. A two-camera PIV technique was used to obtain angle resolved velocity and turbulence data of the flow in a lab-scale stirred tank, equipped with a Rushton turbine.

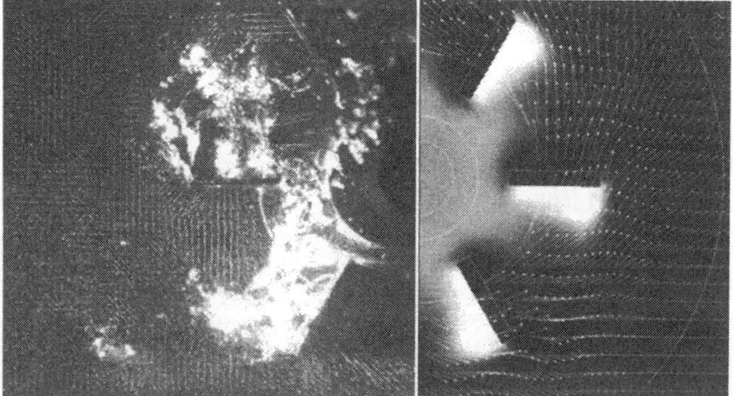

Figure 7-3 Comparisons between measured (left) and computed (right) distribution of gas around the impeller.

Two cases were investigated: a single-phase flow and a gas-liquid flow. In the former case, the classical radial jet flow pattern accompanied by two trailing vortices was observed. In the latter case, the velocity of the radial jet was reduced, and the vortices were diminished by the presence of the gas. Gas cavities clinging to the back of the impeller blades were observed (Figure 7-3).

Both cases were also investigated with the use of three-dimensional transient CFD simulations. The sliding grid technique was used to describe the movement of the impeller. For the single-phase flow the simulations in the impeller region corresponded very well with the experimental data. For the gas-liquid flow both the mean and fluctuating liquid velocities in the impeller region were well predicted (Figure 7-4). This was also the case for the mean radial gas velocities. The largest differences between the simulations and the experiments were found for the mean axial gas velocity.

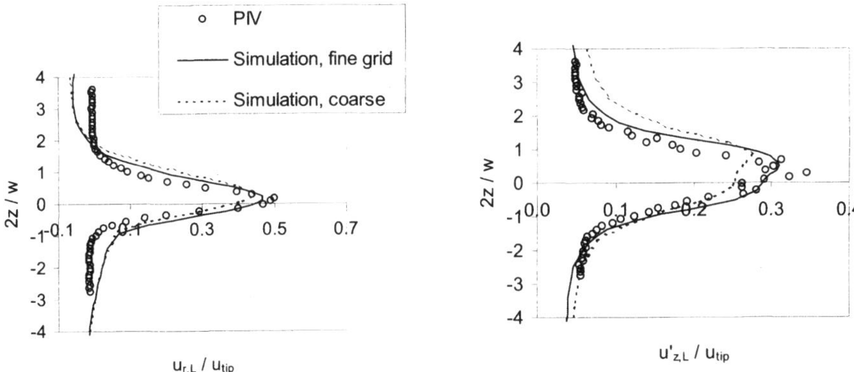

Figure 7-4. Comparisons of measured and predicted radial liquid velocity (left) and axial liquid turbulent velocity (right) close to the impeller region. The gas liquid case using fine or coarse grids.

7.3 Circulating fluidized beds

Cold flow. A computational fluid dynamics simulation of a cold flowing riser fluidized with FCC catalysts has been performed by Hansen et al (2003). The computations were performed using the 3D multiphase computational fluid dynamics code with an Eulerian description of both gas and particle phase, described above. The turbulent motion of the particulate phase was modeled using the kinetic theory for granular flow, and the gas phase turbulence was modeled using a Sub-Grid-Scale model. The complex inlet geometry was approximated using multiple inlet patches. The first results were submitted to a blind-test in connection to the 10th international workshop on two-phase flow prediction held in Merseburg, Germany, 2002. The results were subsequently validated against experimental findings of particle mass flux across the riser and pressure profile along the riser. The calculations showed good agreement with experimental findings of both mass flux and pressure profile, but further improvements were proposed and investigated. A parameter study showed that mesh refinement, choice of particle diameter and choice of drag model are crucial when simulating FCC riser flow. The result of the submitted blind test and best result after the parameter study is given in Figure 7-5.

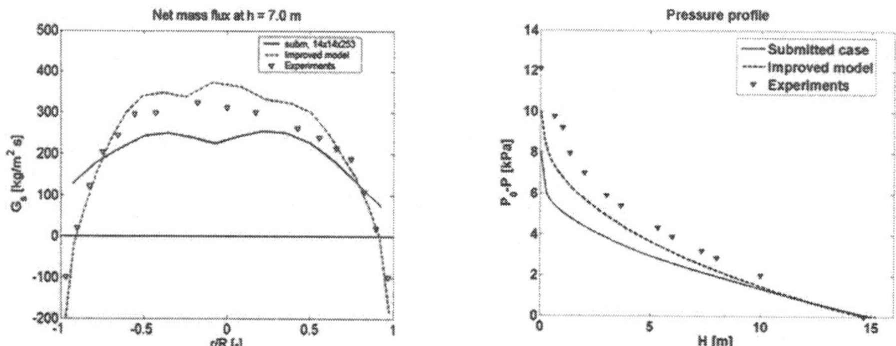

Figure 7-5 (left) shows predicted and measured solids fluxes (left) and (right) pressure profile for blind test and improved cases.

Reactive flow. The isothermal decomposition of ozone has been implemented by Hansen et al. (2004) in the CFD code FLOTRACS-MP-3D described above. The code is a three dimensional (3D) multiphase computational fluid dynamics code with an Eulerian description of both gas and particle phase. The turbulent motion of the particulate phase is modelled using the kinetic theory for granular flow, and the gas phase turbulence is modelled using a sub-grid-scale model,

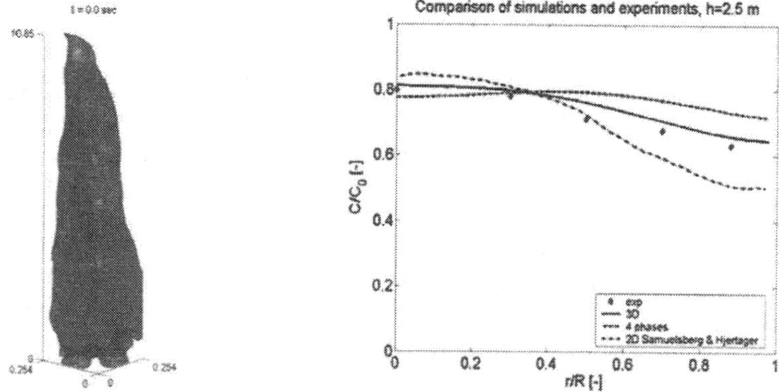

Figure 7-6. (left) Predicted contours of ozone concentration (iso-surface of $C/C_{inlet} = 0.6$); (right) comparison of measured and predicted radial profile of ozone.

The decomposition reaction is studied in a 3D representation of a 0.254m I.D. riser, which has been studied experimentally by Ouyang et al. (1993). These authors obtained profiles of ozone concentration in the 10.85-m high riser by the use of a UV detector system. Furthermore, a pressure drop profile was reported. Comparison between measured and simulated time-averaged ozone concentration at different elevations in the riser shows good agreement. The 3D representation of the reactor geometry gives better predictions of the radial variation in concentration than in a similar 2D simulation, Samuelsberg and Hjertager (1995). Typical results are given in Figure 7-6.

8 Summary

The present Chapter has presented a derivation of the multi-fluid model for multi-phase flows. The mixture model has also been presented. Closure laws for both bubble-liquid and solids-gas systems have been exposed. In particular the kinetic theory of granular flow (KTGF) has been shown for the solids gas system. Several sub-models are presented including bubble size, chemical reactions, heat and mass transfer as well as stirrers. Numerical solution issues are also considered. Finally, examples of results from bubble-liquid and solids-gas simulations are given.

References

K. Agrawal, P.N. Loezos, M. Syamlal and S.Sundaresan (2001). The role of meso-scale structures in rapid gas-solid flows. *Journal of Fluid Mechanics*, 445:151-185.

A. Albrecht, O. Simonin, D. Barthod and D. Vedrine (2001). Multidimensional numerical simulation of the liquid feed injection in an Industrial FCC Riser, in: M. Kwauk, J. Li, W.-C. Yang (Eds.), *Fluidization X.* Engineering Foundation 349-356.

A.A. Aris, (1962). *Introduction to fluid Mechanics*. Springer-Verlag. New York.

S. Benyahia, A,G. Ortiz and J.I.P. Paredes (2003). Numerical analysis of a reacting gas/solid flow in the riser of an industrial fluid catalytic cracking unit. *International Journal of Chemical Reactor Engineering*, 1 (A.41).

G.A. Bokkers, M. van Sint Annaland and J.A.M. Kuipers (2004). Mixing and segregation in a bidisperse gas-solid fluidised bed: a numerical and experimental study. *Powder Technology*, 140:176-186.

S. Bove (2005). *Computational fluid dynamics of gas-liquid flows including bubble population balances*, PhD thesis, Aalborg University Esbjerg.

S. Bove, T. Solberg and B. H. Hjertager (2004). Numerical aspects of bubble column simulations. *International Journal of Chemical Reactor Engineering*, 2:A1.

A. Brucato, M. Ciofalo, F. Grisafi and G. Micale (1994). Complete numerical simulation of flow fields in baffled stirred vessels: the inner-outer approach. *IChemE Symposium Series.* 136:155–162.

P.H. Calderbank (1958). Physical rate processes in industrial fermentation. Part 1. The interfacial area in gas–liquid contacting with mechanical agitation, *Transaction of IChemE*, 36:443-463.

J. Chahed, V. Roig, and L. Masbernat (2003). Eulerian-eulerian two-fluid model for turbulent gas-liquid bubbly flows. *International Journal of Multiphase Flow*, 29:23–49.

S. Chapman and G. Cowling (1970). *The mathematical theory of non-uniform gases.* 3 ed.. Cambridge University Press.

R. Clift, J. R. Grace and M. E. Weber (1978). *Bubbles, drops and particles*. Academic Press. London.

T.L Cook and F.H. Harlow (1986). Vortices in bubbly two-phase flow, *International Journal of Multiphase Flow*, 12:35-61.

J.W. Deardorff (1971). On the magnitude of the subgrid scale eddy coefficient. *Journal of Computaional Physics*, 7:120–133.

N.G. Deen, T. Solberg and B. H. Hjertager (2001). Large eddy simulations of the gas-liquid flow in a square cross-sectioned bubble column. *Chemical Engineering Science*, 56:6341–6349.

N.G. Deen, T. Solberg and B.H. Hjertager (2003). Flow generated by an aerated rushton impeller: two-phase PIV experiments and numerical simulations. *The Canadian Journal of Chemical Engineering*, 80(4):638-652.

J. Derksen and H.E.A. van den Akker (1999). Large eddy simulations on the flow driven by a Rushton turbine. *AIChE Journal* 45:209–221.

J. Ding and D. Gidaspow (1990). A Bubbling fluidization model using kinetic theory of granular flow. *AIChE Journal*, 36:523-538.

D.A. Drew, (1983). Mathematical modeling of two-phase flow. *Annual Review of Fluid Mechanics,* 15:261–291.

D.A. Drew and S.L. Passman (1999). *Theory of multicomponent fluids*. Springer-Verlag. New York.

H. Enwald, E. Peirano and A.-E. Almstedt (1996) Eulerian two-fluid flow theory applied to fluidization. *International Journal of Multiphase Flow*, 22 (1):21–66.

S.O. Enfors, S. George and G. Larson (1992). KTH, Private communications

S. Ergun, (1952). Fluid flow through packed bed columns. *Chemical Engineering Progress*, 48 (2), 89-98.

J. Gao, C. Xu, S. Lin, G. Yang and Y. Guo (1999). Advanced model for turbulent gas-solid flow and reaction in FCC riser reactors. *AIChE Journal*, 45 (5):1095-1113.

D. Geldart, ed. (1986). *Gas fluidization technology*, J. Wiley, Chichester.

L.G. Gibilaro, R. di Felice and S.P. Waldram (1985). Generalized friction factor coefficient correlations for fluid-particle interactions. *Chemical Engineering Science*, 40 (10):1817-1823.

D. Gidaspow (1994). *Multiphase flow and fluidization*, Academic Press, London.

D. Gidaspow, L. Huilin, E. Manger (1996). Kinetic theory of multiphase flow and fluidization. Validation and extension to binaries, In: *Proceedings of XIXth Inernational Congress of theoretical and applied mechanics.*

C. Guenther, T. O'Brien and M. Syamlal (2001). A numerical model of silane pyrolysis in a gas-solids fluidized bed. *Proceedings of the International Conference on Multiphase Flow 2001*, New Orleans.

K.G. Hansen, C.H. Ibsen, T. Solberg and B.H. Hjertager (2003). Eulerian/Eulerian CFD simulation of a cold flowing FCC riser. *International Journal of Chemical Reactor Engineering*, 1:A31.

K.G. Hansen, T. Solberg and B. H. Hjertager (2004). A three-dimensional simulation of gas/particle flow and ozone decomposition the riser of a circulating fluidized bed. *Chemical Engineering Science*,59(22-23):5217-5224.

T. Hibiki, and M. Ishii (2002). Distribution parameter and drift velocity of drift-flux model in bubbly flow. *International Journal of Heat and Mass Transfer*, 45:707–721.

B.H. Hjertager (1986). Three-dimensional modeling of flow, heat transfer and combustion, in: *Handbook of Heat and Mass Transfer*, Gulf Publishing, Houston, 1303-1350

C.H. Ibsen, T. Solberg and B.H. Hjertager (2001). Evaluation of a three-dimensional numerical model of a scaled circulating fluidized bed. *Industrial & Engineering Chemistry Research*, 40 (23):5081-5086.

M Ishii (1975) *Thermo fluid dynamic theory of two-phase flow*. Eyrolles, Paris.

M. Ishii and N. Zuber (2002). Drag coefficient and relative velocity in bubbly, droplet or particulate flows. *AIChE Journal*, 45:707–721.

I. Kataoka and A. Serizawa (1989). Basic equations of turbulence in gas-liquid two-phase flow. *International Journal of Multiphase Flow*, 15(5):843–855.

S.M. Kresta and P.E. Wood (1991). Prediction of the three-dimensional turbulent flow in stirred tanks. *AIChE Journal,* 37:448–460.

H.A. Jakobsen (1993). *On the modelling and simulation of bubble column reactors using a two-fluid Model*, Dr. ing. Thesis, Norwegian Institute of Technology, Trondheim, Norway.

J.T. Jenkins and F. Mancini (1987). Balance laws and constitutive relations for plane flows of a dense mixture of smooth, nearly elastic, circular disks. *Journal of Applied Mechanics,* 54:27–34.

J.T. Jenkins and S. B. Savage (1983) A theory for the rapid flow of identical, smooth, nearly elastic, spherical particles. *Journal of Fluid Mechanics*,130:187–202.

M. Jenne and M. Reuss (1999). A critical assessment of the use of k-ε turbulence models for simulation of the turbulent liquid flow induced by a Rushton-turbine in baffled stirred-tank reactors. *Chemical Engineering Science,* 54:3921–3941.

S.T. Johansen and F. Boysan (1988). Fluid dynamics in bubble stirred ladles. Part II. Mathematical modeling, *Metallurgical Transactions B*, 19B:755-764.

D.D. Joseph, T. S. Lundgren, R. Jackson and D. A. Saville (1990). Ensemble average and mixture theory quations for incompressible fluid-particle suspension. *International Journal of Multiphase Flow*, 16(1):35–42.

D.L. Koch and R. Hill (2001). Inertial effects in suspension and porous-media flows. *Annual Review of Fluid Mechanics*, 619-647.

R.T. Lahey, M. Lopez de Bertodano and O.C. Jones (1993). Phase distribution in complex geometry ducts, *Nuclear Engineering Design*, 141:177-701.

A. Lapin, C. Maul, K. Junghans and A. Lubbert (2001). Industrial-scale bubble column reactors: gas-liquid flow and chemical reaction. *Chemical Engineering Science*, 56:239–246.

M. Lopez de Bertodano (1998). Two fluid model for two-phase turbulent jets. *Nuclear Engineering and Design*, 179:65–74.

M. Lopez de Bertodano, R. T. Lahey and O. C. Jones (1994). Development of a $k-\varepsilon$ model for bubbly two-phase flow. *Journal of Fluids Engineering*, 116:128–134.

J.Y. Luo, A.D, Gosman, R.I. Issa, J.C. Middleton and M.K. Fitzgerald (1993). Full flow field computation of mixing in baffled stirred vessels. *Transactions of IChemE* 71A:342–344.

J.Y. Luo, R.I. Issa, and A.D. Gosman (1994). Prediction of impeller induced flows in mixing vessels using multiple frames of reference. *IChemE Symposium Series,* 136:549–556.

C.K.K. Lun, S. B. Savage, D. J. Jeffrey and N. Chepurniy (1984). Kinetic theories for granular flow: inelastic particles in couette flow and slightly inelastic particles in a general flow field. *Journal of Fluid Mechanics*, 140:223–256.

E. Manger (1996). *Modelling and simulation of gas/solids flow in curvilinear coordinates*, PhD thesis, Telemark Institute of Technology, 1996:64

M. Manninen, and V. Taivassalo (1996). *On the mixture model for multiphase flow*. VTT publications, 288.

V. Mathiesen, T. Solberg, E. Manger and B.H. Hjertager (1996). Modelling and predictions of multiphase flow in a pilot scale circulating fluidized bed, *5th International conference on circulating fluidized beds*, May 28-31, Beijing, China.

V. Mathiesen, T. Solberg and B. H. Hjertager (2000*a*). An experimental and computational study of multiphase flow behavior in a circulating fluidized bed. *International Journal of Multiphase flow,* 26:387–419.

V. Mathiesen, T. Solberg and B.H. Hjertager (2000b). Prediction of gas/particle flow with an Eulerian model including a realistic particle size distribution. *Powder Technology*, 112:34-45.

D. Migdal and V.D. Agosta (1967). A source flow model for continuum gas-particle flow, *Applied Mechanics*, Vol. 35, pp 860-865.

M. Milelli (2002). *A numerical analysis of confined turbulent bubble plumes*. PhD thesis, dissertation eth n. 14799, Swiss Federal Institute of Technology (ETH).

F.J. Moraga, A.E. Larreteguy, D.A. Drew and R.T. Lahey (2003). Assessment of turbulent dispersion models for bubbly flows in the low stokes number limit. *Int. Journal of Multiphase Flow*, 29:655– 673.

B.D. Nichols, C.W. Hirt and R.S. Hotshkiss (1980). *A solution algorithm for transient fluid flow with multiple free boundaries*, LASL Report LA-8355.

M.S. Nigam (2003). Numerical simulation of buoyant mixture flows. *International Journal. of Multiphase Flow*, 29:983–1015.

R.I. Nigmatulin (1979). Spatial averaging in the mechanics of heterogeneous and dispersed systems. *International Journal of Multiphase Flow*, 5:353–385.

S. Ouyang, J. Lin and O.E. Potter (1993). Ozone decomposition in a 0.254 m diameter circulating fluidized bed reactor. *Powder Technology,* 75:73-78.

S.V. Patankar (1980). *Numerical heat transfer and fluid flow*, McGraw-Hill.

S.V. Patankar and Spalding D.B. (1972). A calculation procedure for heat, mass and momentum transfer in three-dimensional parabolic flows. *International Journal of Heat and Mass Transfer*, 15:1787-1800.

V.V. Ranade (1997). An efficient computational model for simulating flow in stirred vessels: a case of Rushton turbine. *Chemical Engineering Science,* 52:4473–4484.

V.V. Ranade, Y. Tayalia and H. Krishnan (2002). CFD predictions of flow near impeller blades in baffled stirred vessels: assessment of computational snapshot approach. *Chemical Engineering Communications* 189(7):895-922.

V.V. Ranade and J.B. Joshi (1990*a*). Flow generated by a disc turbine. part I. - experimental. *Chemical Engineering Research Design,* 68A:19–33.

V.V. Ranade and J.B. Joshi (1990*b*). Flow generated by a disc turbine. part II. - mathematical modelling and comparison with experimental data. *Chemical Engineering Research Design,* 68A:34–50.

J. Revstedt, L. Fuchs and C. Trägårdh (1998). Large eddy simulations of the turbulent flow in a stirred reactor. *Chemical Engineering Science,* 53:4041–4053.

P.N. Rowe (1961). Drag forces in a hydraulic model of a fluidized bed, part ii. *Transaction of IChemE,* 39:175–180.

Y. Sato and K. Sekoguchi (1975). Liquid velocity distribution in two-phase bubble flow. *International Journal of Multiphase Flow.* 2:79–95.

D.B. Spalding (1977). The calculation of free-convection phenomena in gas-liquid mixtures, ICHMT seminar 1976, in: *Turbulent Bouyant Convection*, Hemisphere, 569-586

D.B. Spalding (1980). Numerical computation of multi-phase fluid flow and heat transfer, in *Recent Advances in Numerical Methods in Fluids*, Pineridge Press, 139-168.

D.B. Spalding (1985). Computer simulation of two-phase flows with special reference to nuclear reactor systems, in *Computational Techniques in Heat Transfer*, Ed.: R. W. Lewis, K. Morgan, J.A. Johnson,W. R. Smith, Pineridge Press, 1-44.

A. Samuelsberg and B.H. Hjertager (1995). Simulation of two-phase gas/particle flow and ozone decomposition in a 0.25 I.D. riser. In: *Advances in Multiphase Flow*, 679-688.

A. Samuelsberg and B.H. Hjertager (1996). Computational fluid dynamic simulation of an oxy-chlorination reaction in a full-scale fluidized bed reactor, *5th International conference on circulating fluidized beds*, May 28-31, Beijing, China.

J. Smagorinsky (1963). General circulation experiments with the primitive equations: I. the basic equations. *Monthly Weather Review,* 91:99–164.

M. Syamlal and T.J. O'Brien (1988). Simulation of granular layer inversion in liquid fluidized beds. *International Journal of Multiphase Flow,* 14 (4):473–481.

M. Syamlal and T. O'Brien (2003). Fluid dynamic simulation of O_3 decomposition in a bubbling fluidized bed. *AIChE Journal*, 49 (11):2793-2801.

A. Sokolichin and G. Eigenberger (1999). Applicability of the standard $k - \varepsilon$ turbulence model to the dynamic simulation of bubble columns: Part i. detailed numerical simulations. *Chemical Engineering Science,* 54:2273–2284.

A. Sokolichin, G. Eigenberger and A. Lapin (2004). Simulation of buoyancy driven bubbly flow: established simplification and open questions. *Fluid Mechanics and Transport Phenomena,* 50(1):24–45.

K.N. Theologos, I.D. Nikou, A.I. Lygeros and N.C. Markatos (1996). Simulation and design of fluid-catalytic cracking riser-type reactors, *Computers and Chemical Engineering,* 20(supplement):S757-S763.

A. Tomiyama (2004). Drag, lift and virtual mass forces acting on a single bubble. In: *3rd International Symposium on Two-Phase Flow Modelling and Experimentation*, 22-24 September 2004, Pisa, Italy.

A. Tomiyama, G. P. Celata, S. Hosokawa and S. Yoshida (2002). Terminal velocity of single bubbles in surface tension force dominant regime. *International Journal of Multiphase Flow*, 28:1497–1519.

C. Trägårdh (1988). A hydrodynamic model for the simulation of an aerated agitated fed-batch fermenter, in *Bioreactor Fluid Dynamics*, Elsevier Applied Science Publ., 117-134.

E.I. V. van den Hengel, N. G. Deen, and J. A. M. Kuipers (2005). Application of coalescence and breakup models in a discrete bubble model for bubble columns, *Industrial Engineering Chemistry Research*, 44(14):5233 – 5245.

M.A. van der Hoef, M. Van Sint Annaland and J. A.M. Kuipers (2004). Computational fluid dynamics for dense gas–solid fluidized beds: a multi-scale modeling strategy, *Chemical Engineering Science*, 59(22–23):5157–5165.

C.Y. Wen and Y.H. Yu (1966). Mechanics of fluidization. *Chemical Engineering Progress Symposium Series*, 62:100-111.

The Lattice-Boltzmann Method for Multiphase Fluid Flow Simulations and Euler-Lagrange Large-Eddy Simulations

J.J. Derksen

Multi-Scale Physics Department, Delft University of Technology, Netherlands

Abstract. Cellular automata for mimicking physical systems, the lattice gas and lattice Boltzmann automata, fluid dynamics with the lattice-Boltzmann method, practicalities of the lattice-Boltzmann method, DNS of solid-liquid suspensions, scaling in single-phase turbulence, direct numerical and large-eddy simulations, subgrid-scale modeling in LES, solid-liquid flow: point particles in LES, passive and reactive scalar transport in turbulent flow, filtered density function approach to turbulent reactive flow.

1 Cellular Automata

In computational science, the challenge is to describe and understand complex systems (that can be loosely defined as systems with many interacting components) by means of efficient numerical tools. Cellular automata turn out to be very fruitful in this respect. Von Neumann introduced the concept of cellular automata (CA) in the 1940's. The idea is to represent a physical system in terms of discrete space and time. The physical quantities (the state of the automaton) take only a finite set of values.

Von Neumann was intrigued by the mechanisms that lead to self-reproduction in biology. He wanted to devise a system that has the capability to reproduce another system of similar complexity. For this he devised a fully discrete "universe" made of cells. Each cell is characterized by an internal state, typically consisting of a finite number of information bits. The cells evolve in discrete time like simple automata that only know a simple rule to compute their new internal state. The rule determines the evolution of the system. It is the same for all cells and relates the state of a cell with that of its neighbors. Similarly to what happens in biological systems, the activity of cells takes place simultaneously and synchronously.

The *game of life* (proposed by John Conway in 1970) is an example of a simple rule leading to complex behavior. Conway imagined a two-dimensional, square lattice (like a checkerboard). Each cell can be alive (state 1) or dead (state 0). The updating rule is: a dead cell surrounded by exactly three living cells gets back to life; a living cell surrounded by less than two or more than three neighbors dies (of isolation or over-crowdedness). Here surrounding means the four nearest neighbors (north, south, east, west), and the four next nearest neighbors (along the diagonal).

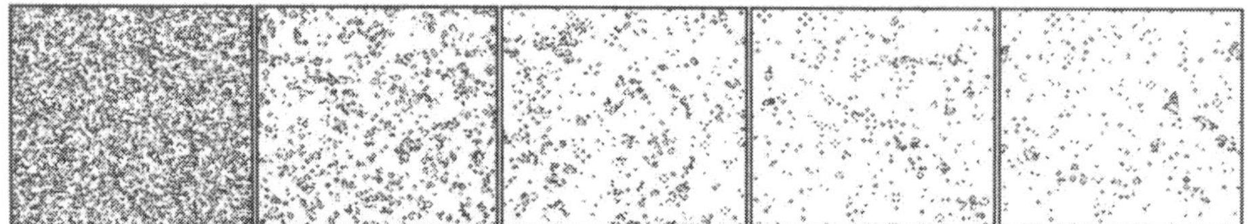

Figure 1. Snapshots of the game of life on a 200x200 grid. Time is running from left to right

This very simple automaton has rich behavior. Complex structures appear and show interesting dynamics. In Figure 1 some single realizations of the lattice are given. More interesting, however, are the dynamics of the system that can be viewed in animations.

Another example along the same lines: We define a two-dimensional square lattice with values 0 or 1 on each node. Starting from some initial condition, we evolve the system as follows: if the sum of the four nearest neighbor values is even, the new state is set to zero, if it is odd it is set to 1. When this rule is iterated, nice geometrical patterns are observed. Figure 2 shows results if we start from the situation given in the left panel. The complexity evolving in time results from spatial organization.

Above were artificial (and in a way esoteric) examples. The route towards describing physical systems in terms of a CA is not straightforward. A natural way is to propose a model of what we think is going on. The "art of modeling" is to retain only those ingredients that are essential; the degree of reality of the model depends on our level of description. If we are interested in global, macroscopic properties (and that is mostly the case in fluid dynamics with variables like fluid velocity and pressure), the microscopic details often are not relevant (as long as we obey symmetries and basic conservation laws).

In 1986, Frisch, Hasslacher and Pomeau (Frisch et al., 1986) announced a striking discovery. They showed that the molecular motion need not to be nearly as detailed as real molecular dynamics (with of the order of 10^{23} molecules, and even more degrees of freedom) to give rise to realistic fluid dynamics. Their fluid was constructed of fictitious particles, each with the same mass and moving with the same speed, and differing only in their velocity directions. Moreover, these directions were constrained to a finite set (in two dimensions only six). This was the so-called lattice-gas automaton, from which later the lattice-Boltzmann method evolved.

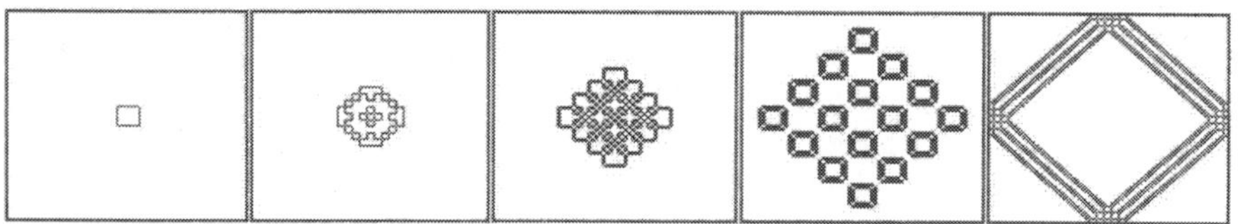

Figure 2. Evolution of a 256x256 cellular automaton.

2 Lattice Gas Automaton

The lattice gas is constructed as a simplified, fictitious molecular dynamic in which space, time, and the particle velocities are all discrete (Chen & Doolen, 1998). From this perspective, the lattice gas method is often referred to as a lattice gas cellular automaton. In general, a lattice gas cellular automaton consists of a regular lattice with particles residing on the nodes. A set of Boolean variables $n_i(\mathbf{x},t)\,(i=1,\cdots,M)$ describing the particle occupation is defined, where M is the number of directions of the particle velocity at each node. The evolution of the LG is as follows

$$n_i(\mathbf{x}+\mathbf{e}_i,t+1)=n_i(\mathbf{x},t)+\Omega_i(n(\mathbf{x},t)) \tag{2.1}$$

where $\mathbf{e}_i$ are the local particle velocities (again $i=1,\cdots,M$), and Ω_i is the collision operator that is a function of all particles $n_i\,(i=1,\cdots,M)$ involved in the collision. Starting from an initial state, the configuration of particles at each time step evolves in two sequential sub-steps: (a) *streaming* in which each particle moves to the nearest node in the direction of its velocity, and (b) *collision*, which occurs when particles arriving at a node interact and change their velocity directions according to scattering rules. For simplicity, the exclusion principle (not more than one particle being allowed at a given time and node with given velocity) is imposed for memory efficiency. This leads to a Fermi-Dirac local equilibrium distribution (Frisch et al., 1987).

3 Lattice-Boltzmann Method

The main feature of the lattice-Boltzmann method (LBM) is to replace the particle occupation variable n_i (Booleans) in Eq. (2.1) by single particle distribution functions (real variables) $f_i=\langle n_i\rangle$ and neglect individual particle motion and particle-particle correlations in the kinetic equations. The brackets $\langle\ \rangle$ denote an ensemble averaging. This procedure largely eliminates noise (which is present (and a problem) in lattice gases). In the LBM, the primitive variables are the averaged particle distributions, which are mesoscopic variables. Because the kinetic form is still the same as the lattice gas automaton, the advantage of locality in the kinetic approach is retained. The locality is essential to parallelism.

4 From Lattice-Boltzmann to Navier-Stokes

Since the lattice-Boltzmann method is a derivative of the lattice gas approach, the LBM will be introduced starting with a discrete kinetic equation for the particle distribution function, which is similar to the kinetic equation in lattice gas automata, see Eq. (2.1):

$$f_i(\mathbf{x}+\mathbf{e}_i\Delta x,t+\Delta t)=f_i(\mathbf{x},t)+\Omega_i(f(\mathbf{x},t)) \quad (i=0,1\cdots,M) \tag{4.1}$$

where f_i is the particle velocity distribution function along the i-th direction (please note that i now runs from 0 to M, in the LBM often a rest particle having zero-velocity and index 0 is introduced); $\Omega_i=\Omega_i(f(\mathbf{x},t))$ is the collision operator that represents the rate of change of f_i as a result of the collision. The collision operator Ω_i depends on all M+1 particles (distribution functions) involved in the collision. The space and time increments are Δx and Δt respectively. If $\Delta x/\Delta t=|\mathbf{e}_i|$, Eqs. (2.1) and (4.1) have the same discretization. In the LBM, space is discretized in a way that is consistent with the kinetic equation, i.e. the coordinates of the neighbors of $\mathbf{x}$ are $\mathbf{x}+\mathbf{e}_i$.

The density ρ and momentum density $\rho\mathbf{u}$ are defined as moments of the distribution function f_i:

$$\rho = \sum_i f_i \quad \rho\mathbf{u} = \sum_i f_i \mathbf{e}_i \tag{4.2}$$

with the sum over i=0...M. The collision operator Ω_i must satisfy mass and momentum conservation at each lattice:

$$\sum_i \Omega_i = 0 \quad \sum_i \Omega_i \mathbf{e}_i = \mathbf{0} \tag{4.3}$$

If only the physics in the long-wavelength and low-frequency limit are of interest, the lattice spacing Δx and time step Δt in Eq. (4.1) can be regarded as small parameters of the same order ε. Performing a Taylor expansion in time and space, we obtain the following continuum form of the kinetic equation, accurate to second order in ε:

$$\frac{\partial f_i}{\partial t} + \mathbf{e}_i \cdot \nabla f_i + \varepsilon\left(\frac{1}{2}\mathbf{e}_i\mathbf{e}_i : \nabla\nabla f_i + \mathbf{e}_i \cdot \nabla \frac{\partial f_i}{\partial t} + \frac{1}{2}\frac{\partial^2 f_i}{\partial t^2}\right) = \frac{\Omega_i}{\varepsilon} \tag{4.4}$$

To derive the macroscopic hydrodynamic equations, we apply the Chapman-Enskog expansion, which essentially is a multi-scale expansion

$$\frac{\partial}{\partial t} = \varepsilon\frac{\partial}{\partial t_1} + \varepsilon^2\frac{\partial}{\partial t_2} \qquad \frac{\partial}{\partial x} = \varepsilon\frac{\partial}{\partial x_1} \tag{4.5}$$

The above assumes the diffusion time scale t_2 to be much slower than the convection time scale t_1. Likewise, the distribution function f_i can be expanded formally about the local equilibrium distribution function f_i^{eq}

$$f_i = f_i^{\text{eq}} + \varepsilon f_i^{\text{neq}} \tag{4.6}$$

Here f_i^{eq} depends on the local macroscopic variables (ρ and $\rho\mathbf{u}$) and should satisfy

$$\sum_i f_i^{\text{eq}} = \rho \quad \sum_i f_i^{\text{eq}}\mathbf{e}_i = \rho\mathbf{u} \tag{4.7}$$

$f_i^{\text{neq}} = f_i^{(1)} + \varepsilon f_i^{(2)} + O(\varepsilon^2)$ is the non-equilibrium distribution function which should satisfy the following constraints

$$\sum_i f_i^{(k)} = 0 \quad \sum_i f_i^{(k)}\mathbf{e}_i = \mathbf{0} \quad k = 1,2 \tag{4.8}$$

Inserting f_i into the collision operator along with a Taylor expansion gives

$$\Omega_i(f) = \Omega_i\left(f^{\text{eq}}\right) + \varepsilon \frac{\partial \Omega_i\left(f^{\text{eq}}\right)}{\partial f_j} f_j^{(1)} + \varepsilon^2 \left(\frac{\partial \Omega_i\left(f^{\text{eq}}\right)}{\partial f_j} f_j^{(2)} + \frac{\partial^2 \Omega_i\left(f^{\text{eq}}\right)}{\partial f_j \partial f_k} f_j^{(1)} f_k^{(1)} \right) + O\left(\varepsilon^3\right) \quad (4.9)$$

(summing of repeated indices is implied in this equation, and in the rest of the text). From Eq. (4.4) in the limit $\varepsilon \to 0$ (which implies $f_i = f_i^{\text{eq}}$) it follows $\Omega_i\left(f^{\text{eq}}\right) = 0$. This teaches us that we can linearize the collision operator:

$$\frac{\Omega_i(f)}{\varepsilon} = \frac{M_{ij}}{\varepsilon}\left(f_j - f_j^{\text{eq}}\right) \quad \text{with} \quad M_{ij} \equiv \frac{\partial \Omega_i\left(f^{\text{eq}}\right)}{\partial f_j} \quad (4.10)$$

The matrix M_{ij} is the collision matrix, which determines the scattering rate between direction i and j. M_{ij} only depends on the angle between the directions i and j. Mass and momentum conservation imply:

$$\sum_{i=1}^{M} M_{ij} = 0 \quad \sum_{i=1}^{M} \mathbf{e}_i M_{ij} = \mathbf{0} \quad (4.11)$$

In the widely used lattice BGK (Bhatnagar-Gross-Krook, see Bhatnagar et al. 1954) collision operator, the distribution function relaxes to an equilibrium state at a single rate with time constant τ:

$$M_{ij} = -\frac{1}{\tau}\delta_{ij} \quad (4.12)$$

Then the collision term reads

$$\frac{\Omega_i}{\varepsilon} = -\frac{1}{\tau} f_i^{\text{neq}} = -\frac{1}{\tau}\left(f_i^{(1)} + \varepsilon f_i^{(2)}\right) \quad (4.13)$$

The lattice-BGK equation

$$f_i\left(\mathbf{x} + \mathbf{e}_i, t+1\right) = f_i\left(\mathbf{x}, t\right) - \frac{f_i - f_i^{\text{eq}}}{\tau} \quad (4.14)$$

forms the heart of many lattice-Boltzmann computer codes.

If we now substitute the LBGK collision operator, and the expansions of the distribution functions as given above in Eq. (4.4), equations to various orders of ε appear. For ε^0:

$$\frac{\partial f_i^{\mathrm{eq}}}{\partial t_1} + \mathbf{e}_i \cdot \nabla_1 f_i^{\mathrm{eq}} = -\frac{f_i^{(1)}}{\tau} \tag{4.15}$$

For order ε^1:

$$\frac{\partial f_i^{(1)}}{\partial t_1} + \frac{\partial f_i^{\mathrm{eq}}}{\partial t_2} + \mathbf{e}_i \cdot \nabla f_i^{(1)} + \frac{1}{2}\mathbf{e}_i\mathbf{e}_i : \nabla\nabla f_i^{\mathrm{eq}} + \mathbf{e}_i \cdot \nabla \frac{\partial f_i^{\mathrm{eq}}}{\partial t_1} + \frac{1}{2}\frac{\partial^2}{\partial t_1^2} f_i^{\mathrm{eq}} = \frac{f_i^{(2)}}{\tau} \tag{4.16}$$

Combining Eqs. (4.15) and (4.16) gives

$$\frac{\partial f_i^{(1)}}{\partial t_2} + \left(1 - \frac{2}{\tau}\right)\left(\frac{\partial f_i^{(1)}}{\partial t_1} + \mathbf{e}_i \cdot \nabla f_i^{(1)}\right) = -\frac{f_i^{(2)}}{\tau} \tag{4.17}$$

From Eqs. (4.15) and (4.17) the continuity equation and momentum balance can be derived:

$$\frac{\partial \rho}{\partial t} + \nabla \cdot \rho\mathbf{u} = 0 \tag{4.18}$$

$$\frac{\partial \rho\mathbf{u}}{\partial t} + \nabla \cdot \mathbf{\Pi} = \mathbf{0} \tag{4.19}$$

These equations are accurate to second order in ε. The momentum flux tensor Π has the form

$$\Pi_{\alpha\beta} = \sum_i e_{i\alpha} e_{i\beta} \left[f_i^{\mathrm{eq}} + \left(1 - \frac{1}{2\tau}\right) f_i^{(1)} \right] \tag{4.20}$$

with $e_{i\alpha}$ the component of the velocity vector $\mathbf{e}_i$ in the α-th coordinate direction. Note that the momentum flux has an equilibrium part and a non-equilibrium part.

To specify the flux tensor, we need to specify the lattice structure and the equilibrium distribution. We consider a two-dimensional, square lattice. This relatively simple case has all the features that also apply to different lattices and number of dimensions. The set of velocity vectors can be written as

$$\mathbf{e}_0 = \begin{bmatrix} 0 \\ 0 \end{bmatrix} \; \mathbf{e}_1 = \begin{bmatrix} 1 \\ 0 \end{bmatrix} \; \mathbf{e}_2 = \begin{bmatrix} 1 \\ 1 \end{bmatrix} \; \mathbf{e}_3 = \begin{bmatrix} 0 \\ 1 \end{bmatrix} \; \mathbf{e}_4 = \begin{bmatrix} -1 \\ 1 \end{bmatrix} \; \mathbf{e}_5 = \begin{bmatrix} -1 \\ 0 \end{bmatrix} \; \mathbf{e}_6 = \begin{bmatrix} -1 \\ -1 \end{bmatrix} \; \mathbf{e}_7 = \begin{bmatrix} 0 \\ -1 \end{bmatrix} \; \mathbf{e}_8 = \begin{bmatrix} 1 \\ -1 \end{bmatrix} \tag{4.21}$$

The requirement for using the nine-velocity model (D2Q9 in the LB jargon: two-dimensions, nine speeds) instead of the simpler five-velocity model comes from considerations of lattice-symmetry. The lattice-Boltzmann equation cannot recover the correct Navier-Stokes equation unless sufficient lattice symmetry is present (Frisch et al 1986).

The Navier-Stokes equation has a second-order non-linearity. According to Chen et al (1992), the general form of the equilibrium distribution can be written up to $O(u^2)$:

$$f_i^{eq} = \rho\left[a + b\mathbf{e}_i \cdot \mathbf{u} + c(\mathbf{e}_i \cdot \mathbf{u})^2 + du^2\right] \tag{4.22}$$

where a, b, c, and d are so-called lattice constants. This expansion of the distribution function only makes physical sense if the velocities are small compared to the (obviously finite) speed of sound of the lattice-Boltzmann system. Using the constraints as given in Eq. (4.7), the lattice constants can be obtained analytically and Eq. (4.22) can be written as

$$f_i^{eq} = \rho w_i\left[1 + 3\mathbf{e}_i \cdot \mathbf{u} + \frac{9}{2}(\mathbf{e}_i \cdot \mathbf{u})^2 - \frac{3}{2}u^2\right] \tag{4.23}$$

The weight factors w_i are

$$w_0 = 4/9 \quad w_1 = w_3 = w_5 = w_7 = 1/9 \quad w_2 = w_4 = w_6 = w_8 = 1/36 \tag{4.24}$$

If we now get back to Eq. (4.20), the equilibrium part can be written as

$$\Pi_{\alpha\beta}^{eq} = \sum_i e_{i\alpha} e_{i\beta} f_i^{eq} = p\delta_{\alpha\beta} + \rho u_\alpha u_\beta \tag{4.25}$$

$$\Pi_{\alpha\beta}^{(1)} = \left(1 - \frac{1}{2\tau}\right)\sum_i e_{i\alpha} e_{i\beta} f_i^{(1)} = \nu\left(\frac{\partial(\rho u_\beta)}{\partial x_\alpha} + \frac{\partial(\rho u_\alpha)}{\partial x_\beta}\right) \tag{4.26}$$

where $p=\rho/3$ is the pressure, and $\nu=(2\tau-1)/6$ is the kinematic viscosity. From the pressure relation the speed of sound c_s can be derived:

$$c_s^2 = \frac{\partial p}{\partial \rho} = \frac{1}{3} \tag{4.27}$$

The momentum balance then reads

$$\rho\left(\frac{\partial u_\alpha}{\partial t} + \frac{\partial u_\alpha u_\beta}{\partial x_\beta}\right) = -\frac{\partial p}{\partial x_\alpha} + \nu\frac{\partial}{\partial x_\beta}\left(\frac{\partial \rho u_\beta}{\partial x_\alpha} + \frac{\partial \rho u_\alpha}{\partial x_\beta}\right) \tag{4.28}$$

The momentum balance for a Newtonian fluid (the Navier-Stokes equation) for a compressible fluid reads

$$\rho\left(\frac{\partial u_\alpha}{\partial t} + \frac{\partial u_\alpha u_\beta}{\partial x_\beta}\right) = -\frac{\partial p}{\partial x_\alpha} + \frac{\partial}{\partial x_\beta}\left(\rho\nu\left(\frac{\partial u_\beta}{\partial x_\alpha} + \frac{\partial u_\alpha}{\partial x_\beta} - \frac{1}{3}\frac{\partial u_\gamma}{\partial x_\gamma}\delta_{\alpha\beta}\right)\right) \tag{4.29}$$

In the limit of constant density Eqs. (4.28) and (4.29) are the same. The limit of constant density can be effectuated in the LB scheme by keeping the fluid velocities well below the speed of sound (low Mach numbers), since $\delta\rho \propto \delta p / c_s^2 \propto \rho|\mathbf{u}|^2 / c_s^2$. That is, one should satisfy

$$\frac{|\mathbf{u}|^2}{c_s^2} \equiv \mathrm{Ma}^2 << 1 \tag{4.30}$$

5 Some Practical Aspects of the Lattice-Boltzmann Method for Single-Phase Flows

5.1 Implementation of the lattice-Boltzmann method in computer code

Implementing the above rules in computer code is fairly straightforward. This will be outlined here briefly in terms of Fortran-like pseudo code for a D2Q9 lattice-Boltzmann scheme (that scheme also was the subject of Section 4).

Define a main real array `f(n,i,j)` containing the distribution functions (or the LB fluid particles) at some specific moment in time. Experience learns that in most cases this can be a single-precision (`real*4`) array. The index `n` relates to the 19 velocity vectors (the numbering being the same as in Eq. (4.21)), and the indices `i,j` to the two coordinate directions that run from `1...nx`, and `1...ny` respectively (flow the domain is a rectangle with size `nx·ny`). The evolution of this system has two major steps: streaming and collision. In the streaming step, particles move to neighboring lattice sites. Computer code could look like this:

Code fragment
```
do j=1,ny
  do i=1,nx
    f(5,i,j)=f(5,i+1,j)
    f(7,i,j)=f(7,i,j+1)
    f(6,i,j)=f(6,i+1,j+1)
    f(8,i,j)=f(8,i-1,j+1)
  enddo
enddo
do j=ny,1,-1
  do i=nx,1,-1
    f(1,i,j)=f(1,i-1,j)
    f(3,i,j)=f(3,i,j-1)
    f(2,i,j)=f(2,i-1,j-1)
    f(4,i,j)=f(4,i+1,j-1)
  enddo
enddo
```
end of code fragment

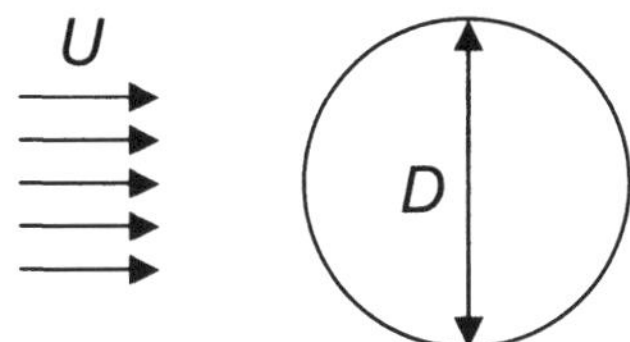

Figure 3. Schematic of the flow past a circular cylinder.

The above structure has been chosen such that we only need one field array `f`, i.e. we overwrite the array before the streaming step with the array after the streaming step.

The LBGK collision step, see Eq. (4.14) requires that we determine the equilibrium distribution According to Eq. (4.28) this needs computing the density and velocity per lattice node. For the latter we use Eq. (4.2).

It is (in my view) convenient to work in lattice-units while setting up an LB simulation. The unit of time is then set to 1, and is the time for an LB fluid particle (or distribution function) to travel to the neighboring lattice site. The unit of length is set to 1 as well and equals the spacing between two nearest-neighbor lattice nodes. The lattice spacing is uniform over the entire lattice and in the coordinate directions (square lattice in two dimensions, cubic in three directions) due to the coupling between velocity-space and physical space: particles have velocities such that they travel to their neighboring sites in exactly one time step.

Translating a flow case defined in physical units to LB units goes via dimensionless numbers. Suppose we would like to determine the flow past a square cylinder at flow and geometrical conditions as given in Figure 3. From Figure 3 a Reynolds number can be deduced: $\mathrm{Re} = UD/\nu$. Suppose we would like to simulate the case Re=100. In the LB simulation, we first choose a reasonable spatial resolution. Say we would like to resolve the flow such that the

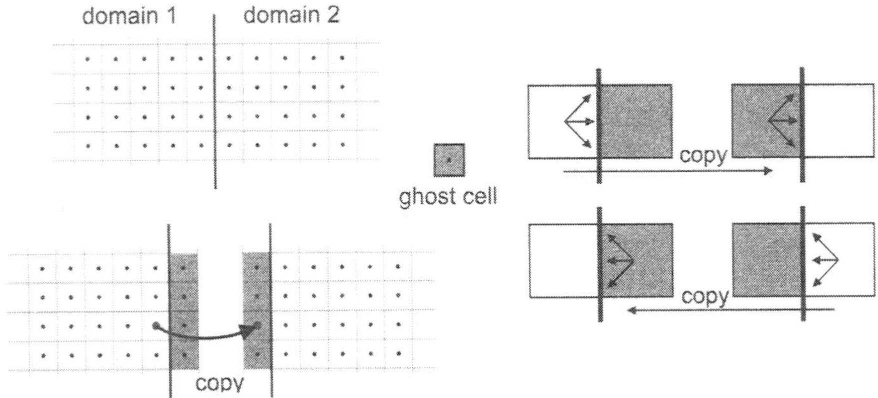

Figure 4. Parallellization through domain decomposition. Left: global view; right: at node level.

cylinder spans 10 lattice spacings. Then D=10 (in LB units). Flow velocities need to be such that $Ma^2 << 1$ (Eq. (4.30)). This can be achieved by setting the free stream velocity U to 0.1 (LB units). For mimicking the physical case that has Re=100, the viscosity then needs to be set to ν=0.01 (again in LB units).

Parallellization of LB computer code is straightforward. The LB nodes only communicate with each other in the streaming step. For parallellization we decompose the flow domain in sub-domains. At the borders of the sub-domains we introduce ghost-cells. Before each streaming step the contents at the edges of each sub-domain is copied in the ghost cells of the neigboring domain (see Figure 4). Once this is done, the streaming step can be carried out in each subdomain according to the procedure presented earlier in this section.

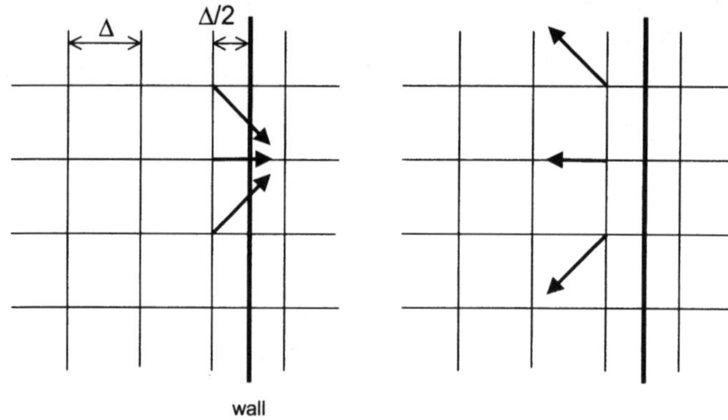

Figure 5. Streaming step for simulating a no-slip wall placed halfway lattice nodes.

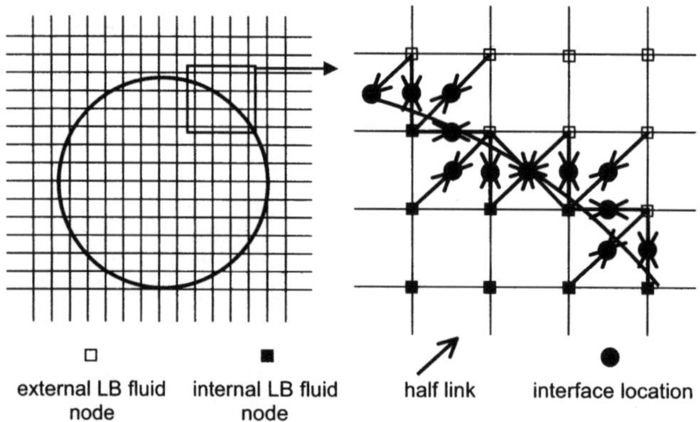

Figure 6. Definition of the halfway bounce-back rule for a circular particle. Right: zoomed-in view.

5.2 Setting boundary conditions

No-slip boundary conditions at solid walls can be set according the so-called bounce-back rule. In the half-way bounce back rule, the wall is located midway half a lattice-spacing away from a lattice node (see Figure 5). The particles propagating in the direction of the wall, bounce back at the wall and after exactly one time step arrive at the lattice node they left at the beginning of the time step. The half-way bounce-back method for no-slip walls is second order accurate (Rohde et al 2002), just as the discretization of the Navier-Stokes equation as given in Section 4. Placing the wall at a different location with respect to the grid reduces the accuracy at the boundary to first order.

If the no-slip wall has a non-zero velocity, the bounce-back method as described above can be generalized such that LB particles that bounce at the wall receive additional momentum due to the wall's motion. The half-way bounce-back rule including additional momentum from moving walls is the basis of Ladd's procedure (Ladd, 1994) of simulating solid-liquid suspensions with spherical particles. Figure 6 shows a representation of a circular object in Ladd's approach. The lattice-Boltzmann fluid particles bounce back halfway the link between two LB nodes, with the bounce back location close to the actual surface of the object. The circle (and sphere in 3D) are represented by a stair-step object. Such a schematic (rough) approach requires calibration. Usually a distinction is made between the input radius of a spherical particle, and the so-called hydrodynamic radius. The input radius relates to the particle as discretized in Figure 6. Of this particle the Stokes drag force is determined in a simulation of the sphere in a periodic domain. The result is compared to the analytical result (Hasimoto, 1959). Then the hydrodynamic radius is the radius of a sphere having a drag force equal to the one of the simulation, but now according to the analytical expression. Usually the hydrodynamic radius is slightly higher than the input radius (some 0.5 lattice spacing), it furthermore depends on the viscosity of the fluid.

Note that in this approach, there also is fluid inside the particle. This internal fluid has no physical meaning. It is there for computational convenience. The particles move relative to the grid and therefore cover and uncover LB nodes on a regular basis. If the particles have internal fluid, in the cover and uncover process fluid simply turns into internal and external fluid respectively and fluid mass is conserved easily. The consequence of having internal fluid is that the density of the particle always has to be at least the density of the fluid; the dynamics of lighter particles cannot be simulated. Also in setting up and solving the equations of motion of the particles, the internal fluid needs to be taken into account. This will be discussed in Section 6 of this chapter.

Aidun and co-workers (Aidun et al 1998, Ding and Aidun 2003) have adapted Ladd's method such that internal fluid is largely avoided. The price is the execution of a dedicated procedure to conserve fluid mass as good as possible.

In recent years, immersed boundary methods are gaining popularity in mainstream CFD based on e.g. the finite volume method, or spectral methods (Goldstein et al 1993). Similar techniques can be employed in an LB context. The idea is to apply body forces on the fluid such that at prescribed locations the fluid has a prescribed velocity (equal to the velocity of a solid wall). An advantage of the method is that the locations can (in principle) be chosen independent of the grid. In the case of defining a moving sphere, the forcing method works as follows: The sphere's surface is defined as a set of M control points $\mathbf{r}_j^{(n)}$ (j=1...M) on their surface, where the superscript (n) now indicates the moment in time. There is no restriction on the position of these points in the flow domain; they do not need to coincide with lattice sites. At these points we require a velocity equal to $\mathbf{w}_j^{(n)} = \mathbf{v}_p + \mathbf{\Omega}_p \times (\mathbf{r}_j^{(n)} - \mathbf{r}_p^{(n)})$, with $\mathbf{v}_p$ the linear velocity and $\mathbf{\Omega}_p$ the angular velocity of the particle.

The above demand can be achieved effectively through a control algorithm (Goldstein et al 1993) that at each time step determines the (interpolated) mismatch between the actual flow velocity and the prescribed flow velocity at the control points, and then adapts the force field in such a way that it suppresses the mismatch. The deviation between the actual and the prescribed velocity ($\mathbf{d}_j^{(n)}$) is determined by a second-order interpolation of the flow velocities at the lattice sites:

$$\mathbf{d}_j^{(n)} = \mathbf{w}_j^{(n)} - \Sigma_k G_k\left(\mathbf{r}_j^{(n)}\right)\mathbf{u}_k^{(n)} \tag{5.1}$$

where the sum is over the lattice sites in the vicinity of $\mathbf{r}_j^{(n)}$, $\mathbf{u}_k$ is the fluid velocity at lattice site k, and G_k are the interpolation coefficients. These coefficients also serve to distribute the forces that reduce the deviation $\mathbf{d}_j^{(n)}$, over the lattice sites:

$$\mathbf{f}_k^{(n)} = \mathbf{f}_k^{(n-1)} + q\Sigma_j G_k\left(\mathbf{r}_j^{(n)}\right)\mathbf{d}_j^{(n)} \tag{5.2}$$

with $\mathbf{f}_k^{(n)}$ the force acting on the fluid at lattice node k and moment n, and q a relaxation factor. Also the forcing method needs calibration of the hydrodynamic diameter. For this we use the same procedure as proposed by Ladd.

Other sorts of boundary conditions can be achieved in the LBM relatively easily and along intuitive lines:

- Zero-shear at flat walls is achieved through specular reflection of the LB particles at the wall.
- Zero gradient conditions (at inflows and outflows) are achieved by copying distribution functions in the direction normal to the boundary; at inflows the forcing scheme (see above) can then be used to impose velocity profiles.
- Periodic boundaries imply copying outgoing distribution functions to the other side of the domain where they enter again.

5.3 Alternative collision operators

The LBGK collision operator, based on a single time constant with which distribution functions relax towards equilibrium gets unstable for low viscosity values. This has led to research towards more stable schemes. Two more stable schemes are mentioned here: the multiple relaxation time (MRT) scheme (Lallemand and Luo, 2000), and a scheme due to Somers (1993).

For a MRT-LB model with M velocities, a set of velocity distribution functions f_i $(i = 0,1\cdots,M)$ is defined. The collision (and this is different from LBGK) is executed in moment space (not in velocity space). Moment space and velocity space are connected via a linear transformation

$$m_i = T_{ij} f_j \quad f_i = \left(T^{-1}\right)_{ij} m_j \tag{5.3}$$

The evolution equation then reads

$$f_i(\mathbf{x} + \mathbf{e}_i \Delta x, t + \Delta t) = f_i(\mathbf{x}, t) + \Omega_i(f(\mathbf{x}, t)) = -\left(T^{-1}\right)_{ij} \Lambda_j \left(m_j - m_j^{eq}\right) \tag{5.4}$$

The moments are related to density, momentum, strain, and energy. Their equilibriums are functions of the conserved quantities which are mass and momentum. The M+1 coefficients Λ_j determine the viscosity (as did the relaxation time) and are used to enhance the stability of the scheme.

The scheme due to Somers has been described in detail by Eggels and Somers (1995). It goes along similar lines as the MRT approach, i.e. its collision operator acts on moments of the velocity distribution function. Furthermore, it uses a staggered discretization in space and time.

6 Direct Numerical Simulations of Solid-Liquid Suspensions

In this section we will discuss a methodology for directly simulating solid-liquid suspensions. In these simulations we resolve the solid-liquid interface and the flow of the interstitial fluid (the latter we do with the LBM), i.e. the spherical solid particles have finite size and the flow around the particles is directly simulated. The forcing method described in Section 5.2 is used for setting the no-slip boundary condition at the sphere's surfaces. At the end of this section we will briefly discuss two examples of such simulations: a turbulent suspension, and solid-liquid fluidization.

Our starting point is a fully periodic, three-dimensional domain containing fluid and solid particles with a spherical shape. In the fluidization cases, the flow is driven by a net gravity force acting on the particles, and a pressure gradient acting on the fluid to balance gravity. In the case of the turbulent suspension, a random body force that can generate turbulence with prescribed properties agitates the suspension.

The fluid flow and the particle motion are coupled by demanding that at the surface of the sphere the fluid velocity matches the local velocity of the solid surface (that is the sum of the linear velocity $\mathbf{v}_p$ and $\mathbf{\Omega}_p \times (\mathbf{r} - \mathbf{r}_p)$ with $\mathbf{\Omega}_p$ being the angular velocity of the particle); in the forcing scheme this is accomplished by imposing additional forces on the fluid at the surface of the solid sphere (which is then distributed to the lattice nodes in the vicinity of the particle surface). The collection of forces acting on the fluid at the sphere's surface and its interior is subsequently used to determine the hydrodynamic force and torque acting on the sphere (action = –reaction).

The effective body force on the fluid (in the fluidization case) mentioned above can be related to the gravitational acceleration, $\mathbf{g} = -g\mathbf{e}_z$, as follows. The net gravity force acting on each spherical particle is $\mathbf{F}_G = -(\rho_s - \bar{\rho})\frac{\pi}{6} d_p^3 g\mathbf{e}_z$, and the force per unit volume acting on the (internal and external) fluid is

$$\mathbf{f}_B = (\bar{\rho} - \rho_f) g\mathbf{e}_z \tag{6.1}$$

with $\bar{\rho} = \bar{\phi}\rho_s + (1-\bar{\phi})\rho_f$ the density of the fluid-solid mixture, and $\bar{\phi}$ the overall (spatially averaged) solids volume fraction in the periodic domain.

The fluid inside the spherical particles is an artefact of the forcing scheme. As long as the density of the solid is higher than the density of the fluid, the effects of the internal fluid can be effectively corrected for: The force $\mathbf{F}_{LB}$ acting on the fluid determined by the forcing method is the sum of the force needed to accelerate the internal fluid and the force of the particle acting on the external fluid. Since the internal fluid largely behaves as a solid body, one can partition $\mathbf{F}_{LB}$ as follows: $\mathbf{F}_{LB} = \mathbf{F}_{LB,1} + \mathbf{F}_{ext}$ where the force $\mathbf{F}_{LB,1}$ is the component that ensures that the internal fluid translates with the particle; $\mathbf{F}_{ext}$ is the force on the external fluid due to the particle. The overall linear momentum balance for the internal fluid can be written as

$$\rho_f \frac{\pi}{6} d_p^3 \frac{d\mathbf{v}_p}{dt} = +\mathbf{F}_{LB,1} + (\bar{\rho} - \rho_f)\frac{\pi}{6} d_p^3 g\mathbf{e}_z = \mathbf{F}_{LB} - \mathbf{F}_{ext} + (\bar{\rho} - \rho_f)\frac{\pi}{6} d_p^3 g\mathbf{e}_z \tag{6.2}$$

where it has been recognized that the internal fluid translates with the particle. The corresponding equation for the particle is then

$$\rho_s \frac{\pi}{6} d_p^3 \frac{d\mathbf{v}_p}{dt} = -\mathbf{F}_{ext} + (\rho_s - \bar{\rho})\frac{\pi}{6} d_p^3 g\mathbf{e}_z \tag{6.3}$$

Lubrication forces which arise because of inadequate resolution of the flow in between neighbouring particles and those arising from direct particle-particle interactions (e.g., collision) will be added to the right hand side later. Combining Eqs. (6.2) and (6.3), we get

$$(\rho_s - \rho_f)\frac{\pi}{6} d_p^3 \frac{d\mathbf{v}_p}{dt} = -\mathbf{F}_{LB} - (\rho_s - \rho_f)\frac{\pi}{6} d_p^3 g\mathbf{e}_z \tag{6.4}$$

Following the same reasoning, we obtain the following angular momentum balance:

$$(\rho_s - \rho_f)\frac{\pi}{60} d_p^5 \frac{d\mathbf{\Omega}_p}{dt} = -\mathbf{T}_{LB} \tag{6.5}$$

with $\mathbf{T}_{LB}$ the torque as determined by the forcing method to impose the no-slip conditions at the sphere's surface.

In order to test if the above procedure represents the dynamics of spheres immersed in liquid properly, we considered the transient motion of a single sphere that is accelerated from rest under the influence of gravity. In the limit of zero Reynolds number in an unbounded fluid the equation of motion of the sphere has been derived by Maxey and Riley (1983). Results obtained by integrating the Maxey and Riley equation showed excellent agreement with the LB simulations combined with the above equations of motion. Ten Cate et al. (2002) compared the results on sedimentation of a single sphere in a limited-size container at higher Reynolds numbers (up to Re = 30) obtained through lattice-Boltzmann simulations and forcing boundary conditions with particle image velocimetry (PIV) data and found good agreement in terms of the sphere's trajectory and the fluid flow field around the moving sphere.

We also take into consideration the interaction between particles through binary, hard-sphere collisions and lubrication forces. For the former, we apply an event-driven collision algorithm: we move the collection of particles until two particles get into contact. At that moment we carry out the collision (i.e. update the velocities of the two particles taking part in the collision). Subsequently, the movements of all particles are continued until the next collision or until the end of a LB time step. The collision model that we apply (described in detail in Yamamoto et al., 2001) has two parameters: a restitution coefficient e and a friction coefficient μ. As the default situation we consider fully elastic, frictionless collisions ($e = 1$, $\mu = 0$).

When two particles are at close proximity, with their separation being of the order of or less than the lattice spacing, the hydrodynamic interaction between them will not be properly resolved in the LB simulations. Therefore, we explicitly impose lubrication forces on the particles, in addition to the hydrodynamic forces stemming from the LBM. The general framework for lubrication forces and torques acting on two particles (1 and 2) as a result of the relative motion of their surfaces can be written in the form of the following vector equation:

$$\begin{bmatrix} \mathbf{F}_{\text{lub},1} \\ \mathbf{T}_{\text{lub},1} \\ \mathbf{T}_{\text{lub},2} \end{bmatrix} = \begin{bmatrix} \mathbf{A}_{11} & -\mathbf{B}_{11} & \mathbf{B}_{22} \\ \mathbf{B}_{11} & \mathbf{C}_{11} & \mathbf{C}_{12} \\ -\mathbf{B}_{22} & \mathbf{C}_{12} & \mathbf{C}_{22} \end{bmatrix} \begin{bmatrix} \mathbf{v}_{\text{p},12} \\ \mathbf{\Omega}_1 \\ \mathbf{\Omega}_2 \end{bmatrix} \tag{6.6}$$

with $\mathbf{F}_{\text{lub},2} = -\mathbf{F}_{\text{lub},1}$ and $\mathbf{v}_{\text{p},12} = \mathbf{v}_{\text{p},1} - \mathbf{v}_{\text{p},2}$ (Kim and Karilla, 1991; Nguyen and Ladd, 2002). In the tensors $\mathbf{A}_{11}$, $\mathbf{B}_{11}$, $\mathbf{B}_{22}$, $\mathbf{C}_{11}$, $\mathbf{C}_{22}$, and $\mathbf{C}_{12}$, we only use the leading order terms in the parameter $\frac{d_\text{p}}{h}$, with h the minimum spacing of the particle surfaces. For the radial lubrication force (contained in the diagonal elements of the $\mathbf{A}_{11}$ matrix in the equation $\mathbf{F}_{\text{lub},1} = \mathbf{A}_{11}\mathbf{v}_{\text{p},12}$,) the leading order is $\frac{d_\text{p}}{h}$, while for the tangential lubrication forces and torques it is $\ln\left(\frac{d_\text{p}}{h}\right)$ (Kim and Karilla, 1991). Two modifications to the above expressions were implemented to tailor them to our numerical needs:

(a) Lubrication only acts if particle separation is less than $\delta = 0.1 d_\text{p}$ (which is equivalent to roughly one lattice spacing in the default resolutions we use). To smoothly switch on/off the lubrication force at $h = \delta$, in the lubrication expressions $\frac{d_\text{p}}{h}$ is replaced by $\frac{d_\text{p}}{h} - \frac{d_\text{p}}{\delta}$, and $\ln\left(\frac{d_\text{p}}{h}\right)$ by $\ln\left(\frac{d_\text{p}}{h}\right) - \ln\left(\frac{d_\text{p}}{\delta}\right)$ (Nguyen and Ladd, 2002).

(b) The lubrication force saturates once the particles are very close (at$10^{-4} d_\text{p}$). The latter restriction we use for numerical reasons (to avoid high force levels and associated instabilities) but also with the surface roughness of the particles and/or the mean-free-path of the fluid in mind.

The time-step-driven (LBM) and the event-driven (collisional) parts of the simulation have been combined by first performing the LBM time step from t to $t+\Delta t$ and subsequently moving the particles until also the particle system has advanced Δt in time. Since in dense systems usually more than one collision occurs during Δt, the particle motion algorithm sets a number of sub-time-steps, the number being equal to one plus the number of (potential, see below) collisions.

At the start of every particle motion sub-time-step, we update the lubrication forces and torques. Then we move the particles over the sub-time-step, i.e. until the next potential collision (or until $t+\Delta t$ is reached). At the new positions of the particles we again determine the lubrication force and torque. The linear and angular velocities of the particles are now updated according to the average of the lubrication forces and torques at the beginning and at the end of the sub-time-step. The velocity update may result in the collision not to occur: in that situation the lubrication forces were sufficiently strong to change the sign of the relative particle velocity so that a hard-sphere collision was prevented.

6.1 Some results for solid-liquid fluidization

Two-dimensional waves and the onset of bubbles have been studied experimentally by Duru and Guazzelli (2002). They used flat, liquid-fluidized beds that could not develop three-dimensional structures, and allowed for good visible observations, well-resolved void fraction measurements, and particle tracking velocimetry measurements. Figure 7 shows a typical result of their experiments: the development of a bubble-like void, starting from a planar wave instability. In this case, steel beads (density $7.8 \cdot 10^3$ kg/m^3) with a diameter of d_p = 1 mm were

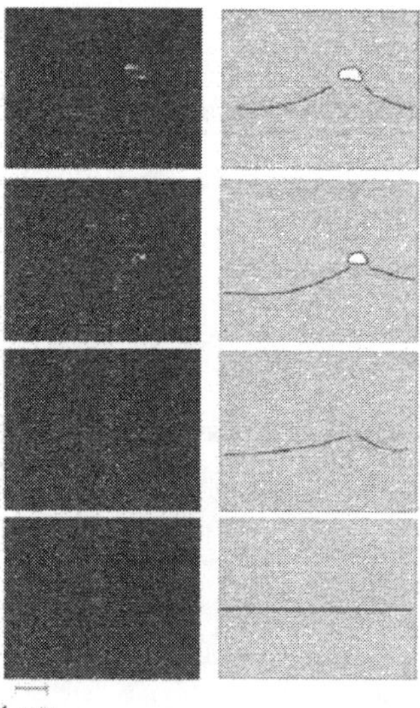

Figure 7. Experiment showing the onset of a bubble in a flat fluidized bed. Left: snapshots of the bed; right: schematization of the observations. Reprinted from Duru and Guazelli (2002).

fluidized with water in a domain that was $120d_p$ wide, $12d_p$ thick and some $2000d_p$ high.

Such two-dimensional structures can indeed be simulated using the approach described above. In order to see the evolution of two-dimensional structures, we have performed simulations in a $24d_p \times 6d_p \times 20d_p$ periodic domain (the $20d_p$ being in the streamwise direction). As an initial condition for the particle positions and velocities (translational and rotational), we juxtaposed four copies of a fully developed planar wave [computed in a $6d_p \times 6d_p \times 20d_p$ periodic domain]. This simulation was performed at a lower resolution such that the hydrodynamic diameter of the particles was set to be 12 lattice units. The density ratio was set to 8. The viscosity and body force were chosen such that the terminal velocity of a single bead was 0.04 (in lattice units) and the Reynolds number based on the terminal velocity matched the value in the experiments (Re = 400). The collisions were smooth and elastic ($\mu = 0,\ e = 1$).

In Figure 8 we show how the numerical system developed a bubble very similar to the experimental observations: the initially plane wave buckles and at its crest forms a bubble-like void. The particle velocity field in the vicinity shows qualitative similarity with the field measured by Duru and Guazzelli (2002); see Figure 9. We again note that non-ideality of bead-bead collisions is not an essential condition for resolving the behavior of void fraction instabilities in

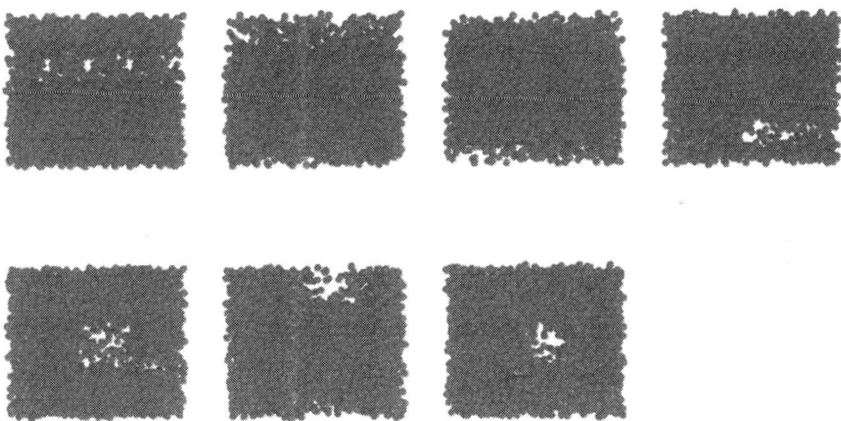

Figure 8. Series of snapshots showing bubble formation. The upper left frame shows the initial condition. The time-spacing between the subsequent (left-to-right, top-to-bottom) frames is

$$0.057\frac{d_p^2}{\nu}.$$

fluidized beds. The domain size in the flow direction is too short for the bubble to behave as an isolated bubble. The periodic system resembles a bubble train, which generally has a higher velocity than an isolated bubble. If we estimate the bubble rise velocity and translate it back to the experimental steel-water system, the bubble rise velocity in the simulation is approximately 15 cm/s. The bubble radius is approximately 0.4 cm. This bubble size and rise velocity combination is at the lower end of the range of rise velocities for bubble trains which is 15 to 22 cm/s according to the experiments (Duru and Guazzelli, 2002).

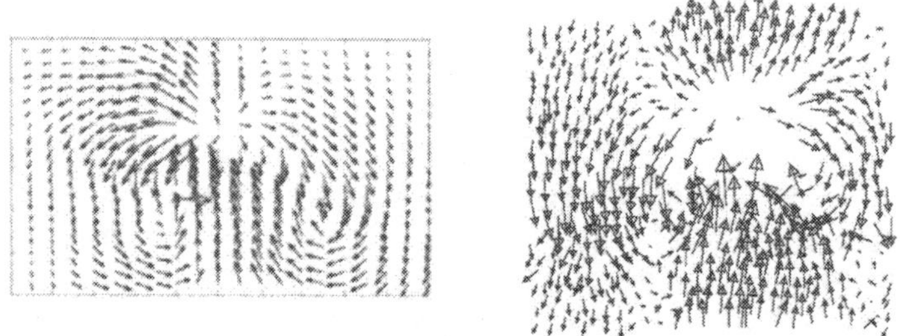

Figure 9. Measured (left, reprinted from Duru and Guazelli 2002), and simulated particle velocities in the vicinity of a bubble.

6.2 DNS of turbulently agitated solid-liquid suspensions

Another example of the application of our approach for directly simulating solid-liquid suspensions is in the field of turbulence and has been reported earlier by ten Cate et al. (2004). We again define a three-dimensional system of solid particles dispersed in a liquid with fully peri-

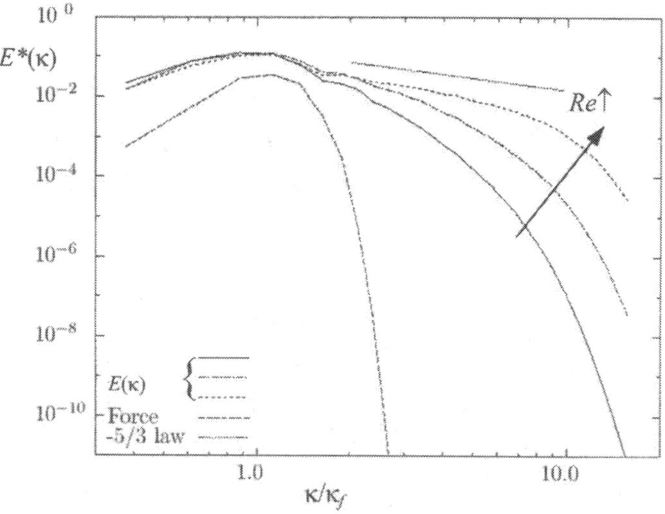

Figure 10. Energy spectrum of the random forcing and the resulting single-phase fluid flow. The wavenumber κ was normalized with the peak-forcing wavenumber κ_f. The energy E is non-dimenionalized with $u_*^2 l_*$ with u_* the rms velocity, and $l_*=2\pi/\kappa_f$.

odic boundary conditions. Instead of driving the system by a uniform body force on the particles and a (opposing) body force on the fluid (mimicking the pressure gradient), we now force the system in a random manner as to generate homogeneous, isotropic turbulence. The procedure for doing this is due to Alvelius (1999); Ten Cate et al (2006) adapted the method so that it could be combined with the lattice-Boltzmann method. In Figure 10 we show results for single-phase turbulence in terms of the energy spectrum. The spectrum of the random forcing is limited to small wavenumbers (large scales). With increasing Reynolds number, the flow develops smaller and smaller scales (the spectrum extends to higher wavenumbers), and develops and inertial subrange characterized by a -5/3 slope in the energy spectrum.

The presence of particles changes the spectrum (see Figure 11): the particles create turbulence at scales comparable to and smaller than the particle diameter. One of our interests in this study was the way particles collide (collision frequencies and intensities). In this respect it was interesting to study the PDF of the time between two collisions of a particle (as given in Figure 12). For "long times" this PDF is exponential indicating Poisson statistics (with a steeper slope for more dense systems). These collisions are uncorrelated events. For "short times" the PDF deviates from exponential and shows a peak towards zero time. These are correlated events: once turbulence has brought two (or more) particles in each others vicinity they tend to cluster due to short range hydrodynamic interaction (lubrication) and undergo many (weak) collisions at short time intervals. Eventually the particles in the cluster are separated when a strong enough eddy comes by.

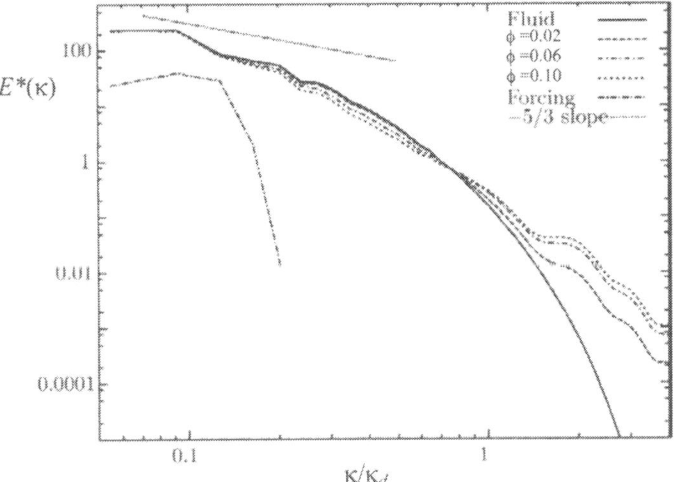

Figure 11. Energy spectra of the two-phase simulations (with solids volume fractions ϕ) compared to the single-phase (fluid) spectrum. The wavenumber κ was normalized with the particle size wavenumber κ_d=$2\pi/d_p$. The energy E is non-dimenionalized with $\varepsilon^{2/3}\eta^{5/3}$ with ε the average energy dissipation rate, and η the Kolmogorov lenght scale.

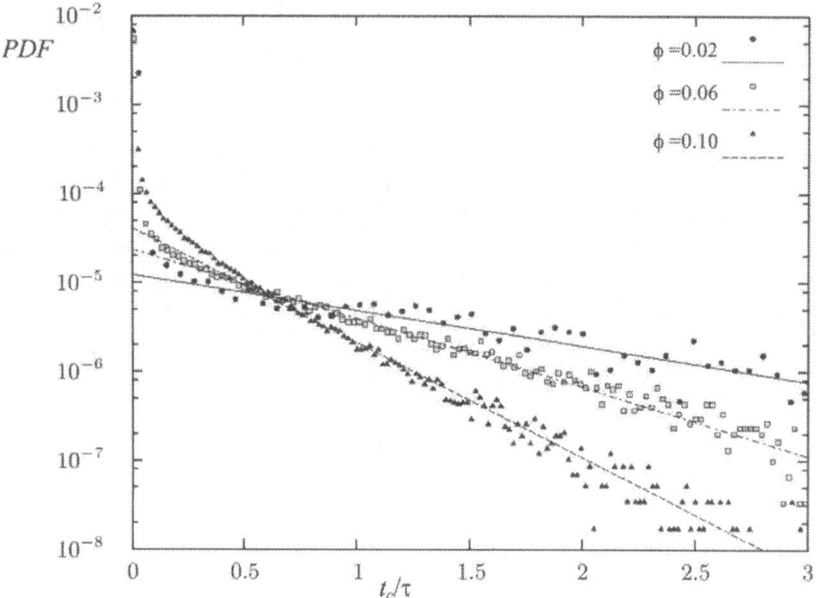

Figure 12. PDF of the time between two collisions of a particle for three solids volume fractions. The lines are linear fits of the tails of the distributions. The collision time has been made dimensionless with the Kolmogorov time scale τ.

7 Single Phase Turbulence

A turbulent flow exhibits an irregular behavior in space and time. A typical example of a time signal corresponding to a turbulent flow quantity is shown in Figure 13 where the streamwise velocity recorded in a turbulent pipe flow is shown as a function of time. At first glance, the velocity may seem to behave randomly. Detailed studies, however, have shown that turbulent flows are not completely random in space and time. They contain spatial (coherent) structures that evolve in time. These structures are often referred to as eddies, as they are usually associated with rotating motions of fluid flow. A fundamental result of turbulence theory is that these eddies are not all of a particular size, but that an often broad continuous range of large to small eddies exists. If we return to Figure 13, and carefully study the temporal evolution of the turbulence signal shown there, we see that in this signal both "fast" and "slow" temporal variations occur, that might be associated to small and large eddies respectively. In general, the size of the largest eddies in a turbulent flow is determined by the geometry of the flow configuration. Here it is characterized by a length scale l. Typical values of l for wall-bounded, shear-driven turbulence are $l \approx 0.1L$ with L a length scale corresponding to the flow geometry (e.g. the pipe diameter). Besides a length scale, these large eddies also have a velocity scale denoted by u. From this we can deduce that large-scale eddies have a typical time-scale proportional to l/u, and a turbulent kinetic energy (per unit mass) proportional to u^2. This kinetic energy is extracted from the mean flow by interaction of the mean flow and the turbulent fluctuations.

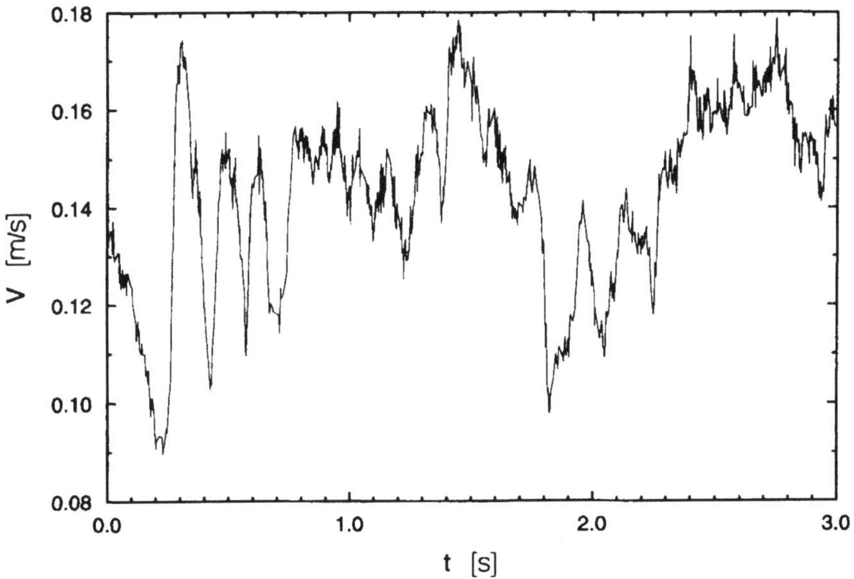

Figure 13. Evolution of the streamwise velocity as a function of time measured in fully developed turbulent pipe flow (reprinted from Eggels, 1994; courtesy J. den Toonder).

The smaller eddies do not extract their kinetic energy directly from the mean flow but are fed by a continuous decay of (unstable) large eddies which break up into smaller ones. These smaller ones in turn decay to even smaller eddies until this cascade reaches the smallest scales of turbulent motion (the energy cascade). The length and velocity scales of these smallest eddies are determined by the amount of kinetic energy that is being transformed along the energy cascade from large to small eddies, and by the molecular viscosity of the fluid that eventually (at the smallest dynamical scales) dissipates the energy.

The loss of kinetic energy of the large-scale eddies is represented by the dissipation rate ε (energy per unit time and unit mass). In turbulence theory, it is assumed that the dissipation rate is independent of the turbulent micro-structure (or the small eddies) since it is fully determined by what happens at the large scales. This is expressed by the following:

$$\varepsilon = \frac{u^3}{l} \tag{7.1}$$

This relation can be interpreted as the kinetic energy of the macro-structure eddies (u^2) being transferred to smaller scales (break-up of bigger eddies into smaller eddies) during their lifetime (l/u).

The smallest eddies are fully controlled by the energy transfer (or, equivalently, by the dissipation rate ε), and by the molecular viscosity ν, they are e.g. decoupled from the flow geometry. By means of dimensional analysis, the length, velocity, and time scales of these smallest eddies can be determined:

$$\eta = \left(\frac{\nu^3}{\varepsilon}\right)^{1/4} \qquad \upsilon = (\nu\varepsilon)^{1/4} \qquad \tau = \left(\frac{\nu}{\varepsilon}\right)^{1/2} \tag{7.2}$$

These are the celebrated Kolmogorov scales. Since the dissipation rate is known in terms of macro-structure properties, we can easily deduce relations between the various scales of macro-structure and micro-structure. Substituting Eq. (7.1) into Eq. (7.2) yields

$$\frac{\eta}{l} = \mathrm{Re}_l^{3/4} \qquad \frac{\upsilon}{u} = \mathrm{Re}_l^{-1/4} \qquad \frac{\tau}{l/u} = \mathrm{Re}_l^{-1/2} \qquad \text{with} \quad \mathrm{Re}_l = \frac{ul}{\nu} \tag{7.3}$$

For large Reynolds numbers Re_l, the scales of the micro-structure become much smaller compared to those of the macro-structure. In other words, the energy cascade process determines the scales of the micro-structure in such a way that the smallest eddies can transform their kinetic energy into internal energy (heat) by means of molecular viscosity. If e.g. the kinetic energy of the macro-structure is increased (i.e. if Re_l becomes larger), then the scales of the micro-structure become smaller (compared to l and u) in order to more effectively transform the increased amount of kinetic energy into internal energy.

The appearance of a broad range of scales in a turbulent flow with the macro-structure characterized by its large eddies on one hand and the micro-structure with its small-scale eddies on the other, is our point of departure to illustrate the principles of the simulation techniques described here.

8 Numerical Simulation of Fluid Flow

For an incompressible, Newtonian fluid, the conservation of mass and momentum are as follows

$$\nabla \cdot \mathbf{v} = 0 \tag{8.1}$$

$$\frac{\partial \mathbf{v}}{\partial t} + \mathbf{v} \cdot \nabla \mathbf{v} = -\frac{1}{\rho}\nabla p + \nu\nabla^2 \mathbf{v} \tag{8.2}$$

with $\mathbf{v}$ the velocity vector, ρ the fluid density, and p the pressure. The Reynolds number (introduced in Section 7), can be interpreted as the ratio of the advection term $\mathbf{v} \cdot \nabla \mathbf{v}$ and the viscous term $\nu\nabla^2\mathbf{v}$ both scaled by the macro-scales u and l:

$$\frac{|\mathbf{v} \cdot \nabla \mathbf{v}|}{|\nu\nabla^2 \mathbf{v}|} \propto \frac{u\frac{1}{l}u}{\nu\frac{1}{l^2}u} = \frac{ul}{\nu} = \mathrm{Re}_l \tag{8.3}$$

For large Re_l, the nonlinear advection term dominates over the viscous term and (in general) the flow will be turbulent. The nonlinearity of the advection term is the reason for the appearance of a broad range of scales in turbulent flows. If the viscous term dominates over the advection term (small Re_l), the flow is laminar and has a regular flow pattern.

Since the equations which describe the flow field are known, it should be possible (in principle) to solve them in a discretized form by means of a computer. Such a numerical simulation should resolve the spatial and temporal evolution of the flow field in all detail to capture all relevant flow phenomena. For laminar flow with low Re_l, the equations of motion can be discretized and solved straightforwardly from a computational point of view. For turbulent flows, the situation is different and more complicated.

8.1 Direct numerical simulation (DNS)

Discretized versions of the partial differential equations involve a grid (spatial discretization) and time steps. Let the distance between two sequential point in space and time be denoted by Δx and Δt respectively. Turbulent flows are characterized by a broad range of length and time scales that should be resolved in all detail by the simulation. Hence, Δx and Δt should be proportional to the smallest length and time scales respectively: $\Delta x \propto \eta$ and $\Delta t \propto \tau$. A more restrictive criterion, however, must sometimes be applied for Δt in which $\Delta t \propto \tau_a$, with τ_a the time scale associated with a small-scale eddy passing a fixed point when being advected by the macro-structure velocity, i.e. $\tau_a \propto \eta / u$.

In Eq. (7.3) we related the macro and micro-structure. Using $l \approx 0.1L$:

$$N_L = \frac{L}{\Delta x} \propto \frac{10l}{\eta} = 10\,\mathrm{Re}_l^{3/4} \tag{8.4}$$

For the ratio T and Δt with T a time interval during which the flow field is monitored, a similar expression can be obtained. As for the length l and L, we have to relate T to the time-scale of the macro-structure l/u. To obtain a correct impression of the turbulent flow, it should be at least monitored for several time scales l/u, say 50 times. Then we find

$$N_T = \frac{T}{\Delta t} \propto \frac{50l/u}{\tau_a} = 50\,\mathrm{Re}_l^{3/4} \tag{8.5}$$

in which we used the (in general) most restrictive criterion to the time step.

The computational time T_C for direct numerical simulations scales at least as the total number of grid points (being N_L^3) and the number of time steps N_T. Memory requirements M_C scale with N_L^3. As a result

$$T_C \propto \mathrm{Re}_l^3 \quad M_C \propto \mathrm{Re}_l^{9/4} \tag{8.6}$$

(for wall bounded flows with boundary layers the above scalings are even on the low side since generally more resolution is required close to walls). Both M_C and T_C are proportional to Re_l raised to a positive power larger than 1. With increasing Re_l, the computational effort rapidly increases. It is obvious that for strongly turbulent flows (say $\mathrm{Re}_l = 10^4$) DNS cannot be performed due to the limited capacity of present-day (super)computers. For relatively modest Re_l, DNS is a very useful technique for studying turbulence in all its detail.

8.2 Large-eddy simulation (LES)

A remedy to overcome the limitations of DNS with increasing Re_l is to reduce the range of scales that are resolved on the numerical grid. A viable concept is to remove the small-scale eddies by a spatial filtering procedure, and to resolve the large-scale eddies only. This approach is called large-eddy simulation (LES). The separation of large and small scales is inspired by observations that the large eddies of the macro-structure are mostly anisotropic. Furthermore, they depend on the geometry of the flow considered. On the other hand, the small eddies of the micro-structure can be considered to be closer to isotropic. They are much less dependent on the flow geometry, as they are fed by the energy cascade in which the geometry information present in the larger eddies gets lost due to break-ups. Therefore, the micro-structure-eddies may tentatively be regarded more or less universal. Since in LES the large eddies are resolved explicitly on the spatial grid, only the effect of the removed small scales remains to be modeled. The more isotropic and (perhaps) universal nature of the small scales is a favorable point of departure for the modeling of the small-scale eddies. This modeling is referred to as subgrid-scale (SGS) modeling.

Before turning to the subject of SGS modeling, let us first consider the equations that need to be solved in an LES and estimate the associated computational effort. In an LES, a flow variable ϕ is decomposed in a grid-scale (GS) component $\overline{\phi}$ and a SGS component ϕ'. The GS component is defined by the moving average filter operation:

$$\overline{\phi}(x,t) = \int_V G(\mathbf{x}-\mathbf{x}')\phi(\mathbf{x}',t)d\mathbf{x}' \tag{8.7}$$

in which G is a spatial filter function that depends on the separation between **x** and **x'**, and V is the total volume of the computational domain. An often applied filter is the top-hat filter:

$$G(\mathbf{x}-\mathbf{x}') = \begin{cases} \dfrac{1}{\Delta V} & \text{if } |\mathbf{x}-\mathbf{x}'| \in \Delta V \quad \text{with} \quad \Delta V \propto l_f^3 \\ 0 & \text{elsewhere} \end{cases} \tag{8.8}$$

Here l_f is a characteristic filter length. Application of the filter (Eq. 8.7) to the flow equations (Eqs. (8.1) and (8.2)) yields equations for the GS variables

$$\nabla \cdot \overline{\mathbf{v}} = 0 \tag{8.9}$$

$$\frac{\partial \overline{\mathbf{v}}}{\partial t} + \nabla \cdot (\overline{\mathbf{v}}\,\overline{\mathbf{v}}) = -\frac{1}{\rho}\nabla \overline{p} + \nu \nabla^2 \overline{\mathbf{v}} - \nabla \cdot \boldsymbol{\tau} \tag{8.10}$$

where the continuity equations has been used in the momentum equation to reformulate the advective term in its conservative form. The stress $\boldsymbol{\tau}$ tensor represents the influence of the SGS motion on the GS motion. It reads

$$\boldsymbol{\tau} = \left(\overline{\overline{\mathbf{v}}\,\overline{\mathbf{v}}} - \overline{\mathbf{v}}\,\overline{\mathbf{v}}\right) + \left(\overline{\overline{\mathbf{v}}\,\mathbf{v}'} + \overline{\mathbf{v}'\,\overline{\mathbf{v}}}\right) + \overline{\mathbf{v}'\,\mathbf{v}'} \tag{8.11}$$

In a later stage, Eq. (8.11) will be discussed in more detail; for the moment it suffices to realize that $\boldsymbol{\tau}$ contains the unknown $\mathbf{v}'$ and requires modeling.

In LES, the macro-structure with length-scale l is still resolved on the spatial grid. The filter length l_f associated to the filter function $G(\mathbf{x}-\mathbf{x}')$ is now the characteristic length of the smallest resolved (GS) motions, as well as of the largest SGS motions. For LES to be realistic, the ratio l/l_f representing the range of length-scales of the GS motion should be $l/l_f>1$ (smaller than one would imply that even the large-scale eddies of the macro structure are filtered out). The filter length l_f can be associated with the grid-spacing Δx where one usually assumes $l_f=2\Delta x$ (Nyquist criterion). Larger values of l_f with unchanged Δx yield a more accurate numerical representation of the smallest resolved scales but simultaneously reduce the range of GS motions without any reduction of the computational effort. In LES one aims at keeping the range of GS motion as large as possible and therefore retains $l_f=2\Delta x$. Thus, the linear grid-size N_L now becomes

$$N_L = \frac{L}{\Delta x} = \frac{L}{l}\frac{l}{l_f}\frac{l_f}{\Delta x} \approx 20\frac{l}{l_f} \tag{8.12}$$

In contrast to Eq. (8.4), Eq. (8.12) is not related to Re_l anymore, but to the ratio l/l_f. As a result, LES is not restricted to low Re_l. The coefficient "20" in Eq. (8.12) appears due to the product L/l and $l_f/\Delta x$. The parameter L/l depends on the type of flow. For shear driven flow it typically is 10. Apparently the ratio l/l_f is essential in LES. The choice of the ratio depends on the flow type under consideration. Roughly speaking, an LES with a value larger than 10 is considered to be well resolved.

Let us turn to the temporal resolution in LES. We have seen that the smallest resolved motions in LES are characterized by the filter length l_f. On similar grounds as before for the DNS, the time step Δt should thus be proportional to the time scale l_f/u. In practice, however, Δt in LES computations is determined by criteria for numerical stability that generally lead to small time steps such that the time-scale l_f/u is well-resolved.

8.3 Subgrid-scale modeling in LES

Due to the filtering operation, a stress tensor is introduced in the momentum equations which represents the effect of the SGS motion on the GS motion, see Eqs. (8.10) and (8.11). Here we repeat the expression for the stress tensor:

$$\boldsymbol{\tau} = \tau_{ij}\mathbf{e}_i\mathbf{e}_j = \left(\overline{\overline{\mathbf{v}}\,\overline{\mathbf{v}}} - \overline{\mathbf{v}}\,\overline{\mathbf{v}}\right) + \left(\overline{\overline{\mathbf{v}}\,\mathbf{v}'} + \overline{\mathbf{v}'\,\overline{\mathbf{v}}}\right) + \overline{\mathbf{v}'\,\mathbf{v}'} = \mathbf{L} + \mathbf{C} + \mathbf{R} \tag{8.13}$$

with τ_{ij} the components of $\boldsymbol{\tau}$, and **L**, **R** and **C** the Leonard stress, cross-terms, and SGS stress respectively. In most SGS modeling approaches, modeling applies to the tensor $\boldsymbol{\tau}$ as a whole. In that context $\boldsymbol{\tau}$ is simply referred to as the SGS stress. The tensor $\boldsymbol{\tau}$ is usually decomposed in an isotropic part and a non-isotropic part

$$\tau_{ij} = \frac{1}{3}\tau_{kk}\delta_{ij} + \left(\tau_{ij} - \frac{1}{3}\tau_{kk}\delta_{ij}\right) \tag{8.14}$$

with δ_{ij} the Kronecker delta. The summation convention applies to the repeated index k. Usually, the isotropic part is combined with the resolved pressure. The non-isotropic part is denoted by $\boldsymbol{\tau}'$ and is thus given by

$$\boldsymbol{\tau}' = \tau'_{ij}\mathbf{e}_i\mathbf{e}_j = \left(\tau_{ij} - \frac{1}{3}\tau_{kk}\right)\mathbf{e}_i\mathbf{e}_j \tag{8.15}$$

Let us now consider modeling of $\boldsymbol{\tau}'$. Various approaches can be followed. Prognostic differential equations for all components τ'_{ij} can be derived form the momentum equations. In these equations, however, several terms appear which need to be parametrized, as they contain higher-order correlations of small scale fluctuations. Using this so-called second-order modeling would require solving 7 additional partial differential equations. Deardorff (1973) performed second-order LES for atmospheric turbulence. It yielded results superior to first-order approaches at increased computational cost. It is noteworthy that in later papers Deardorff reverted to much simpler gradient hypothesis approaches. A general simplification of second-order modeling is the reduction of partial differential equations into algebraic equations to parametrize the SGS stresses. It is believed that this approach retains the advantage of second-order modeling while it avoids an increase of computing time and memory usage. Applications of algebraic SGS models have been reported by Schem and Lipps (1976) and Schmidt and Schumann (1989) for studies of atmospheric turbulence. Of particular concern in second-order models is the attention that must be paid to the determination of the various coefficients involved (see also Schmidt and Schumann, 1989).

In view of the above remarks, one usually turns to first-order modeling in LES, where the SGS stresses are directly related to the GS velocity field in an algebraic way. To mimic the net energy transfer from large to small scales, the stress tensor $\boldsymbol{\tau}'$ is formulated in such a way that the SGS stresses reduce the kinetic energy of the GS velocity field. In this respect one introduces the so-called SGS eddy viscosity ν_e in analogy to the molecular viscosity to write $\boldsymbol{\tau}'$ as

$$\boldsymbol{\tau}' = -\nu_e\left(\nabla\overline{\mathbf{v}} + \left(\nabla\overline{\mathbf{v}}\right)^T\right) \tag{8.16}$$

Application of Eq. (8.16) requires a specification of ν_e. The most widely used model is the one proposed by Smagorinsky (1963) in which ν_e is related to the deformation of the resolved velocity field as

$$\nu_e = l_{\text{mix}}^2\sqrt{\overline{S}^2} \quad \text{with} \quad \overline{S}^2 = \frac{1}{2}\left(\nabla\overline{\mathbf{v}} + \left(\nabla\overline{\mathbf{v}}\right)^T\right):\left(\nabla\overline{\mathbf{v}} + \left(\nabla\overline{\mathbf{v}}\right)^T\right) \tag{8.17}$$

The length-scale l_{mix} represents the mixing length of the SGS motions and will be specified below.

The Smagorinsky model originates from the assumption of local equilibrium between production and dissipation in the equation governing the SGS kinetic energy E_{SGS} By definition:

$$E_{SGS} = \int_{2\pi/l_f}^{\infty} E(\kappa)\mathrm{d}\kappa \tag{8.18}$$

with $E(\kappa)$ the three-dimensional energy spectrum, κ the wavenumber and $2\pi/l_f$ the smallest wavenumber corresponding to SGS motions. Local equilibrium between production and dissipation of E_{SGS}.is expressed by

$$-(\boldsymbol{\tau}' : \nabla\overline{\mathbf{v}}) = \varepsilon \tag{8.19}$$

which can be written with Eq. (8.16) as

$$\nu_e \overline{S}^2 = \varepsilon \tag{8.20}$$

We now would like to relate the mixing length l_{mix} with the filter length l_f. For this reason the deformation rate $\overline{S}^2$ is formulated in terms of the energy spectrum $E(\kappa)$:

$$\overline{S}^2 = 2\int_0^{2\pi/l_f} \kappa^2 E(\kappa)\mathrm{d}\kappa \tag{8.21}$$

This equation states that the resolved deformation rate $\overline{S}^2$ equals the dissipation spectrum $\kappa^2 E(\kappa)$ integrated over the resolved scales. For high Reynolds number turbulent flows and wavenumbers κ in the inertial subrange, the energy spectrum $E(\kappa)$ can be written as (e.g. Tennekes and Lumley, 1972)

$$E(\kappa) = \alpha_K \varepsilon^{2/3} \kappa^{-5/3} \tag{8.22}$$

known as the $\kappa^{-5/3}$-law. The constant (attributed to Kolmogorov) α_K is approximately 1.6. Substitution of Eq. (8.22) in Eq. (8.18) and Eq. (8.21) respectively yields:

$$E_{SGS} = \frac{2}{3}\alpha_K \varepsilon^{2/3}\left(\frac{2\pi}{l_f}\right)^{-2/3} \tag{8.23}$$

$$\overline{S}^2 = \frac{2}{3}\alpha_K \varepsilon^{2/3}\left(\frac{2\pi}{l_f}\right)^{4/3} \tag{8.24}$$

where we assumed Eq. (8.22) also for small wavenumbers ($\kappa \to 0$). This is not a valid assumption. However, $\kappa^2 E(\kappa)$ increases with increasing κ, and the contribution of the lower wavenumbers to the integral is relatively small. Substituting the expression for ν_e (Eq. 8.17), and the expression for ε that can be derived from Eq. (8.24) into Eq. (8.20) yields an expression for the ratio of mixing length and filter length:

$$\frac{l_{\text{mix}}}{l_{\text{f}}} = \frac{\left(\frac{3}{2}\alpha_{\text{K}}\right)^{-3/4}}{2\pi} = 0.0825 \tag{8.25}$$

Let us now specify the mixing length. Similar to the filter length, the mixing length can be related to the grid spacing Δ (that for convenience is considered uniform and the same in all directions; a cubic grid). The ratio l_{mix} and Δ is denoted by the Smagorinsky constant c_{S}.:

$$c_{\text{S}} = \frac{l_{\text{mix}}}{\Delta} \tag{8.26}$$

If we assume $l_{\text{f}}=2\Delta$, and use Eq. (8.25) we find $c_{\text{S}} = 0.165$.

This value of the Smagorinsky constant was based on the equilibrium assumption, and on the existence of an inertial subrange ($\kappa^{-5/3}$-law). Especially in wall bounded flows one finds that $c_{\text{S}} = 0.165$ leads to overly damped simulations in which turbulence sometimes cannot be sustained. The remedy is to reduce c_{S} to values of typically 0.1. In principle this implies that the simulations are underresolved: $c_{\text{S}} < 0.165$ can be interpreted as a filter length smaller than 2Δ.

The Smagorinsky model is a simple and robust model. It has two major drawbacks: it does not allow for backscatter (i.e. transfer of energy from the SGS to the GS), and it yields unrealistic results in the vicinity of walls. Since ν_{e} is a positive quantity, the effect of the SGS stresses in the momentum equations will always cause an energy transfer from GS to SGS, mimicking a cascade process. In real turbulence, there indeed is a net energy transfer from large to small scales. This net transfer, however, is the sum of forward scatter (GS to SGS) and backscatter (SGS to GS), where forward scatter dominates. In DNS it has been observed, however, that backscatter occurs in significant parts of a flow.

The Smagorinsky model has an isotropic ν_{e}. In particular near walls, ν_{e} should presumably be taken anisotropic which is not included in the standard Smagorinsky model. Also related to walls is the unphysical behavior of the SGS stress near walls. Precisely at walls, the SGS stress should vanish (no-slip condition), which is not guaranteed in the SGS model governed by Eqs. (8.16), (8.17), (8.26). A way to repair this is by means of wall-damping functions that reduce the mixing length towards the wall. An often-used damping function is due to Van Driest (1965) which relates l_{mix} to the (dimensionless) distance to the wall y^+:

$$l_{\text{mix}} = c_{\text{S}}\Delta\left(1 - e^{-y^+/A^+}\right) \tag{8.27}$$

The parameter

$$y^+ = \frac{yu_*}{\nu} \tag{8.28}$$

is the wall unit or Reynolds number based on the distance to the wall y and the wall-shear velocity u_*. A^+ is a constant equal to 26.

Dynamic subgrid-scale models (Germano et al, 1991) address the issues of walls and back-scatter in the Smagorinsky model (see also the review article by Lesieur and Métais, 1996). In dynamic models the idea of an eddy viscosity related to the deformation rate, i.e. (Eqs. (8.17) and (8.26))

$$\nu_e = c_S^2 \Delta^2 \sqrt{\bar{S}^2} \tag{8.29}$$

is retained. The c_S in Eq. (8.29), however, is not a constant but is determined locally. It can become negative, thereby allowing for backscatter. In dynamic modeling, two filters are applied with different filter widths: *e.g.* Δ_1 and Δ_2 with $\Delta_1 > \Delta_2$. Two eddy viscosities can now be determined:

$$\nu_{e1} = c_S^2 \Delta_1^2 \sqrt{\left(\bar{S}\right)_1^2} \quad \nu_{e2} = c_S^2 \Delta_2^2 \sqrt{\left(\bar{S}\right)_2^2} \tag{8.30}$$

with $\left(\bar{S}\right)_1$ and $\left(\bar{S}\right)_2$ the resolved deformation rates according to the two filters. The energy transfer (dissipation rate) at the two filter wavenumbers κ_1 and κ_2 is the same if the filter wavenumbers are chosen within the inertial subrange. Therefore (see Eq. 8.20)

$$\nu_{e1} \left(\bar{S}\right)_1^2 = \nu_{e2} \left(\bar{S}\right)_2^2 \tag{8.31}$$

The set of three equations (Eqs. 8.30 and 8.31) allows us to determine the unknowns ν_{e1}, ν_{e2} and c_S. There are quite some practical issues in applying dynamic models. The most important is stability. Negative values of c_S (can) lead to negative viscosities and as a result unstable behavior of numerical schemes. Since the work of Germano et al. (1991) there have been extensive efforts in improving dynamic models.

8.4 Examples of single-phase LES by means of the lattice-Boltzmann method

The implementation of the concept of an eddy viscosity in the lattice-Boltzmann scheme is fairly straightforward. Instead of the molecular (constant) viscosity we substitute the sum of the molecular viscosity and the eddy viscosity in Eq. (4.28). The eddy viscosity varies in space and time and is determined via a SGS model.

If the viscosity is based on the local deformation rate, as is the case for the Smagorinsky model (see Eqs. (8.16) and (8.17)), an advantage of applying the LBM is that the deformation rate is readily available from the LB distribution functions (see Eq. (4.26)). This means that we do not need to calculate spatial derivatives of the velocity field (based on e.g. finite differences) to determine deformation rates.

What now follows are a few examples of LB/LES of flow systems typically encountered in (chemical) engineering. Such systems have complexly shaped (and sometimes moving) boundaries. In these situations, lattice-Boltzmann discretization of the flow equations can be a favorable approach in view of computational efficiency; in order to do realistic LES we need to have a sufficiently fine grid and numerical (parallel) efficiency is then of key importance. The examples are the flow in a mixing tank, and the flow in a swirl tube. In the latter case some aspects of SGS modeling will be highlighted.

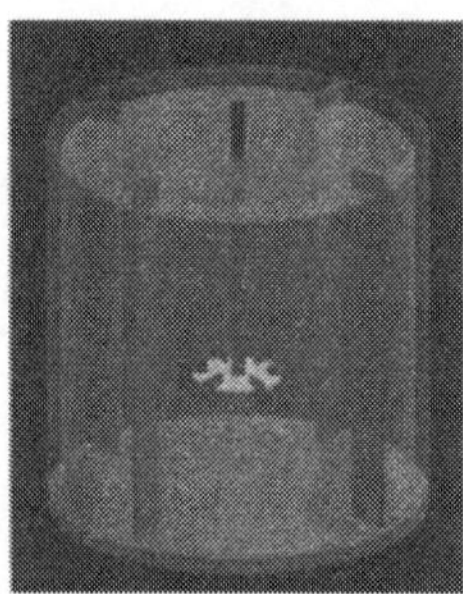

Figure 14. Typical mixing tank geometry: a baffled tank with a disk impeller (Rushton turbine) placed one-third of the tank height from the bottom.

Mixing tank simulation. Turbulently agitated tanks are used in various industries to perform mixing tasks in order to e.g. stimulate chemical reactions, bring species into contact, or disperse gases or solids into liquid. Predicting the single-phase flow in a mixing tank therefore is a relevant task. In order to prepare such a single-phase simulation for inclusion of chemical reactions, or solid particle dispersion, an LES approach is highly desirable. For instance: the dynamics of solid particles suspended in the turbulent flow is governed by their direct (hydrodynamic) surroundings that can be largely provided by an LES, not by the (time) averaged flow in the tank that would follow from a RANS-type of simulation. Here we report on simulations in a tank geometry as given in Figure 14. A revolving turbine with six vertical blades drives the flow. The vertically placed baffles at the perimeter of the tank enhance mixing as they largely prevent a solid-body rotation of the fluid. The Reynolds number of this flow is traditionally based on the impeller diameter D, and the angular velocity of the impeller N [in rev/s]: Re=D^2N/ν. Here, Re=$1\cdot 10^5$. This flow has been simulated by means of lattice-Boltzmann discretization and a Smagorinsky subgrid-scale model with c_S=0.1. The uniform cubic mesh had a size of 240^3. The off-grid, no-slip boundary conditions at the tank wall, baffles, and revolving impeller were imposed by means of an immersed boundary technique that has been briefly described in Section 5.2. Further details of the simulations shown here can be found in e.g. Derksen and Van den Akker (1999), and Derksen (2003).

In Figure 15, a snapshot and the average flow in a vertical cross section are compared. The average flow visualizes the action of the impeller: fluid is pumped in radial direction. Once it hits the outer wall, the fluid stream emerging from the impeller splits and drives two large recirculations, one above and one below the impeller. The slight upward inclination of the impeller stream is due to the impeller being placed closer to the bottom than to the top. This makes the lower recirculation stronger and pushes the impeller stream slightly upwards. A solid particle dispersed in the tank does not “see” the average flow. It is moved around by the instantaneous flow structures that are to a large extent resolved by the LES.

Validation of these simulations by means of experimental data is essential. Figure 16 shows a simulation-experiment comparison in a critical region of the flow: the wake of an impeller blade. The trailing vortex structure is well resolved by the simulations. Also the fluctuation levels (in terms of the turbulent kinetic energy) are well predicted by the LES (not shown here, see Derksen and Van den Akker, 1999).

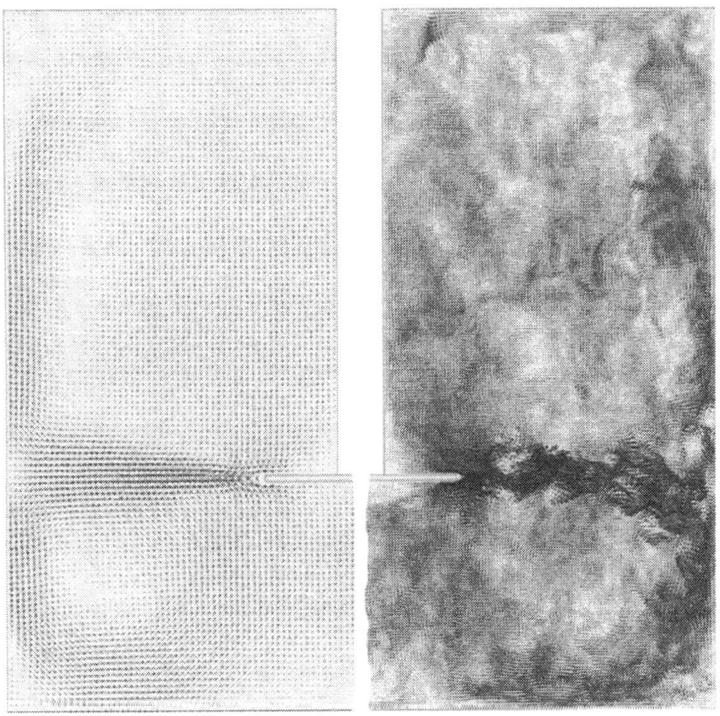

Figure 15. Flow field (in terms of velocity vectors) in a vertical plane through the center. Left: time-averaged flow; right: single realization.

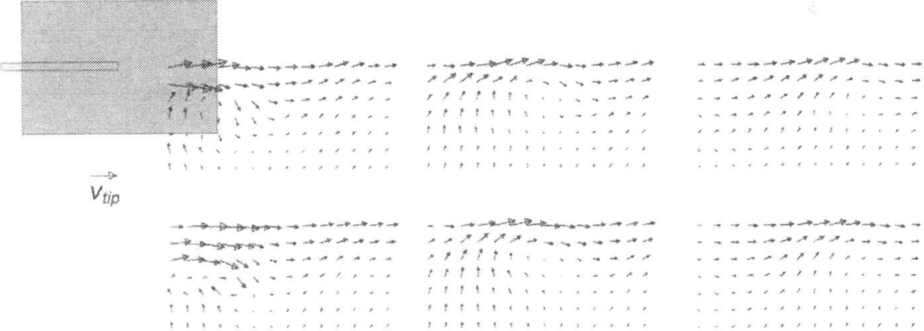

Figure 16. Comparison between experimental velocity data (top row, from Derksen et al 1999) and LES results (bottom row) at three angular positions in the wake of an impeller blade. From left to right the observation plane and the impeller blade make an angle of 10°, 31° and 49° respectively.

Swirl tube with tangential inlet. Turbulent swirling flows form a good testing ground for assessing simulation techniques and turbulence modeling approaches. This is because the turbulence is swirling flow is known to be highly anisotropic. Furthermore, swirl tends to laminarize parts of turbulent flows. Swirling flows have practical significance in separation devices and combustion (flames stabilized by swirl). A classical experimental data set on turbulent swirling flow is due to Escudier and co-workers (see Escudier et al, 1980). The flow geometry is given in Figure 17.

The complexity of the turbulence can be appreciated from Figure 18, which shows the flow visualized in an axial cross-section. A laminar-like vortex core can be distinguished from a turbulent outer region. Furthermore, vortex breakdown is observed (not shown here). Large-eddy simulations (Derksen, 2005) were able to represent these specific flow features very well (see Figure 18).

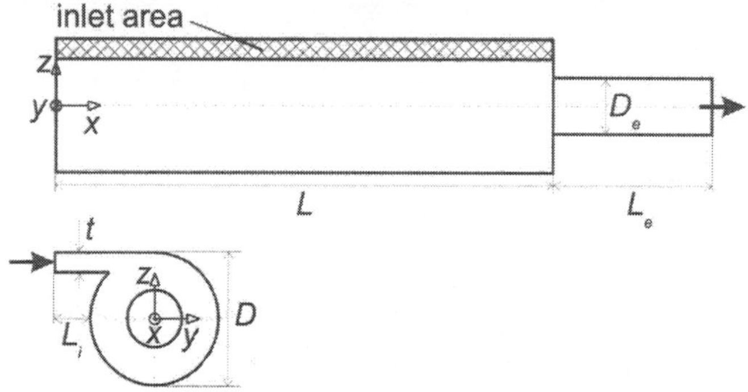

Figure 17. Flow geometry of the experimental setup due to Escudier et al' (1980).

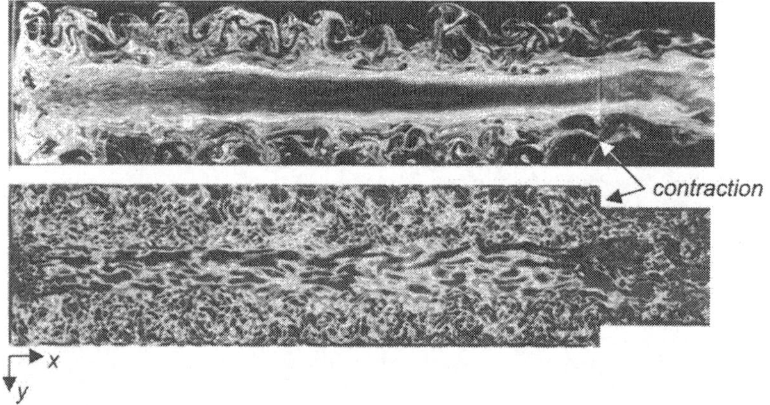

Figure 18. Top: (experimental) flow visualization due to Escudier et al' (1980) at Re=410, D_e/D=0.58. Bottom: LES snapshot at Re=2,100, D_e/D=0.73 in terms of the vorticity.

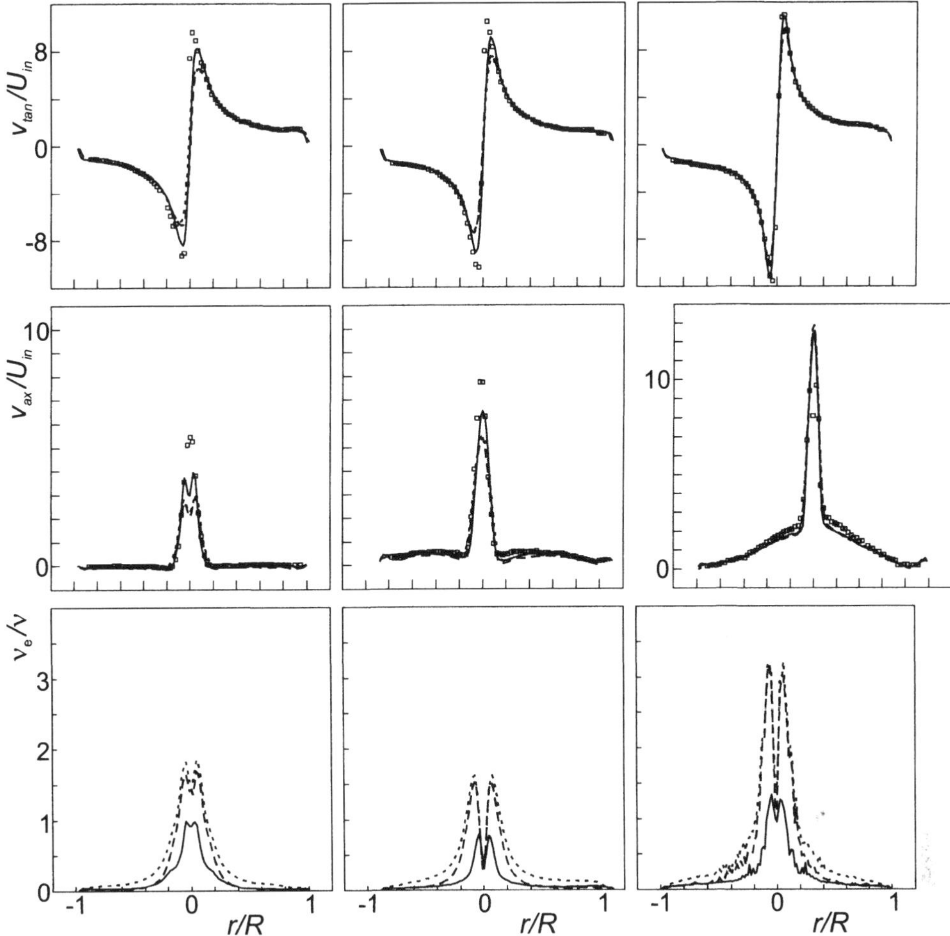

Figure 19. Average velocity and eddy-viscosity profiles for D_e/D=0.33. Top: tangential velocity; middle: axial velocity; bottom: eddy viscosity. From left to right: three axial positions x/D=0.15, 2.15, and 3.61. The symbols denote experimental data. The short-dashed line is the LES result with the Smagorinsky model, the long-dashed line with the model due to Voke, and the solid line with the mixed-scale model.

A quantitative comparison was made based on velocity profiles measured by Escudier et al (1980) by means of laser-Doppler anemometry (LDA). It was interesting to see the impact of the grid resolution, and the choice of the SGS model on the quality of the LES predictions. Next to the Smagorinsly model supplemented with Van Driest wall damping functions (Eq. (8.27)), we also applied a mixed-scale model due to Sagaut et al (2000), and a low-Reynolds number version of the Smagorinsky model due to Voke (1996). Some sample results are shown in Figure 19. They show that Sagaut et al's model is better capable of resolving the large gradients present in the flow. In relation to that, it can be seen that Sagaut et al's model has much lower eddy viscosities. Apparently the Smagorinsky model (and Voke's variant) produce relatively high eddy viscosities that smear out (dampen) steep flow gradients.

9 Point Particles in LES

In the previous section, two single-phase LES examples were discussed. In typical applications, however, mixing tanks and swirl separators are operated with multiple phases. Here we will discuss how to incorporate solid particles in LES flow fields. We will limit ourselves to a "point" particle approach. With this we mean that we do not resolve the flow around the particles with our lattice-Boltzmann discretization; the particles are smaller than the lattice spacing Δ (in Section 6, an approach is discussed where we do resolve the finite size of the particles; per particle such an approach requires much more computer resources than a point-particle approach and only small systems containing typically 10^3 particles can be simulated).

In a point-particle approach, the motion of the particles is determined by integrating the equation of motion of each individual particle (Lagrangian type simulations). The underlying fluid flow is based on an Eulerian (fixed-grid) method. Such Euler-Lagrange flow simulations are mostly limited to relatively dilute systems, i.e. systems with a low dispersed phase volume fraction. This is because of computational reasons: we can only track a limited number of particles that have relatively simple interactions; and because of reasons related to physical modeling: in dense suspensions the finite extent of the particles becomes relatively important and a point-particle approach cannot represent the physics very well. As an example: in fluidized beds (with typically 50% solids volume fractions) the fluid flow in between the solid particles is akin to the flow through a porous medium (i.e. flow through the channel-like structures in between particles) that is mimicked poorly by viewing the particles as points in space.

In the dilute limit amenable to an Euler-Lagrange approach, various subdivisions can be made, one of the most important being the choice of taking or not taking into account particle-particle collisions. In my view there is no general rule for taking into account collisions, especially in inhomogeneous flows with regions preferentially occupied by particles. In what follows we discuss a mixing tank containing a mixture of liquid and solid particles. The fluid flow is solved based on an LES approach and LB discretization (see Section 8.4). The geometry and fluid properties are the same as in the example discussed in that section.

In order to better appreciate the agitated solid-liquid system, we will now relate to dimensional numbers. The vessel volume was 10^{-2} m^3. This implies D=7.78·10^{-2} m. The continuous phase was water (with $\nu=1\cdot10^{-6}$ m^2/s, and $\rho_l=1\cdot10^3$ kg/m^3). A set of 6,705,623 spherical particles was released in the tank. The particles have a diameter of d_p=0.30 mm, and density ratio $\frac{\rho_p}{\rho_l}=2.5$ (with ρ_p the density of the particles), typical for glass beads in water. As a result, the solids volume and mass fractions amounted to Φ_V=0.95% and Φ_m=2.37% respectively. The impeller was set to revolve with N=16.5 rev/s (Re=1·10^5). The Stokes number of the particles was Stk=1.2 (with $\mathrm{Stk}=\frac{\rho_p}{\rho_l}\frac{d_p^2}{18\nu}6N$; the ratio of the Stokesian particle relaxation time, and the time of one impeller blade passage). The Stokes number is of the order of 1, and we may expect appreciable effects of particle inertia (i.e. the particles neither follow the flow (Stk<<1), nor move around in a ballistic manner (Stk>>1) and hardly feel the turbulent flow variations).

Without much discussion it can be anticipated that particle inertia and fluid inertia (added mass), gravity, and drag need to be part of the equations of motion of the solid particles. Since a stirred tank flow is very inhomogeneous, it is difficult to estimate *a priori* if more exotic forces like lift and history forces play an important role. For instance, an estimate of the ratio between lift (Magnus and/or Saffman force) and drag forces is $0.2\sqrt{\frac{d_p^2|\hat{\omega}|}{\nu}}$, with $|\hat{\omega}|$ the vorticity in the liquid phase or the angular (slip) velocity of the particle. In the impeller region of single-phase stirred tank flow, vorticity easily exceeds $10N$ (Derksen and Van den Akker, 1999). With N=16.5 rev/s, and d_p=0.3 mm the ratio amounts to 0.8, indicating the potential relevance of lift forces.

Each particle dispersed in the stirred tank has six degrees of freedom associated to it: three linear coordinates, and three angles. Since we consider the particles to be spherical, the particle's orientation has no physical consequence. As will be demonstrated below, the angular velocity has physical significance. For the linear motion, the following set of equations will to be solved:

$$\frac{d\mathbf{x}_p}{dt} = \mathbf{v}_p \tag{9.1}$$

$$\frac{\pi}{6}d_p^3\left(\rho_p + \frac{1}{2}\rho_l\right)\frac{d\mathbf{v}_p}{dt} = \frac{\pi}{8}d_p^2\rho_l C_D|\mathbf{v}-\mathbf{v}_p|(\mathbf{v}-\mathbf{v}_p) + \mathbf{F}_{Saffman} + \mathbf{F}_{Magnus} + \mathbf{F}_{stress} + \frac{\pi}{6}d_p^3(\rho_p - \rho_l)\mathbf{g} \tag{9.2}$$

with $\mathbf{x}_p$ the center position of the particle, $\mathbf{v}_p$ and $\mathbf{v}$ the velocity of the particle, and the velocity of the liquid at $\mathbf{x}_p$ respectively, and C_D the drag coefficient. The latter depends on the particle Reynolds number $Re_p = \frac{|\mathbf{v}-\mathbf{v}_p|d_p}{\nu}$ according to Eq. (A.1), given in the *Appendix* (Sommerfeld, 2001). The solids volume fractions are considered to be sufficiently low not to include a dependency of C_D on Φ_V. Added mass is accounted for by the additional particle inertia $\frac{\pi}{12}d_p^3\rho_l$ (Maxey and Riley, 1983). The influence of the Saffman force ($\mathbf{F}_{Saffman}$), Magnus force ($\mathbf{F}_{Magnus}$), and the force due to stress gradients ($\mathbf{F}_{stress}$) will be discussed below where simulations with and without these forces will be compared.

The Basset history force (Odar and Hamilton, 1964) may have some impact in the impeller region, with its strong velocity fluctuations at frequencies of the order of $6N$. The ratio between the Basset history force and Stokes drag in a time-varying flow field with frequency f is of the order of $0.1\sqrt{\frac{d_p^2 f}{\nu}}$. If we take f=6N, the ratio is 0.3. It will be demonstrated, however, that in the impeller region, Re_p is of the order 10^2. As a result, the drag force is one order of magnitude higher than estimated from Stokes drag, and the Basset force becomes small compared to the drag force. For this reason, and for computational reasons (inclusion of the force would add appreciably to the computational effort), the Basset force has been neglected.

The non-Stokes expressions we use for the Saffman and Magnus force respectively are (Mei, 1992; Oesterlé and Bui Dinh, 1998)

$$\mathbf{F}_{\text{Saffman}} = \frac{\pi}{4} d_{\text{p}}^{3} \frac{\rho_{\text{l}}}{2} C_{\text{S}} \left(\left(\mathbf{v} - \mathbf{v}_{\text{p}} \right) \times \boldsymbol{\omega} \right) \tag{9.3}$$

$$\mathbf{F}_{\text{Magnus}} = \frac{\pi}{4} d_{\text{p}}^{2} \frac{\rho_{\text{l}}}{2} C_{\text{M}} \left| \mathbf{v} - \mathbf{v}_{\text{p}} \right| \frac{\left(\boldsymbol{\omega} - 2\boldsymbol{\omega}_{\text{p}} \right) \times \left(\mathbf{v} - \mathbf{v}_{\text{p}} \right)}{\left| \boldsymbol{\omega} - 2\boldsymbol{\omega}_{\text{p}} \right|} \tag{9.4}$$

with $\boldsymbol{\omega}$ the vorticity of the liquid, and $\boldsymbol{\omega}_{\text{p}}$ the angular velocity of the particle. The lift coefficients C_{S} and C_{M} depend on Re_{p}, and on the rotational Reynolds numbers $\text{Re}_{\text{S}} = \frac{|\boldsymbol{\omega}| d_{\text{p}}^{2}}{\nu}$, and $\text{Re}_{\text{R}} = \frac{\left| \frac{1}{2}\boldsymbol{\omega} - \boldsymbol{\omega}_{\text{p}} \right| d_{\text{p}}^{2}}{\nu}$ according Eqs. (A.2) and (A.3). The force due to stress gradients has a pressure and a viscous stress part:

$$\mathbf{F}_{\text{stress}} = \frac{\pi}{6} d_{\text{p}}^{3} \left(-\nabla p + \rho_{\text{l}} \nu \nabla^{2} \mathbf{v} \right) \tag{9.5}$$

In order to determine the Magnus force (Eq. (9.4)), the angular velocity of the particles needs to be solved. This is done by solving the following dynamic equation:

$$\frac{\text{d}\boldsymbol{\omega}_{\text{p}}}{\text{d}t} = \frac{60}{d_{\text{p}}^{2}} \frac{\rho_{\text{l}}}{\rho_{\text{p}}} \nu \left(\frac{1}{2}\boldsymbol{\omega} - \boldsymbol{\omega}_{\text{p}} \right) \tag{9.6}$$

which is valid for $\text{Re}_{\text{R}} \leq 30$ (Dennis et al., 1980). The particles' angular velocity may also be relevant from a practical point of view. Mass transfer between solid particles and continuous phase liquid depends on the motion of the solid surface relative to the liquid. Apart from linear velocities, particle rotation might play a role in mass exchange.

The fluid's velocity $\mathbf{v}$, vorticity $\boldsymbol{\omega}$, pressure p and viscous stress $\rho_{\text{l}} \nu \nabla \mathbf{v}$ contained in the above equations all consist of a resolved and a subgrid-scale (SGS) part. For reasons of simplicity, the SGS parts have been discarded, except when the drag force is involved. For determining the drag force, the local fluid velocity is considered to be the sum of the resolved velocity and a Gaussian random process with standard deviation $u_{\text{sgs}} = \sqrt{\frac{2}{3} k_{\text{sgs}}}$ representing the SGS motion. The SGS kinetic energy k_{sgs} was estimated based on isotropic, local-equilibrium mixing-length reasoning according to

$$k_{sgs} = C_k c_s^2 \Delta^2 \overline{S}^2 \tag{9.7}$$

with C_k a constant amounting to 5 (Mason and Callen, 1986). To have temporal coherency in the SGS motion, a new random velocity was picked after the elapse of a SGS eddy lifetime

$$t_{sgs} = C_L \frac{k_{sgs}}{\varepsilon} \tag{9.8}$$

with the constant C_L=0.15 (Weber et al. 1984), and ε the energy dissipation rate.

The resolved part of the liquid velocity was determined by linearly interpolating the velocities on the lattice-Boltzmann grid to the particle position (although a higher order interpolation scheme may be needed for accurate evaluation of single- and, especially, two-particle Lagrangian statistics; see, for example Kontomaris et al. 1992). The vorticity, and the pressure and viscous stress gradients felt by the solid particles were taken uniform over a grid cell. The stresses are directly contained in the solution vector of the lattice-Boltzmann scheme. Their gradients, as well as the velocity gradients contained in the vorticity were determined from central finite differencing. Note that the determination of the force due to stress-gradients differs from how it is usually done, i.e. by means of the material time derivative of the fluid velocity (see e.g. Crowe et al. 1998). Since in a lattice-Boltzmann scheme the stresses are readily available, it is not expensive to directly determine the stress gradients.

Two types of collisions need to be distinguished: particle-wall, and particle-particle collisions. Collisions of all types were considered to be fully elastic and frictionless (the latter implies that in a particle-wall collision the wall parallel components of the velocity of the particle *surface* are unchanged after a collision; in a particle-particle collision the rotation of the particles does not play a role in the collisional process). For the particle-wall collisions with the wall being part of the impeller, only a collision with one of the (six) impeller blades adds momentum to a particle since only the impeller blades have a velocity component in their wall-normal direction.

The method for detecting and handling particle-particle collisions was similar to the one proposed by Chen et al. (1998). In their method, they make use of a collision detection algorithm that anticipates collisions in the upcoming time step. Subsequently, the path of two particles that are bound to collide is integrated in a three-step-process: the pre-collision step, the collision step (in which the particles exchange momentum), and the post-collision step. In order to limit the computational effort spent in handling the particle-particle collisions (which in principle is an M^2 process, with M the number of particles) we have grouped the particles in each other's vicinity in a so-called link-list (Chen et al., 1998). The extent of the vicinity of a particle in which potential collision partners are sought is the lattice cell in which the particle under consideration resides, and the 26 neighboring cells. The distance traveled by a particle during one time step was at most 0.2Δ. This reduces the number of possible collisions partners to a few for a specific particle during a specific time-step. The collision algorithm assumes that one particle can only collide once during one time step. The reason is purely practical: taking into account multiple collisions in one time step would lengthen the computations to an unfeasible extent (e.g. allowing for the possibility to have two collisions per particle per time step would make an M^3 process). The assumption either limits the time step, or the particle volume fraction. In any case, in the simulations there is a finite chance that the collision detection algorithm misses a collision. This is reflected in the situation that at the next time step, two approaching particles have a mutual distance less than d_p. If this occurs, a so-called missed collision procedure is executed: directly at the start of the time-step, the particles involved are given their post-collision velocities (making that they now are moving apart). During the time step, the particles are displaced as a pair according to their average velocity, and they move apart with their relative velocity until they have a mu-

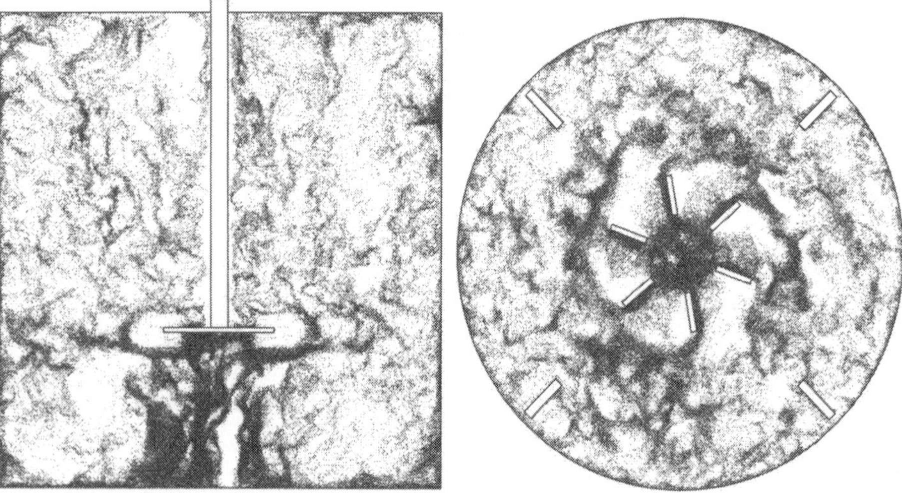

Figure 20. Instantaneous realization of the particle distribution in the tank. Left: vertical cross section through the center of the tank midway between two baffles; bottom: horizontal cross section at z/T=0.308 (i.e. just below the impeller disk). The impeller rotates in the counter-clockwise direction. In both graphs, the particles in a slice with thickness 0.0083T have been

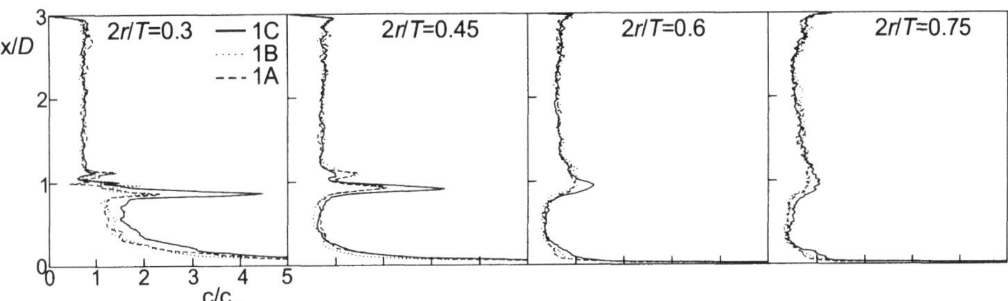

Figure 21. Phase-averaged particle concentration profiles in the plane midway between two baffles as a function of the vertical position in the tank at four different radial positions. Comparison between case #1A (only drag and gravity), #1B (plus lift forces), and #1C (plus particle-particle collisions).

tual separation of at least d_p.

The particle-particle collision algorithm has been tested by (numerically) releasing a set of particles with random initial velocity (according to a uniform distribution) in a periodic vacuum box. The velocity distribution should relax to a Maxwellian distribution, which it did. In the same setup, the algorithm described above to repair missed collisions was tested. Ignoring missed collisions led to one order of magnitude more overlapping particles at any moment in time compared to a situation in which the missed collision procedure was applied.

In the simulations, solids volume fractions are such that it is expected that two-way coupling effects are relevant (Elgobashi, 1994). Two-way coupling was achieved by feeding the force that the fluid exerts on the particle back to the fluid. Since the center position of a particle in general does not coincide with a grid point, the back-coupling force needs to be distributed over the (lattice-Boltzmann) grid nodes in the vicinity of the particle (particle-source-in-cell (PSIC) method, see Crow et al, 1996). For this extrapolation we used the same coefficients as were used for the linear interpolation of the velocity at the grid nodes to the particle location.

Some impressions of these Euler-lagrange two-phase flow simulations are given in Figures 20 and 21. In Figure 20 we see the strongly inhomogeneous distribution of particles throughout the tank. In the vertical cross-section, a highly concentrated region underneath the impeller can be observed. Here gravity and an upwardly directed flow somehow balance the particle motion. This gives rise to long residence times there. The streaky patterns are due to particles collecting at the edges of eddies. This is a typical phenomenon for particles with Stokes numbers of the order of one. Smaller particles would show a much more homogeneous distribution. Particle inertia is also apparent from the horizontal cross-section: particles collect in front of the impeller blades while the wakes of the blades are almost void of particles. In Figure 21 particle concentration profiles are presented obtained with different modeling assumptions. The most striking feature is the impact of taking into account particle-particle collisions has on the concentration profiles. A more detailed analysis reveals that it is the volume exclusion effect brought about by the collision algorithm that makes the difference. If particle-particle collisions are not taken into account, unrealistically high particle concentrations closely above the bottom of the tank are observed. Volume exclusion reduces these concentrations strongly. The profiles obtained with particle-particle collisions fairly well agree with experimental data (Derksen, 2003); the strong peak at the impeller level has also been observed in experiments.

10 Passive Scalar Transport

In order to describe passive scalar transport in a laminar or turbulent flow, the convection-diffusion-equation needs to be solved

$$\frac{\partial \phi}{\partial t} + \mathbf{v} \cdot \nabla \phi = -\nabla \cdot \mathbf{J} \tag{10.1}$$

with ϕ the scalar concentration, and $\mathbf{J}$ the diffusive mass flux. In many cases Fick's law applies:

$$\mathbf{J} = -\Gamma \nabla \phi \tag{10.2}$$

with Γ the scalar's diffusivity. The scalar transport has its own micro length and time-scales. Of prime importance in this respect is the Batchelor length-scale:

$$\eta_{\mathrm{B}} = \left(\Gamma^2 \frac{\nu}{\varepsilon} \right)^{1/4} = \eta \mathrm{Sc}^{-1/2} \quad \text{with} \quad \mathrm{Sc} = \frac{\nu}{\Gamma} \quad \text{the Schmidt number} \tag{10.3}$$

The Batchelor scale can be interpreted as the diffusion distance during one Kolmogorov time scale. Now an essential difference between gases and liquids can be appreciated. In gases the Schmidt number is of the order 1 (momentum and species diffuse at approximately the same pace); in liquids the Schmidt number is $O(10^3)$ (species diffuse much slower than momentum). In a direct numerical simulation of turbulent flow and associated scalar transport, the grid needs to be fine enough to resolve all scales. If the medium is a gas, the grid that was used to resolve the gas motion is sufficiently fine to also resolve the concentration field. If the medium is a liquid, a DNS resolving the all scales of the concentration field would require a grid that (in linear terms) is of the order of $\sqrt{1000} \approx 30$ times finer than the grid required for flow simulations.

Suppose we perform an LES, and would like to represent the scalar concentration fields on the same grid as the velocity and pressure field. This implies filtering of Eq. (10.1):

$$\frac{\partial \bar{\phi}}{\partial t} + \nabla \cdot \left(\bar{\mathbf{v}} \bar{\phi} \right) = \Gamma \nabla^2 \bar{\phi} - \nabla \cdot \boldsymbol{\sigma} \tag{10.4}$$

with $\boldsymbol{\sigma}$ the equivalent of $\boldsymbol{\tau}$ in Eq. (8.10). Usually $\boldsymbol{\sigma}$ is closed in a manner similar to the closure of $\boldsymbol{\tau}$: we assume that the subgrid-scales merely act in a diffusive manner on the concentration field. The eddy diffusion coefficient is then taken proportional to the eddy-viscosity:

$$\Gamma_{\mathrm{e}} = \frac{\nu_{\mathrm{e}}}{\mathrm{Sc}_{\mathrm{e}}} \tag{10.5}$$

with Sc_{e} the turbulent (eddy) Schmidt number. For Sc_{e} a value in the range of 0.6 to 1.0 is usually taken. The rationale behind $\mathrm{Sc}_{\mathrm{e}} < 1$ (at least in liquid systems) is that the unresolved part of the scalar spectrum is larger (it runs up to $\kappa \approx 2\pi/\eta_{\mathrm{B}}$) than the unresolved part of the turbulent kinetic energy spectrum.

Eggels and Somers (1995) have performed scalar transport calculations on free convective cavity flow with the lattice-Boltzmann discretization scheme (i.e. they solved the fluid flow equations, and the convection diffusion equation with the lattice-Boltzmann method). This approach, however, is more memory intensive than using a finite volume formulation for the convection-diffusion equation. In a finite volume discretization we only need to store two or three (depending on the time integrator) double precision concentration fields, whereas in the lattice-Boltzmann discretization typically 18 single-precision variables need to be stored. In a time-explicit approximation of the discretized convection-diffusion equation we have (as in the lattice-Boltzmann scheme) fully local operations: The communication between the nodes of the grid defining the concentration field does not go beyond the stencil that is used for discretizing

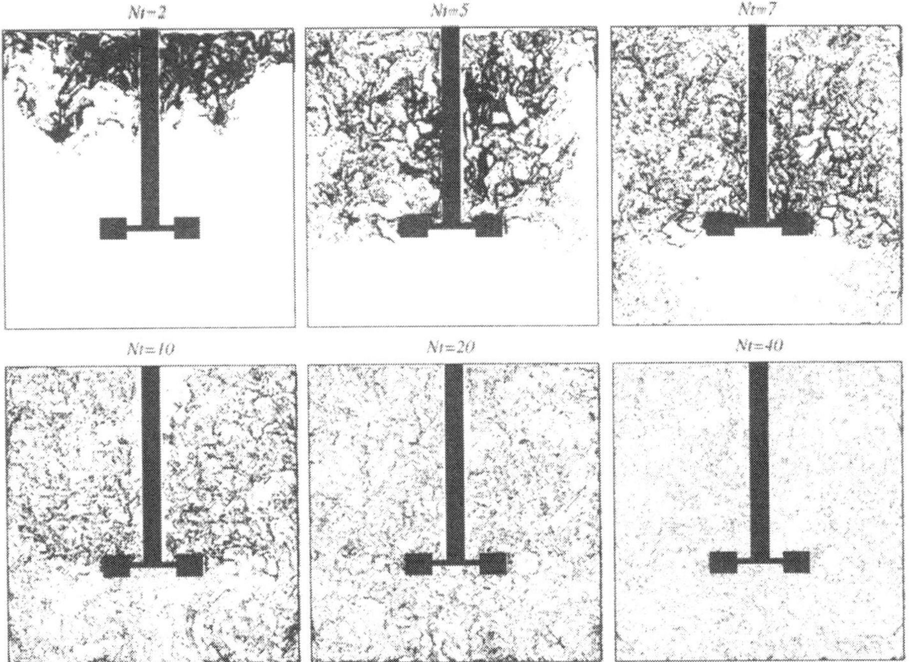

Figure 22. Snapshot of the particle distribution in a vertical plane midway between two baffles.

the convection and diffusion term in Eq. (10.4). An additional advantage of using finite volume discretization for the convection diffusion equation is that we can make use of the various methods to suppress numerical diffusion and still retain stability, such as TVD (total variation diminishing) schemes introduced by Harten (1983).

As an example of LES including scalar transport and solid particle dynamics we briefly discuss here work presented earlier by Hartmann et al (2006) on a dissolution process in a mixing tank. In this example we combine the solid-liquid LES based on the point particle approach (see Section 9) with a finite-volume scalar transport solver for keeping track of the scalar concentration as a result of dissolution. The coupling between the scalar field and the solids is established via a mass transfer coefficient *k*, which is a function of the particle size, the relative velocity of particle and fluid v_{slip}, and the material properties (viscosity and diffusivity) of the fluid. In dimensionless form this implies that the Sherwood number $\mathrm{Sh} = kd_p / \Gamma$ is a function of the particle Reynolds number and the Schmidt number. We use the correlation due to Ranz and Marshall (1952):

$$\mathrm{Sh} = 2.0 + \mathrm{Re}_p^{1/2}\,\mathrm{Sc}^{1/3} \tag{10.6}$$

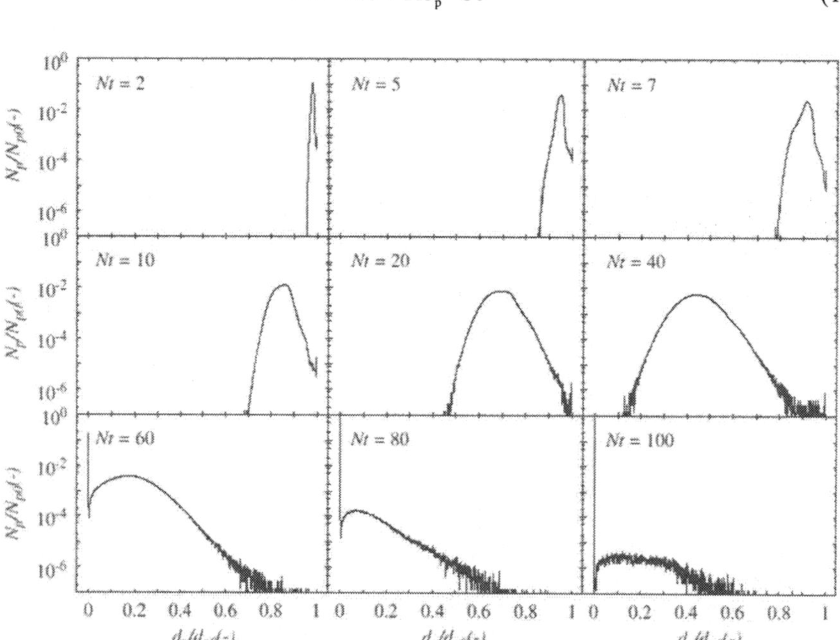

Figure 23. Instantaneous realizations of the particle size distribution throughout the tank.

The mass transfer from solid to liquid is then modeled as

$$\dot{m} = k\pi d_p^2 \left(c_{sat} - c\right) \tag{10.7}$$

Since this relation is linear in the solute concentration, micro-mixing effects are considered not important (there is, however, a velocity-concentration correlation since k is a function of Re that in principle requires SGS closure; here this correlation is neglected).

The tank that Hartmann et al (2006) used was the same as the one discussed in the previous sections. Particles of uniform size were released in the top part of the tank in a fully developed flow field. They spread through the tank and reduce in size (see Figure 22). As they reduce in size they go through a spectrum of Stokes numbers, and the particle field gradually loses its streaky structure; particles get more homogeneously dispersed over the entire tank (Figure 22). The simulations allow for an estimate of the dissolution time. The evolution of the particle size distribution (Figure 23) reflects the diversity in the history of the dissolution process per particle, and thereby the inhomogeneous flow conditions in the tank.

11 Filtered Density Functions for Reactive Flows

In the final section of this chapter, we consider reactive scalar transport. In that case, the species transport equation (the convection diffusion Eq. (10.1)) is adapted in two ways. In the first place, instead of a single scalar ϕ we now write the vector $\boldsymbol{\varphi}$, reflecting that we have a number of species involved in the reactions that can be organized in a vector. In the second place, concentrations can now change as a result of reactions: we need to add a reaction term $\boldsymbol{\omega}$ that (in the general case) depends on all species involved (i.e. the vector $\boldsymbol{\varphi}$):

$$\frac{\partial \boldsymbol{\varphi}}{\partial t} + \mathbf{v} \cdot \nabla \boldsymbol{\varphi} = -\nabla \cdot \mathbf{J} + \boldsymbol{\omega}(\boldsymbol{\varphi}) \tag{11.1}$$

In the case of second order reactions, $\boldsymbol{\omega}$ depends on products of the elements of the vector $\boldsymbol{\varphi}$. In an LES context, Eq. (11.1) could be filtered just as Eq. (10.1). The result is (again we have assumed Fickian diffusion for all species involved)

$$\frac{\partial \overline{\boldsymbol{\varphi}}}{\partial t} + \nabla \cdot \left(\overline{\boldsymbol{\varphi}}\,\overline{\mathbf{v}}\right) = \Gamma \nabla^2 \overline{\boldsymbol{\varphi}} + \overline{\boldsymbol{\omega}}(\boldsymbol{\varphi}) - \nabla \cdot \boldsymbol{\sigma} \tag{11.2}$$

where $\boldsymbol{\sigma}$ now is a tensor. The filtered reaction term $\overline{\boldsymbol{\omega}}(\boldsymbol{\varphi})$ needs specific care. It would be too gross a simplification to write $\overline{\boldsymbol{\omega}}(\boldsymbol{\varphi}) = \boldsymbol{\omega}(\overline{\boldsymbol{\varphi}})$. Remember that we are performing an LES. It only makes sense to perform an LES if the grid is courser than the Kolmogorov scale. This implies that also the Batchelor scale (the smallest scale of scalar transport) is not resolved. As a result, the concentration field shows details finer than the grid spacing. If we would state $\overline{\boldsymbol{\omega}}(\boldsymbol{\varphi}) = \boldsymbol{\omega}(\overline{\boldsymbol{\varphi}})$, this would imply the assumption that at the grid level species concentrations are uniform, and this is not the case.

The filtered reaction term $\overline{\boldsymbol{\omega}}(\boldsymbol{\varphi})$ is known once the filtered version of the probability density function $P_{\mathrm{L}}(\boldsymbol{\varphi};\mathbf{x},t)$ (the fdf, filtered density function) of the vector $\boldsymbol{\varphi}$ at the nodes of the computational domain is known. For example, for a second order reaction $A+B \rightarrow C$ with reaction rate k

$$\overline{\omega}(\varphi) = \int k\psi_A\psi_B P_L(\psi_A,\psi_B)d\psi_A d\psi_B \tag{11.3}$$

where the integration is over the entire (in this case two-dimensional) composition space.

This would imply that, instead of solving transport equations for the species concentrations $\boldsymbol{\varphi}$, we would need to solve transport equations for $P_L(\boldsymbol{\varphi};\mathbf{x},t)$. This may seem a quite impossible task since the dimensionality of the system of equations to be solved increases rapidly. Apart from the three spatial dimensions the composition space adds to the dimensionality. With each species the dimension of the problem increases with 1. A way around this is to solve the fdf-transport equations by means of a Monte-Carlo (MC) method, i.e. to release computational particles in the flow domain and track their position in spatial and compositional space, the major advantage being that adding a reactant increases the computational load approximately linearly.

The idea is to release MC particles randomly in the computational domain. Each particle represent the scalar composition $\boldsymbol{\varphi}$ at its current position **x**(t). The MC particle position and composition are evolved according to the following stochastic differential equations

$$\mathrm{d}\mathbf{x} = \mathbf{D}(\mathbf{x}(t),t)\mathrm{d}t + E(\mathbf{x}(t),t)\mathrm{d}\mathbf{W}(t) \quad \text{and} \quad \mathrm{d}\boldsymbol{\varphi} = \mathbf{B}(\boldsymbol{\varphi}(t),t)\mathrm{d}t \tag{11.4}$$

where **D** and E are the drift (convection) and diffusion coefficients of the particles in the physical domain. The random process $\mathrm{d}\mathbf{W}$ is a Wiener process; $\mathrm{d}\mathbf{W} = \mathrm{d}W_i = \sqrt{\mathrm{d}t}\zeta_i$ with ζ_i a random variable with Gaussian pdf. The drift **B** in the scalar domain is due to micro-mixing and chemical reactions. The various processes in Eq. (11.4) can be related to physical quantities:

$$E = \sqrt{2(\Gamma + \Gamma_e)}, \quad \mathbf{D} = \overline{\mathbf{v}} + \nabla(\Gamma + \Gamma_e) \quad \text{and} \quad \mathbf{B} = -\Omega_m(\boldsymbol{\varphi} - \overline{\boldsymbol{\varphi}}) + \boldsymbol{\omega}(\boldsymbol{\varphi})$$

where as a micro-mixing model we have substituted the interaction-by-exchange-with-the-mean (IEM) model. The SGS mixing frequency Ω_m can be related to the total (molecular plus eddy) diffusivity $\Gamma + \Gamma_e$:

$$\Omega_m = \frac{C_\Omega(\Gamma + \Gamma_e)}{\Delta^2} \tag{11.5}$$

with Δ the filter width, and C_Ω a constant equal to 3 (Colucci et al 1998).

As an example of the application of the LES/FDF approach to a semi-practical system, some results earlier presented by Van Vliet et al (2005) are shown here. Van Vliet et al performed an LES in a straight tube with a deeply protruding feed pipe. The Reynolds number of the main flow was 4,000. Upstream of the feed pipe, the main flow contained species *B* and *C* in the same amount. Component *A* was fed through the feed pipe. *A* could react either with *B* or *C* according to a second-order reaction to form product *P* or *Q* respectively. The reaction rate to form *P*, however, was 10^3 higher than the one to form *Q*. If the chemical kinetics would control the system, the amount of *Q* formed (Φ_Q) would be a thousand times smaller than the amount of *P* formed (Φ_P). Non-ideal mixing, however, will generally increase the ratio Φ_Q/Φ_P: The reaction between *A* and *B* locally depletes the flow of *B*. If mixing cannot bring *A* into contact with fresh *B* quickly enough, the slow reaction will get a chance.

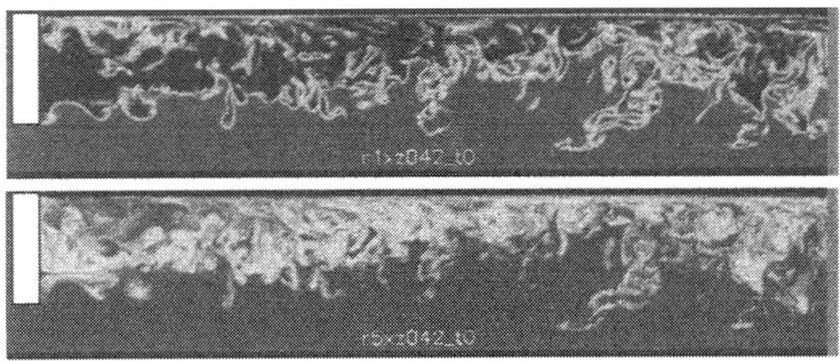

Figure 24. Instantaneous realizations of c_B in the tube reactor. Top: at Da=$2.5 \cdot 10^3$ (poor mixing); bottom: Da=$2.5 \cdot 10^{-5}$ (intense mixing).

The strength of mixing has been quantified in terms of a Damkohler number, which we define as the ratio of the integral hydrodynamic time-scale D/U (with D and U the tube diameter and bulk velocity respectively), and the chemical time scale $1/\sqrt{k_2 c_{A0} c_{C0}}$ with k_2 the rate constant of the slow reaction, and c_{A0} and c_{C0} the inlet concentrations:

$$\mathrm{Da} = \frac{D\sqrt{k_2 c_{A0} c_{C0}}}{U} \tag{11.6}$$

Some results of the simulations are shown in Figures 24 and 25. In Figure 24 we see that in the case of poor mixing (high Da) component B is not very well able to penetrate the reaction zone which in turn allows for the formation of Q. Figure 25 illustrates this point further. Here it is shown how the product ratio varies with Da between its theoretical limits (10^{-3} for Da→0; 1 for Da→∞).

Appendix: Coefficients in hydrodynamic force relations

Drag force:

$$C_D = \frac{24}{\mathrm{Re}_p}\left(1.0 + 0.15\,\mathrm{Re}_p^{0.687}\right) \quad \mathrm{Re}_p < 1{,}000$$

$$C_D = 0.44 \quad \mathrm{Re}_p \geq 1{,}000 \tag{A.1}$$

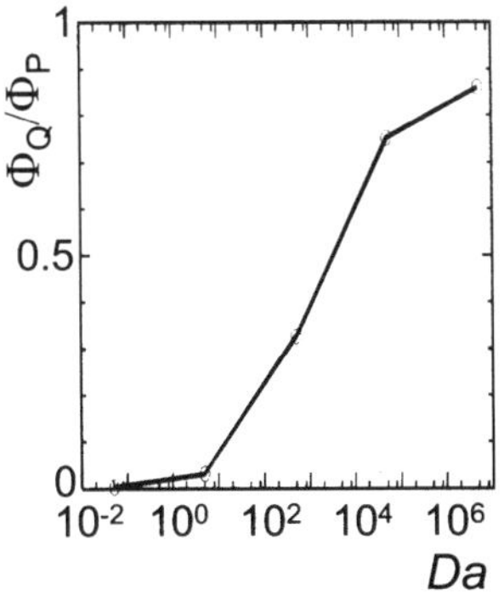

Figure 25. The product ratio Φ_Q/Φ_P as a function of the Damkohler number.

Saffman force:

$$C_S = \frac{4.1126}{\mathrm{Re}_S^{0.5}}\left[\left(1.0 - 0.234\left(\frac{\mathrm{Re}_S}{\mathrm{Re}_p}\right)^{0.5}\right)e^{-0.1\mathrm{Re}_p} + 0.234\left(\frac{\mathrm{Re}_S}{\mathrm{Re}_p}\right)^{0.5}\right] \quad \mathrm{Re}_p < 40$$

$$C_S = 0.1524 \quad \mathrm{Re}_p \geq 40 \tag{A.2}$$

Magnus force:

$$C_M = 0.45 + \left(\frac{\mathrm{Re}_R}{\mathrm{Re}_p} - 0.45\right)e^{-0.05684\,\mathrm{Re}_R^{0.4}\,\mathrm{Re}_p^{0.3}} \tag{A.3}$$

References

C.K. Aidun, Y. Lu, and E.J. Ding (1998). Direct analysis of particle suspensions with inertia using the discrete Boltzmann equation. *Journal of Fluid Mechanics*, 373: 287-311.

K. Alvelius (1999). Random forcing of three-dimesnional homogeneous turbulence. *Physics of Fluids*, 11: 1880-1889.

P.L Bhatnagar, E.P. Gross and M. Krook (1954). A model for collision processes in gases. I: small amplitude processes in charged and neutral one-component system. *Physical Review*, 94: 511-525.

M. Chen, K. Kontomaris and J.B. McLaughlin (1998). Direct numerical simulation of droplet collisions in a turbulent channel flow. Part I: collision algorithm. *International Journal of Multiphase Flow*, 24: 1079-1103.

S. Chen, Z. Wang, X.W. Shan and G.D. Doolen (1992). Lattice-Boltzmann computational fluid dynamics in three dimensions. *Journal of Statistical Physics*, 68: 379-400.

S. Chen and G.D. Doolen (1998). Lattice-Boltzmann method for fluid flows. *Annual Review Fluid Mechanics*, 30: 329-364.

P.J. Colucci, F.A. Jaberi, P. Givi and S.B. Pope (1998). Filtered density functions for large eddy simulation of turbulent reacting flows. *Physics of Fluids*, 10: 499-515.

C. Crowe, M. Sommerfeld and Y. Tsuji (1998). *Multiphase flows with droplets and particles*, CRC Press, Boca Raton.

J.W. Deardorff (1973). The use of subgrid transport equations in a three-dimensional model of atmospheric turbulence. *Journal of Fluids Engineering*, 95: 429-438.

S.C.R. Dennis, S.N. Singh and D.B. Ingham (1980). The steady flow due to rotating sphere at low and moderate Reynolds numbers. *Journal of Fluid Mechanics*, 101: 257-279.

J.J. Derksen and H.E.A. van den Akker (1999). Large eddy simulations on the flow driven by a Rushton turbine. *AIChE Journal*, 45: 209-221.

J.J. Derksen (2003). Numerical simulation of solids suspension in a stirred tank. *AIChE Journal*, 49: 2700-2714.

J.J. Derksen (2005). Simulations of confined turbulent vortex flow. *Computers & Fluids*, 34:301-318.

E.J. Ding and C.K. Aidun (2003). Extension of the lattice-Boltzmann method for direct simulation of suspended particles near contact. *Journal of Computational Physics*, 112: 685-708.

P. Duru and E. Guazelli (2002). Experimental investigations on the secondary instability of liquid-fluidized beds and the formation of bubbles. *Journal of Fluid Mechanics*, 470: 359-382.

J.G.M. Eggels (1994). *Direct and large eddy simulation of turbulent flow in a cylindrical pipe geometry*. PhD Thesis, Delft University of Technology.

J.G.M. Eggels and J.A. Somers (1995). Numerical simulation of free convective flow using the lattice-Boltzmann scheme. *International Journal of Heat and Fluid Flow*, 16: 357-364.

S. Elghobashi (1994). On predicting particle-laden turbulent flows. *Applied Scientific Research*, 52: 309-329.

M.P. Escudier, J. Bornstein and N. Zehnder (1980). Observations and LDA measurements of confined turbulent vortex flow. *Journal of Fluid Mechanics*, 98: 49-63.

U. Frisch, B. Hasslacher and Y. Pomeau (1986). Lattice-gas automata for the Navier-Stokes equations. *Physical Review Letters*, 56: 1505-1508.

U. Frisch, D. d'Humieres, B. Hasslacher, Y. Pomeau and J.P. Rivet (1987). Lattice gas hydrodynamics in two and three dimensions. *Complex Systems*, 1: 649.

M. Germano, U. Piomelli, P. Moin and W. Cabot (1991). A dynamic subgrid-scale eddy-viscosity model. *Physics of Fluids A*, 3: 1760-1765.

D. Goldstein, R. Handler and L. Sirovich (1993) Modeling a no-slip flow boundary with an external force field. *Journal of Computational Physics*, 105: 354-366.

A. Harten (1983). High resolution schemes for hyperbolic conservation laws. *Journal of Computational Physics*, 49: 357-393.

H. Hartmann, J.J. Derksen and H.E.A. van den Akker (2006). Numerical simulation of a dissolution process in a stirred tank reactor. *Chemical Engineering Science*, 61: 3025-3032.

H. Hasimoto (1959). On the periodic fundamental solutions of the Stokes equation and their application to viscous flow past a cubic array of spheres. *Journal of Fluid Mechanics*, 5: 317-328.

S. Kim and S.J. Karrila (1991). *Microhydrodynamics: Principles and selected applications*. Butterworth-Heinemann.

K. Kontomaris, T.J. Hanratty and J.B. McLaughlin (1992). An algorithm for tracking fluid particles in a spectral simulation of turbulent channel flow. *Journal of Computational Physics*, 103: 231-242.

A.J.C. Ladd (1994). Numerical simulations of particle suspensions via a discretized Boltzmann equation. Part I: Theoretical Foundation. *Journal of Fluid Mechanics*, 271: 285-309.

P. Lallemand and L.S. Luo (2000). Theory of the lattice-Boltzmann method: dispersion, dissipation, isotropy, Galilean invariance, and stability. *Physical Review E*, 61: 6546-6562.

M. Lesieur and O. Métais (1996). New trends in large-eddy simulations of turbulence. *Annual Review Fluid Mechanics*, 28: 45-82.

P.J. Mason and N.S. Callen (1983). On the magnitude of the subgrid-scale eddy coefficient in large-eddy simulations of turbulent channel flow. *Journal of Fluid Mechanics*, 162: 439-462.

M.R. Maxey and J.J. Riley (1983). Equation of motion for a small rigid sphere in a nonuniform flow. *Physics of Fluids*, 26: 883-889.

R. Mei (1992). An approximate expression for the shear lift force on a spherical particle at finite Reynolds numbers. *International Journal of Multiphase Flow*, 18: 145-147.

N.-Q. Nguyen and A.J.C. Ladd (2002). Lubrication corrections for lattice-Boltzmann simulations of particle suspensions. *Physical Review, E* 66: 046708.

F. Odar and W.S. Hamilton (1964). Forces on a sphere accelerating in a viscous fluid. *Journal of Fluid Mechanics*, 18: 302-314.

B. Oesterlé and T. Bui Dinh (1998). Experiments on the lift of a spinning sphere in the range of intermediate Reynolds numbers. *Experiments in Fluids*, 25: 16-22.

W.E. Ranz and W.R. Marshall (1952). Evaporation from drops. *Chemical Engineering Progress*, 48: 141-146.

M. Rohde, J.J. Derksen and H.E.A. van den Akker (2002). Volumetric method for calculating the flow around a moving object in lattice-Boltzmann schemes. *Physical Review E*, 65: 056701.

P. Sagaut, P. Comte and F. Ducros (2000). Filtered subgrid-scale models. *Physics of Fluids*, 12: 233.

C.E. Schem and F.B. Lipps (1976). Some results from a simplified three-dimensional numerical model of atmospheric turbulence. *Journal of Atmospheric Science*, 33: 1021-1041.

H. Schmidt and U. Schumann (1989). Coherent structure of the convective boundary layer from large-eddy simulations. *Journal of Fluid Mechanics*, 200: 511-562.

J. Smagorinsky (1963). General circulation experiments with the primitive equations: 1. The basic experiment. *Monthly Weather Review*, 91: 99-164.

J.A. Somers (1993) Direct simulation of fluid flow with cellular automata and the lattice-Boltzmann equation. *Applied Scientific Research*, 51: 127-133.

M. Sommerfeld (2001). Validation of a stochastic Lagrangian modelling approach for inter-particle collisions in homogeneous isotropic turbulence. *International Journal of Multiphase Flow*, 27: 1829-1858.

S. Succi (2001). *The lattice Boltzmann equation for fluid dynamics and beyond.* Clarondon Press, Oxford.

A. ten Cate, C.H. Nieuwstad, J.J. Derksen and H.E.A. van den Akker (2002). PIV experiments and lattice-Boltzmann simulations on a single sphere settling under gravity. *Physics of Fluids*, 14: 4012-4025.

A. ten Cate, J.J. Derksen, L.M. Portela and H.E.A. van den Akker (2004). Fully resolved simulations of colliding spheres in forced isotropic turbulence. *Journal of Fluid Mechanics*, 519: 233-271.

A. ten Cate, E. van Vliet, J.J. Derksen, and H.E.A. van den Akker (2006). Application of spectral forcing in lattice-Boltzmann simulations of homogeneous turbulence. *Computers & Fluids*, 35:1239-1251.

H. Tennekes and J.L. Lumley (1972). A first course in turbulence. MIT Press, Cambridge MA.

E.R. van Driest (1965). On turbulent flow near a wall. *Journal of Aerodynamic Science*, 23: 1007-1011.

E. van Vliet, J.J. Derksen and H.E.A. van den Akker (2005). Turbulent mixing in a tubular reactor: assessment of an FDF/LES approach. *AIChE Journal*, 51: 725-739.

P.R. Voke (1996). Subgrid-scale modelling at low mesh Reynolds number. *Theoretical and Computational Fluid Dynamics*, 8: 131-143.

R. Weber, F. Boysan, W.H. Ayers and J. Swithenbank (1984). Simulation of dispersion of heavy particles in confined turbulent flows. *AIChE Journal*, 30: 490-492.

Y. Yamamoto, M. Potthoff, T. Tanaka, T. Kajishima Y. Tsuji (2001). Large-eddy simulation of turbulent gas-particle flow in a vertical channel: effect of considering inter-particle collisions. *Journal of Fluid Mechanics*, 442: 303-334.

Direct Numerical Simulation of Sprays: Turbulent Dispersion, Evaporation and Combustion

Julien Reveillon
CORIA CNRS, University of Rouen, Rouen, France

Abstract Numerical procedures to describe the dispersion, evaporation and combustion of a polydisperse liquid fuel in a turbulent oxidizer are presented. Direct Numerical Simulation (DNS) allows one to describe accurately the evolution of the fully compressible gas-phase coupled with a Lagrangian description in order to describe two-phase flows. Standard coupling is used for the Eulerian/Lagrangian system while some practical issues related to the reactive source terms are addressed by suggesting a fast single-step Arrhenius law allowing one to capture the main fundamental properties of the flame whatever the local equivalence ratio. Then some basic procedures to describe spray preferential segregation in a turbulent reactor are described. Eventually spray combustion is addressed by first demonstrating the complex interactions caused by the presence of an evaporating liquid phase: definition of various equivalence ratios, apparition of flame instabilities for a unit Lewis number, etc. Then a history of the development of the existing spray combustion diagrams is presented to display the possible flame structures and combustion regimes encountered in spray combustion.

1 Introduction

It is generally admitted that the Navier-Stokes (NS) equations offer an accurate description of fluid motion. The basis for these equations is that the fluid under consideration is a continuum. Numerical resolution of the NS equations on a fine computational mesh allows one to capture all the macroscopic structures since all the considered length scales are considerably larger than the molecular length and time scales. Such NS resolution is defined as Direct Numerical Simulation (DNS). DNS solves all the characteristic scales of a turbulent flow from the Kolmogorov 'dissipative' length scale up to the integral 'energy-containing' length scale. However, if a two-phase flow is considered (gas/liquid for example), the apparition of an interface and a strong variation of density jeopardizes the possibility to achieve a complete DNS of the flow. It is especially true if fundamental physical phenomena, like evaporation or heat transfer, are present at the interface. Then the computational cost of the DNS of the whole flow, including both phases, would sky rocket unless some major assumptions were made.

A first possibility is to adopt an interface-tracking approach like the 'volume of fluid' (VOF) method developed by Hirt and Nichols (1981). It is based on the reconstruction of the gas/liquid interface from the time and space evolution of the local volume fraction of liquid. This mass-conservative procedure is complex and time consuming as far as the interface reconstruction is concerned. Another possibility is to use the level-set procedure of Osher and Sethian (1988), which follows the motion of an iso-surface of a specific scalar function that maintains algebraic distances. Even if these simulations are still designed as DNS because no turbulence model is used, from a strict point of view, the results are not an exact resolution of the complete NS equations and some approximations are often necessary. For example, an incompressible formulation is generally used. Then evaporation, heat transfer or even combustion phenomena are difficult to account for. Nevertheless, these methods are very promising as demonstrated very recently by Tanguy and Berlemont (2005) who simulated for the first time the complete atomization of a liquid jet.

A second possibility is to give up the idea of a complete DNS of the flow while trying to maintain highly accurate results. This objective seems rather difficult to reach as far as dense flows are concerned. On the other hand, when a dispersed liquid or solid phase is embedded in a gaseous carrier phase, some solutions exist. The principle is to carry out a DNS of the gaseous continuous phase and to model the dispersion of the liquid phase through a Lagrangian or an Eulerian formulation (Gouesbet and Berlemont, 1999). In the framework of DNS, where accurate results are more important than computational cost, Lagrangian modeling of the spray is preferable because every particle (or group of particles) is followed in space and time by the solver, whereas statistical integration of the information is obtained when a Eulerian model is used. It appears however that both procedures are complementary: Lagrangian formulations have to be used for the accurate description the dispersion of few (some millions per processor) particles while, on the other hand, an Eulerian formulation seems to be particularly adapted to complex dispersed or dense flows involving large-scale computations.

DNS was first introduced 35 years ago by Orszag and Patterson (1972) and then Rogallo (1981) and Lee et al. (1991) for the simulation of inert gaseous flows. It has since been used in a large range of applications. During the last two decades, DNS of reactive flows has been carried out to study non-premixed, partially premixed and premixed turbulent combustion of purely gaseous fluids (Givi, 1989; Poinsot et al., 1996; Vervisch and Poinsot, 1998; Poinsot and Veynante, 2001; Pantano et al., 2003). DNS has been extended to two-phase flows starting with the pioneering work of Riley and Patterson (1974). Most of the first numerical studies were dedicated to solid particle dispersion (see for instance Samimy and Lele, 1991; Squires and Eaton, 1991; Elgobashi and Truesdell, 1992; Wang and Maxey, 1993, and Ling et al., 1998). More recently, Mashayek et al. (1997), Reveillon et al. (1998) and Miller and Bellan (1999) conducted the first DNS with evaporating droplets in turbulent flows. Since then, DNS of two-phase flows have been extended to incorporate two-way coupling effects, multicomponent fuels, etc., and to deal with spray evaporation and combustion phenomena (Mashayek, 1998; Miller and Bellan, 1999, 2000; Reveillon and Vervisch, 2005).

The objective of this text is to offer a full description of the numerical procedures used to carry out DNS of two-phase dispersed flows. Note that a similar methodology may be used for large-eddy simulations if subgrid turbulence, mixing and dispersion models are added to the general balance equations. Starting from the classical NS equations, specific source terms are added to account for the presence of a dispersed liquid phase whose evolution is described by a Lagrangian solver. The modeling of the liquid phase embedded in a full DNS of the gas phase implies that the droplets are not resolved by the Eulerian solver. They are considered as local point sources of mass, momentum and energy. On the other hand, these source terms are obtained thanks to a fine-scale description of the evolution of droplets that are considered individually by the Lagrangian solver. Another major task concerns the chemical source terms describing chemical reactions between the various components. Two major solutions emerge: first, detailed chemistry may be used and leads to an accurate description of the combustion phenomena and the flame structures; however, it increases significantly the computational cost. The other possibility, often adopted in single-phase DNS as well, is to use a one-step Arrhenius law to describe the impact of combustion kinetics from a global point of view. In that case, two major conditions have to be respected: an accurate resolution of the basic features of the flame (propagation velocity, thickness, heat release) with respect to the local properties of the flow (equivalence ratio, stretch). However, basic single-step Arrhenius laws show major drawbacks especially when partially premixed combustion is concerned, which is mainly the case with turbulent two-phase flows. Thus, an adapted single-step kinetics has to be used.

In the following, two major parts are developed: first, details concerning the coupling of a complete DNS of a gaseous turbulent flow with a Lagrangian model of the dispersed phase are given. This association is especially useful to study fundamental physical phenomena and to carry out the preliminary development of two-phase flow models. Coupling with the dispersed spray is detailed as well as the specific Arrhenius law allowing one to capture properly the combustion phenomena whatever the local equivalence ratio. Then some applications and analysis procedures are briefly described to demonstrate the ability of DNS to capture every major phenomena present in two-phase flows: dispersion, evaporation and combustion.

2 Direct Numerical Simulation

DNS is a powerful way to study turbulent flows. Indeed, the NS equations are resolved with highly accurate and non-dissipative numerical methods that are able to capture all the time and length scales of the flow. No models are necessary to observe the development and the evolution of turbulent structures and all results may be considered to be as close as possible to reality (if the simulations are conducted properly). Occasionally, DNS is even referred to as a 'numerical experiment'. The grid mesh has to be fine enough to capture the smallest scales of the flow. This leads to severe conditions based on the turbulent Reynolds number and the Damk'öhler number, which is the ratio of the fluid motion time scale to the characteristic reaction time. A large Damköhler number indicates a rapid chemical reaction compared to all other processes. DNS is thus very limited from a technological point of view by the capacity of the current supercomputers. Only configurations with very small dimensions and small turbulent Reynolds numbers may be considered at the present time. Nevertheless, because of the accuracy of the results, it is a useful tool to analyze some specific physical phenomena.

One has to be careful as far as the 'DNS' term is concerned. Indeed in many works what is called DNS is a 'direct numerical resolution' of the considered equations without any model but not necessarily with adapted numerical methods and grids. This may lead to severe numerical dissipation and approximations, in which case speaking of DNS is abusive because the results are not a direct outcome of the equations for the physics. Indeed, an implicit numerical filtering exists in between the two. For a complete introduction to DNS of turbulent reactive single-phase flows and the resolved equations, readers may refer to the recent book of Poinsot and Veynante (2005).

Many formulations may be derived for the NS equations. The fully compressible equations allow for a complete description of the physics including acoustic phenomena. Apart from having a very small time step mainly based on the sound velocity, the main difficulties lie in a good formulation of the initial and boundary conditions that must account for entering and exiting acoustic waves. If acoustic phenomena are negligible, it is possible to select a low Mach number (LMN) formulation with either a constant- or variable-density flow. Then, boundary conditions are straightforward to prescribe. However, an elliptic solver is needed to close the momentum balance equation. For non-reactive flow or by using an implicit scheme for the temperature/species equations, the computational cost may be reduced by a factor of five to ten compared to the corresponding compressible formulation.

In the next section the complete compressible equations are given, and then a low Mach number formulation is derived. These sets of equations describe the evolution of the gas phase. They are rapidly introduced because they are widely documented in the CFD literature. Then a derivation of the Lagrangian solver to be resolved simultaneously with the DNS is proposed with a detailed description of the coupling between the carrier phase and the dispersed evaporating droplets. This part ends with a detailed description of a single-step Arrhenius law adapted to the full range of fuel and equivalence ratios.

2.1 Compressible formulation

The carrier phase is a compressible Newtonian fluid following the equation of state for an ideal gas. The instantaneous balance equations describe the evolution of mass ρ, momentum $\rho\mathbf{U}$, total (except chemical) energy E_t and species mass fraction. Y_F denotes the mass fraction of gaseous fuel resulting from spray evaporation and Y_O is the oxidizer mass fraction. The following set of balance equations are solved where the usual notation is adopted:

$$\frac{\partial \rho}{\partial t} + \frac{\partial \rho U_k}{\partial x_k} = \dot{m} , \tag{2.1}$$

$$\frac{\partial \rho U_i}{\partial t} + \frac{\partial \rho U_i U_k}{\partial x_k} = -\frac{\partial P}{\partial x_i} + \frac{\partial \sigma_{ik}}{\partial x_k} + \dot{v}_i , \tag{2.2}$$

$$\frac{\partial \rho E_t}{\partial t} + \frac{\partial (\rho E_t + P) U_k}{\partial x_k} = \frac{\partial}{\partial x_k}\left(\lambda \frac{\partial T}{\partial x_k}\right) + \frac{\partial \sigma_{ik} U_k}{\partial x_i} + \rho\dot{\omega}_e + \dot{e} , \tag{2.3}$$

$$\frac{\partial \rho Y_F}{\partial t} + \frac{\partial \rho Y_F U_k}{\partial x_k} = \frac{\partial}{\partial x_k}\left(\rho D \frac{\partial Y_F}{\partial x_k}\right) + \rho\dot{\omega}_F + \dot{m} , \tag{2.4}$$

$$\frac{\partial \rho Y_O}{\partial t} + \frac{\partial \rho Y_O U_k}{\partial x_k} = \frac{\partial}{\partial x_k}\left(\rho D \frac{\partial Y_O}{\partial x_k}\right) + \rho\dot{\omega}_O , \tag{2.5}$$

with

$$\sigma_{ij} = \mu\left(\frac{\partial U_i}{\partial x_j} + \frac{\partial U_j}{\partial x_i}\right) - \frac{2}{3}\mu\frac{\partial U_k}{\partial x_k}\delta_{ij} ,$$

together with the equation of state for ideal gases:

$$P = \rho r T .$$

Source terms are present, the $\dot{\omega}_i$ terms are related to the chemical reaction processes and $\dot{m}$, $\dot{\mathbf{v}}$ and $\dot{e}$ result from a two-way coupling between the carrier phase and the spray. These terms will be discussed in detail later.

The sixth-order Pade scheme from Lele (1992) and the Navier-Stokes characteristics boundary conditions (NSCBC) of Poinsot and Lele (1992) or Baum et al. (1994) are usually employed to solve the gas-phase transport equations on a regular mesh. The time integration of both the spray and gas-phase equations is done with a third-order explicit Runge-Kutta scheme with a minimal data storage method (Wray, 1990). A third-order interpolation is employed when gas-phase properties are needed at the droplet positions.

2.2 Low Mach number approximation

The fully compressible set of equations presented above may be normalized by reference physical quantities. Among them the reference velocity will be defined by $u_0 = \mathrm{M}\sqrt{\gamma r T_0}$ where M is the Mach number. The normalized compressible NS equations may be written as

$$\frac{\partial \rho}{\partial t} + \frac{\partial \rho U_i}{\partial x_i} = 0\,, \tag{2.6}$$

$$\frac{\partial \rho U_i}{\partial t} + \frac{\partial \rho U_i U_j}{\partial x_j} = -\frac{1}{\gamma \mathrm{M}^2}\frac{\partial P}{\partial x_i} + \frac{\partial \sigma_{ij}}{\partial x_j}\,, \tag{2.7}$$

$$\frac{\partial \rho E_{\mathrm{I}}}{\partial t} + \frac{\partial \rho E_{\mathrm{I}} U_i}{\partial x_i} = -\frac{(\gamma-1)}{\gamma}P\frac{\partial U_i}{\partial x_i} + \frac{\partial}{\partial x_i}\left(\lambda\frac{\partial T}{\partial x_i}\right) + \mathrm{M}^2(\gamma-1)\sigma_{ij}\frac{\partial U_j}{\partial x_i}\,, \tag{2.8}$$

$$\frac{\partial \rho Y_{\mathrm{F}}}{\partial t} + \frac{\partial \rho Y_{\mathrm{F}} U_i}{\partial x_i} = \frac{\partial}{\partial x_i}\left(\rho D\frac{\partial Y_{\mathrm{F}}}{\partial x_i}\right)\,. \tag{2.9}$$

The various source terms have been dropped in this intermediary expression and the internal energy ($E_{\mathrm{I}} = E_{\mathrm{t}} - U^2/2$) has been selected to simplify the low Mach number system.

A new variable $\epsilon = \gamma\mathrm{M}^2$ is introduced. If the low Mach number hypothesis is adopted, $\epsilon \lll 1$ and the transport variables may be developed in truncated power series such as $A = A^{(0)} + \epsilon A^{(1)}$. If, as a first approach, only the zero-order terms are conserved when developing the NS equations, the following relations are obtained (by letting $A^{(0)} = A$ for all variables except for pressure):

$$\frac{\partial \rho}{\partial t} + \frac{\partial \rho U_i}{\partial x_i} = \dot{m}\,, \tag{2.10}$$

$$\frac{\partial P^{(0)}}{\partial x_i} = 0\,, \tag{2.11}$$

$$\frac{\partial \rho E_{\mathrm{I}}}{\partial t} + \frac{\partial \rho E_{\mathrm{I}} U_i}{\partial x_i} = -P^{(0)}\frac{(\gamma-1)}{\gamma}\frac{\partial U_i}{\partial x_i} + \frac{\partial}{\partial x_i}\left(\lambda\frac{\partial T}{\partial x_i}\right)\rho\dot{\omega}_e + \dot{e}\,, \tag{2.12}$$

$$\frac{\partial \rho Y_{\mathrm{F}}}{\partial t} + \frac{\partial \rho Y_{\mathrm{F}} U_i}{\partial x_i} = \frac{\partial}{\partial x_i}\left(\rho D\frac{\partial Y_{\mathrm{F}}}{\partial x_i}\right) + \rho\dot{\omega}_{\mathrm{F}} + \dot{m}\,, \tag{2.13}$$

and the equation of state gives $P^{(0)} = \rho T$.

The term $P^{(0)}$ is the thermodynamic pressure, which is constant in space in our newly simplified system (Eq. 2.11). To describe a complete system, the description of the fluid momentum needs to be derived from the momentum equation (Eq. 2.7) developed up to first order.

$$\frac{\partial \rho U_i}{\partial t} + \frac{\partial \rho U_i U_j}{\partial x_j} = -\frac{\partial P^{(1)}}{\partial x_i} + \frac{\partial \sigma_{ij}}{\partial x_j} + \dot{v}_i\,. \tag{2.14}$$

The term $P^{(1)}$ is the dynamic pressure linked to the fluid motion. It does not participate in the thermodynamical processes. Energy and momentum have thus been decoupled.

Note that if an open system is considered the pressure $P^{(0)}$ is equal to the external pressure. On the other hand, if the system is closed (periodical or adiabatic boundaries) it is possible to show that

$$\frac{\partial P^{(0)}}{\partial t} = \frac{1}{\mathcal{V}} \int\int\int \rho \gamma \dot{\omega}_e \, \mathrm{d}v \,, \tag{2.15}$$

where the integral is over the volume of the domain. Spatial derivative and temporal integrations schemes similar to the ones used for the compressible formulation may be used. Boundary conditions are straightforward: a given value or gradient is prescribed on all the domain boundaries for the considered variables. The major stumbling block of the LMN formulation concerns the determination of the first-order pressure term $P^{(1)}$. If a derivative operator is applied to Eq. (2.14), a Poisson equation appears with $P^{(1)}$ as the unknown. By using an adapted elliptic solver, the pressure gradient may be determined and Eq. (2.14) is closed. For more details about the low Mach number formulation for reacting single- and two-phase flows, readers may consult the works of McMurtry et al. (1986) or Wang and Rutland (2005) and references therein.

3 Dispersed-Phase Lagrangian Description

As described by Reeks (1991), is it possible to account for many forces to characterize the droplet dynamics. However, the purpose of this text is to present a basis to carry out DNS of two-phase flows. Because of the high density ratio between the liquid and gas phases, only the drag force, which is prevalent, has been selected. Moreover, several usual assumptions have to be formulated. First, the spray is dispersed and each droplet is unaware of the existence of the others. Any internal liquid circulation or droplet rotations are neglected and an infinite heat conduction coefficient is assumed. Therefore, the liquid core temperature remains uniform in every droplet, although it may vary as a function of time. The spray is then composed of local sources of mass following the saturation law and modifying the momentum and gaseous fuel topology, depending on the local temperature, pressure and vapor mass fraction.

3.1 Position and velocity

By letting $\mathbf{V}_k$ and $\mathbf{X}_k$ denote the velocity and position vectors of droplet k, the relations

$$\frac{\mathrm{d}\mathbf{V}_k}{\mathrm{d}t} = \frac{1}{\beta_k^{(\mathrm{V})}} \left(\mathbf{U}\left(\mathbf{X}_k, t\right) - \mathbf{V}_k \right) \,, \tag{3.1}$$

$$\frac{\mathrm{d}\mathbf{X}_k}{\mathrm{d}t} = \mathbf{V}_k \,, \tag{3.2}$$

are used to track their evolution throughout the computational domain. The vector $\mathbf{U}\left(\mathbf{X}_k, t\right)$ represents the gas velocity at the droplet position $\mathbf{X}_k$. The right-hand side of Eq. (3.1) represents a drag force applied to the droplet and $\beta_k^{(\mathrm{V})}$ is a kinetic relaxation time. It may be obtained from the k droplet dynamics:

$$m \frac{\mathrm{d}\mathbf{V}_k}{\mathrm{d}t} = \mathbf{D}_k \,, \tag{3.3}$$

where $\mathbf{D}$ is the drag force applied to a sphere. A summation of all the forces on the droplet surface gives

$$\mathbf{D}_k = 3\pi a_k \mu C_k \left(\mathbf{U}\left(\mathbf{X}_k, t\right) - \mathbf{V}_k \right) \,, \tag{3.4}$$

where a_k is the diameter of droplet k. A corrective coefficient $C_k = 1 + \mathrm{Re}_k^{2/3}/6$ (Crowe et al., 1998) is introduced to allow for the variation of the drag factor according to the value of the droplet Reynolds number $\mathrm{Re}_k = \rho|\mathbf{U}(\mathbf{X}_k, t) - \mathbf{V}_k|a_k/\mu$. Thus from Eqs. (3.3) and (3.4) and with $m_k = \rho_\mathrm{d}\pi a_k^3/6$, the following relation is obtained:

$$\frac{\mathrm{d}\mathbf{V}_k}{\mathrm{d}t} = \frac{(\mathbf{U}(\mathbf{X}_k, t) - \mathbf{V}_k)}{a_k^2}\frac{18\mu C_k}{\rho_\mathrm{d}}, \tag{3.5}$$

and leads directly to the kinetic relaxation time of droplet k:

$$\beta_k^{(\mathrm{V})} = \frac{\rho_\mathrm{d} a_k^2}{18 C_k \mu}. \tag{3.6}$$

3.2 Heating and evaporation

The heating and evaporation of each droplet in the flow may be described through a normalized quantity B_k, called the "mass-transfer number". B_k is the normalized flux of gaseous fuel between the droplet surface, where the fuel mass fraction takes the value Y_k^s, and the surrounding gas at the droplet position, where the fuel mass fraction is $Y_\mathrm{F}(\mathbf{X}_k)$. It may be written

$$B_k = \frac{Y_k^\mathrm{s} - Y_\mathrm{F}(\mathbf{X}_k)}{1 - Y_k^\mathrm{s}}. \tag{3.7}$$

By solving the mass and energy balances at the surface of a vaporizing droplet in a quiescent atmosphere (Kuo, 1986), the following relations for the surface and temperature evolution of the kth droplet are found:

$$\frac{\mathrm{d}a_k^2}{\mathrm{d}t} = -\frac{a_k^2}{\beta_k^{(\mathrm{a})}}, \tag{3.8}$$

$$\frac{\mathrm{d}T_k}{\mathrm{d}t} = \frac{1}{\beta_k^{(\mathrm{T})}}\left(T(\mathbf{X}_\mathrm{d}) - T_k - \frac{B_k L_\mathrm{v}}{C_\mathrm{p}}\right). \tag{3.9}$$

Again, characteristic relaxation times appear and are defined by

$$\beta_k^{(\mathrm{a})} = \frac{\mathrm{Sc}}{4\mathrm{Sh_c}}\frac{\rho_\mathrm{d}}{\mu}\frac{a_k^2}{\ln(1 + B_k)}, \tag{3.10}$$

$$\beta_k^{(\mathrm{T})} = \frac{\mathrm{Pr}}{6\mathrm{Nu_c}}\frac{C_\mathrm{d}}{C_\mathrm{p}}\frac{\rho_\mathrm{d} a_k^2}{\mu}\frac{B_k}{\ln(1 + B_k)}. \tag{3.11}$$

Normalized gas and liquid heat capacities are denoted by C_p and C_d, respectively, and as L_v is the latent heat of evaporation; all of which are considered as constant in the present set of equations. Sc and Pr are the Schmidt and Prandtl numbers, respectively. $\mathrm{Sh_c}$ and $\mathrm{Nu_c}$ are the convective Sherwood and Nusselt numbers, respectively. They are both equal to 2 in a quiescent atmosphere, but a correction has to be applied in a convective environment. In this context, the empirical expression of Faeth and Fendell (Kuo, 1986), identical either for $\mathrm{Sh_c}$ or $\mathrm{Nu_c}$ ($\mathrm{Sh_c} \mid \mathrm{Nu_c}$), may be used:

$$(\mathrm{Sh_c} \mid \mathrm{Nu_c})_k = 2 + \frac{0.55\mathrm{Re}_k(\mathrm{Sc} \mid \mathrm{Pr})_k}{\left(1.232 + \mathrm{Re}_k(\mathrm{Sc} \mid \mathrm{Pr})_k^{4/3}\right)^{1/2}}. \tag{3.12}$$

One of the most accurate models to describe the evaporation process is to consider a phase equilibrium at the interface thanks to the Clausius-Clapeyron relation:

$$\frac{\mathrm{d}\ln(P_k^{\mathrm{s}})}{\mathrm{d}T} \approx \frac{L_{\mathrm{v}}}{r_{\mathrm{F}}T^2}, \tag{3.13}$$

where r_{F} is the ideal gas constant in the gaseous fuel. This leads to the following expression for the partial pressure P_k^{s} of fuel vapor at the surface of a droplet:

$$P_k^{\mathrm{s}} = P_{\mathrm{ref}} \exp\left(-\frac{L_{\mathrm{v}}}{r_{\mathrm{F}}}\left(\frac{1}{T_k^{\mathrm{s}}} - \frac{1}{T_{\mathrm{ref}}}\right)\right), \tag{3.14}$$

where P_{ref} and T_{ref} are two reference parameters. A fuel boiling temperature T_{ref} corresponding to a reference pressure P_{ref} has been introduced. T_k^{s} is the gas temperature at the droplet surface. The liquid temperature is uniform in all droplets. Thus, it is equal to the temperature of the gas at the interface $T_k = T_k^{\mathrm{s}}$.

The gaseous fuel mass fraction at the surface of the droplet may be determined from

$$Y_k^{\mathrm{s}} = \left[1 + \frac{W_{\mathrm{O}}}{W_{\mathrm{F}}}\left(\frac{P(\mathbf{X}_{\mathrm{d}})}{P_k^{\mathrm{s}}} - 1\right)\right]^{-1}, \tag{3.15}$$

where W_{O} and W_{F} are the molecular weights of the considered oxidizer and fuel, respectively. Once the gaseous fuel mixture fraction at the droplet surface is known, the mass-transfer number B_k is determined by introducing relation (3.15) into Eq. (3.7). Consequently, Eqs. (3.8) and (3.9) describing the evolution of droplet surface are and temperature are closed.

4 Eulerian/Lagrangian Coupling

The terms $\dot{m}$, $\dot{\mathbf{v}}$ and $\dot{e}$ modify the gas-phase mass, momentum and temperature owing to a distribution of the Lagrangian quantities on the Eulerian grid. Every droplet has positive or negative source terms to be distributed over the Eulerian nodes and the specification of an accurate projection of these Lagrangian sources onto the Eulerian mesh is not an easy task. In real spray flows, this distribution is not instantaneous and further assumptions are needed to perform simulations. Every Lagrangian source has to be distributed over the Eulerian nodes by adding the volumetric contributions from droplets. This induces a numerical dispersion that remains weak because of the small size of the DNS grid (Reveillon and Vervisch, 2000). This procedure is not really satisfactory but this stumbling block remains an open problem at the present time.

For every Eulerian node, a control volume $\mathcal{V}$ is defined by the mid-distance between the neighboring nodes. If an isotropic Cartesian grid is considered ($\Delta = x_{i+1} - x_i = \Delta_x = \Delta_y = \Delta_z$), then $\mathcal{V} = \Delta^3$. The mass source term applied to an Eulerian node n is denoted $\dot{w}_{\mathrm{v}}^{(n)}$:

$$\dot{w}_{\mathrm{v}}^{(n)} = -\frac{1}{\mathcal{V}} \sum_k \alpha_k^{(n)} \frac{\mathrm{d}m_k}{\mathrm{d}t}, \tag{4.1}$$

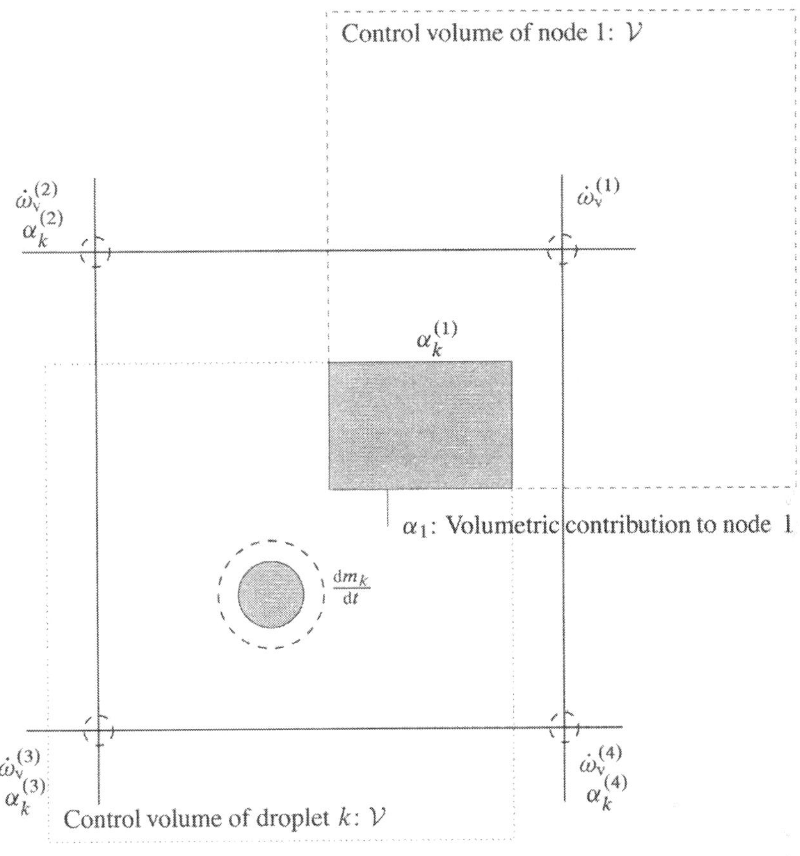

Figure 1. 2-D Sketch of the repartition of droplet source terms onto the closest Eulerian node. A 3-D repartition must be done in the simulations.

where $\sum_k$ is the sum over all droplets affecting node n. $\alpha_k^{(n)}$ is the distribution coefficient of the kth droplet source term on node n. Considering all the nodes affected by droplet k, it is necessary to have $\sum_n \alpha_k^{(n)} = 1$ to conserve mass, momentum and energy during the Lagrangian/Eulerian coupling. In fact, $\alpha_k^{(n)}$ is the portion of the control volume of node n intersecting the control volume of droplet k (see Figure 1):

$$\alpha_k^{(n)} = \frac{1}{\mathcal{V}} \prod_{i=1}^{3} \left(\Delta - \mid x_i^{(n)} - x_{ki} \mid \right), \tag{4.2}$$

where $x_i^{(n)}$ and x_{ki} are the coordinates along the ith direction of node n and droplet k, respectively.

The mass of the considered droplet k in the neighborhood of node n is $m_k = \rho_{\mathrm{d}} \pi a_k^3/6$ and using Eqs. (3.8) and (4.1) one may write

$$\dot{w}_{\mathrm{v}}^{(n)} = \rho_{\mathrm{d}} \frac{\pi}{4} \frac{1}{\mathcal{V}} \sum_k \alpha_k^{(n)} a_k^3 / \beta_k^{(\mathrm{a})} . \tag{4.3}$$

Similarly, the following relation:

$$\dot{\mathbf{v}}^{(n)} = -\frac{1}{\mathcal{V}} \sum_k \alpha_k^{(n)} \frac{\mathrm{d} m_k \mathbf{V}_k}{\mathrm{d}t} , \tag{4.4}$$

leads to the expression for the momentum source term:

$$\dot{\mathbf{v}}^{(n)} = -\rho_{\mathrm{d}} \frac{\pi}{4} \frac{1}{\mathcal{V}} \sum_k \alpha_k^{(n)} a_k^3 \left(\frac{2}{3} \frac{\mathbf{U}(\mathbf{X}_{\mathrm{d}}, t) - \mathbf{V}_k}{\beta_k^{(\mathrm{V})}} - \frac{\mathbf{V}_k}{\beta_k^{(\mathrm{a})}} \right) . \tag{4.5}$$

The energy variation of the gaseous flow induced by the droplets inside volume $\mathcal{V}$ may be written as

$$\dot{e}^{(n)} = -\frac{1}{\mathcal{V}} \sum_k \alpha_k^{(n)} \frac{\mathrm{d} m_k C_{\mathrm{d}} T_k}{\mathrm{d}t} , \tag{4.6}$$

and it can be rewritten as

$$\dot{e}^{(n)} = -C_{\mathrm{d}} \rho_{\mathrm{d}} \frac{\pi}{4} \frac{1}{\mathcal{V}} \sum_k \alpha_k^{(n)} a_k^3 \left(\frac{2}{3} \frac{T(\mathbf{X}_{\mathrm{d}}, t) - T_k - B_k L_{\mathrm{v}} / C_{\mathrm{p}}}{\beta_k^{(\mathrm{T})}} + \frac{T_k}{\beta_k^{(\mathrm{a})}} \right) . \tag{4.7}$$

This last equation details how energy fluxes reaching the droplet surface are distributed (evaporation and liquid core heating) and the loss of energy due to the loss of liquid mass.

5 Reaction Rates

First it may be useful to sketch the usual combustion regimes. Initially segregated gaseous fuel and oxidizer streams are injected into the combustion chamber. Two possible combustion scenarios are shown in Figure 2. In the first case, reaction takes place before any mixing between fuel and oxidizer occurs. Then a diffusion flame emerges piloted by a triple flame. Otherwise if fuel and oxidizer are mixed before ignition takes place, a premixed flame propagates in the domain. These premixed and diffusion flames have very different structures and behaviors, but they are characterized through the same chemical law. Indeed, it is possible to consider their common origin in the chemical kinetics resulting from the presence in the same region of molecules of each species.

As mentioned in the Introduction, detailed chemistry may be used to determine the chemical source terms $\dot{\omega}_e$ and $\dot{\omega}_Y$ appearing in the energy and species balance equations, respectively. The Eulerian/Lagrangian procedure may be coupled with a solver dedicated to the computation of detailed chemistry. An open-source package like Cantera from Caltech (Goodwin, 2003) is an ideal candidate. However detailed kinetics are time consuming. Added to the fact that the

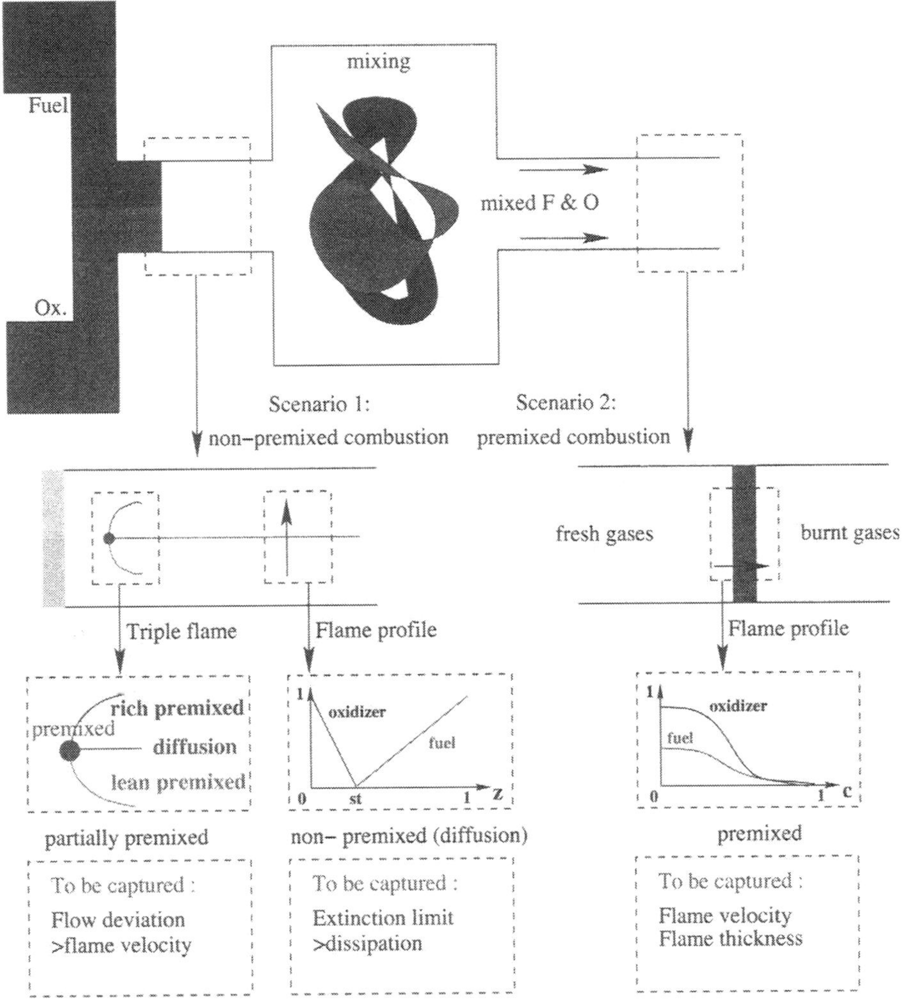

Figure 2. Combustion regimes and basic properties that must be captured by any chemical kinetic scheme.

evolution of millions of droplets has to be computed as well, the association DNS/Lagrangian solver/detailed kinetics is overly expensive in most cases.

Detailed chemical kinetics are not inevitably required in combustion studies. Very often one needs to estimate only the global impact of the presence of a flame within a flow without knowing the reaction details. It is within this framework that a global reaction scheme can be used. In this case, chemical kinetics are reduced to a unique irreversible single-step reaction:

$$\text{Fuel} + \text{Oxidizer} \longrightarrow \text{Products}\,. \tag{5.1}$$

If the reaction rate resulting from this reaction captures the basic properties of the real flames, then the global effects of the combustion on the gas flow will be taken into account. It is possible to briefly enumerate some of these fundamental properties:

- **Combustion heat release** to account for gas dilatation in the flow motion and to determine the correct output temperature,
- **Flame velocity** to capture the burning rate corresponding to the local equivalence ratio,
- **Flame thickness** to estimate correctly the scale of the vortices able to cross the flame front in combustion/turbulence studies and thus the flame wrinkling and turbulent velocity,
- **Extinction limit** to establish local extinction of the flame because of the local dissipation rate,
- **Ignition delay/lift-off height** to capture the flame ignition delay or the distance between the injector and the flame. Single-step chemistry can not capture ignition delays, which are directly linked to the rate of creation of intermediate species.

The most basic model for global kinetics concerns non-premixed combustion. An infinitely fast reaction between fuel and oxidizer occurs because both of them react instantaneously as soon as they meet and the flame is then controlled by the mixing of the flow. Only the heat release (first property) is captured by this model that considers only very high Damköhler numbers. However, even if its global properties and behavior are wrong, this instantaneous and simple method may be useful as a development procedure to test the ability of numerical and physical models to cope with combustion heat release and gas dilatation.

On the other hand, the Arrhenius law, which is commonly used to describe global kinetics, proved to be able to capture most of the fundamental properties. However, as it will be developed later, problems appear as far as realistic stoichiometric coefficients and non-stoichiometric (or partially premixed) combustion are concerned. To overcome this issue, which is fundamental for two-phase flows, a pre-exponential correction for the Arrhenius law is proposed in the following, which takes into account a significant number of the fundamental properties and phenomena resulting from the combustion. This method may be used for any stoichiometric ratio between the fuel and the oxidizer without the usual numerical and physical problems met with standard methods. Moreover, it allows for a proper computation of the velocity and the thickness of the flame, the heat release and the extinction limit whatever the local or global equivalence ratio. The method is called GKAS (Global Kinetics, Any Stoichiometry) in this work.

The evolution of a chemical system is usually described by the quantities of fuel and oxidizer present in the domain. It is possible to add to this conventional representation a new phase space based on the intensity and the progress of any reaction. The definition of the conventional parameters (mixture fraction, reaction rate, etc.) will help us to introduce a new set of parameters allowing us to modify the Arrhenius law.

5.1 Definitions

As it has been mentioned above, it is possible to study two initially segregated gaseous flows injected into the same domain (see Figure 2). It is possible to consider diluted fuel and oxidizer; however, for the sake of clarity, we will not do so here. The first flow consists of a pure fuel stream such that $Y_F = Y_{F,0} = 1$. The second flow consists of an oxidizer stream such that $Y_O = Y_{O,0} = 1$ before injection into the combustion chamber.

Reaction between fuel and oxidizer is then described by single-step kinetics:

$$\nu_F Y_F + \nu_O Y_O \longrightarrow \nu_P Y_P . \tag{5.2}$$

where Y_F, Y_O and Y_P are the mass fraction of fuel, oxidizer and products, respectively. Their corresponding stoichiometric molar coefficients are ν_F, ν_O and ν_P. The reaction rate is denoted by $\dot{\omega}_r$ and the evolution of mass fractions and energy is described by the following expressions:

$$\rho\dot{\omega}_F = -\nu_F W_F \dot{\omega}_r\ , \ \rho\dot{\omega}_O = -\nu_O W_O \dot{\omega}_r\ , \ \rho\dot{\omega}_e = q_0 \nu_F W_F \dot{\omega}_r\ , \tag{5.3}$$

where W_F and W_O are the molar weights of the fuel and oxidizer, respectively, and q_0 is the energy released per unit of mass of fuel. Defining the mass stoichiometric coefficient $s = \nu_O W_O / \nu_F W_F$ allows us to introduce the Schwab-Zeldovitch variable $\varphi = Y_F - Y_O/s$, which may be normalized by its extremum values in the pure fuel and oxidizer streams. The mixture fraction Z is then derived from the normalization of φ, such that $Z = 0$ in the oxidizer stream and $Z = 1$ in the fuel stream. Thus the mixture fraction expression is

$$Z = \frac{sY_F - Y_O + 1}{1+s}\ , \tag{5.4}$$

and allows us to define the local equivalence ratio Φ in fresh gases by

$$\Phi = s\frac{sY_F - Y_O + 1}{s\,(1 - Y_F) + Y_O}\ . \tag{5.5}$$

5.2 Arrhenius law

Many analytical and numerical works have been dedicated to the determination of an accurate law for the reaction rate. A detailed description of this law may be found in Poinsot and Veynante (2005) or Williams (1985) and references therein. Among them, the Arrhenius law is based on the product of an exponential expression and a pre-exponential function:

$$\dot{\omega}_r = K\rho^{n_\rho} {Y_F}^{n_F} {Y_O}^{n_O} \exp\left(-T_a/T\right)\ , \tag{5.6}$$

with K a constant pre-exponential factor. It is possible to express the exponential term in a more convenient form by introducing α, the heat-release coefficient, and β, the coefficient allowing us to measure the activation temperature T_a. By defining T_u the unburnt gas temperature and T_b the burnt gases temperature of a stoichiometric mixture, α and β are expressed as

$$\alpha = \frac{T_b - T_u}{T_b} \quad \text{and} \quad \beta = \alpha\frac{T_a}{T_b}\ . \tag{5.7}$$

For a dimensionless expression of temperature that has been divided by the fresh gas temperature T_u, the Arrhenius law may be expressed as a function of α and β:

$$\dot{\omega}_r = K\rho^{n_\rho} {Y_F}^{n_F} {Y_O}^{n_O} \exp\left(\frac{\beta}{\alpha} - \frac{\beta}{\alpha(1-\alpha)T}\right)\ , \tag{5.8}$$

where the term $\exp(-\beta/\alpha)$ has been included in the K constant to balance the orders of magnitude. The reaction exponents n_F and n_O are not necessarily equal to ν_F and ν_O, and n_ρ is set to $n_\rho = n_F + n_O$. Equation (5.8) allows us to model the kinetics of a single-step reaction. It is widely used to determine asymptotically or numerically the reaction rate. However a lot of physical or numerical restrictions may appear. As summarized by Poinsot and Veynante (2005), with reaction exponents n_F and n_O of the order of unity, asymptotic and numerical estimations of the flame speed as a function of the equivalence ratio provide good estimates on the lean side ($\Phi < 1$), but fail on the rich side ($\Phi > 1$). In fact, a constant growth of the velocity may be observed even for equivalence ratios greater than unity, whereas it should reach a maximum value in the vicinity of $\Phi = 1$. This problem is linked to the pre-exponential function defined by

$$P_e(Y_F, Y_O, n_F, n_O) = Y_F{}^{n_F} Y_O{}^{n_O} . \tag{5.9}$$

By looking for the maximum value of the pre-exponential function P_e for a varying equivalence ratio ($\partial P_e/\partial \Phi = 0$), it is possible to show that the maximum value of the pre-exponential function P_e for a stoichiometric mixture $\Phi = 1$, is obtained only for the following ratio:

$$\frac{n_O}{n_F} = s . \tag{5.10}$$

For combustion of classical fuels (methane, propane, any hydrocarbon, etc.) with air, the stoichiometric coefficient s is greater than 10 and often close to 15. In this last case, exponents between 0.0625 and 15 appear if the standard pre-exponential function is utilized. For example, if $n_F = 0.0626$ and $n_O = 0.9375$, numerical noise may be burnt ($(1.10^{-14})^{0.0625} = 0.13$) and numerical difficulties exist as soon as high values of n_O/n_F are prescribed. Therefore using a realistic value for s is not possible. Otherwise, if other exponents are used, the bell-shape curve (BSC) will not be maximum for a stoichiometric mixture.

For analytical studies based on heat release analysis, this problem may be circumvented by working with a non-realistic fuel such as $s = 1$. In this case, n_O/n_F is equal to unity, and a maximum value of the flame velocity is found for $\Phi = 1$. However, even in this simplified case, another problem remains. Experiments show that the flame speed as a function of the equivalence ratio is almost symmetrical in the vicinity of its maximum value. But the numerical estimation of this curve shows a substantial asymmetry. This is due to the fact that the amount of burnt fuel as a function of the equivalence ratio is not symmetrical around $\Phi = 1$. Therefore, the heat release and flame velocity are more sustained on the rich side.

The new method will have to override two major stumbling blocks: first, the flame velocity as a function of the equivalence ratio has to be maximum in the vicinity of $\Phi = 1$, for all values of s. Second, the (BSC) has to be symmetrical around stoichiometry.

5.3 GKAS procedure

The GKAS procedure may be divided into two parts:

1. **Remapping**: The first stage consists of modifying the expression for the pre-exponential factor so that the new function has a maximum value at $\Phi = 1$ whatever the value of s. This new pre-exponential factor will induce 'naturally' a correct global shape for the BSC.
2. **Scaling**: A correction factor depending on the mixture fraction Φ is added to the Arrhenius law so that the numerical BSC matches the experimental one.

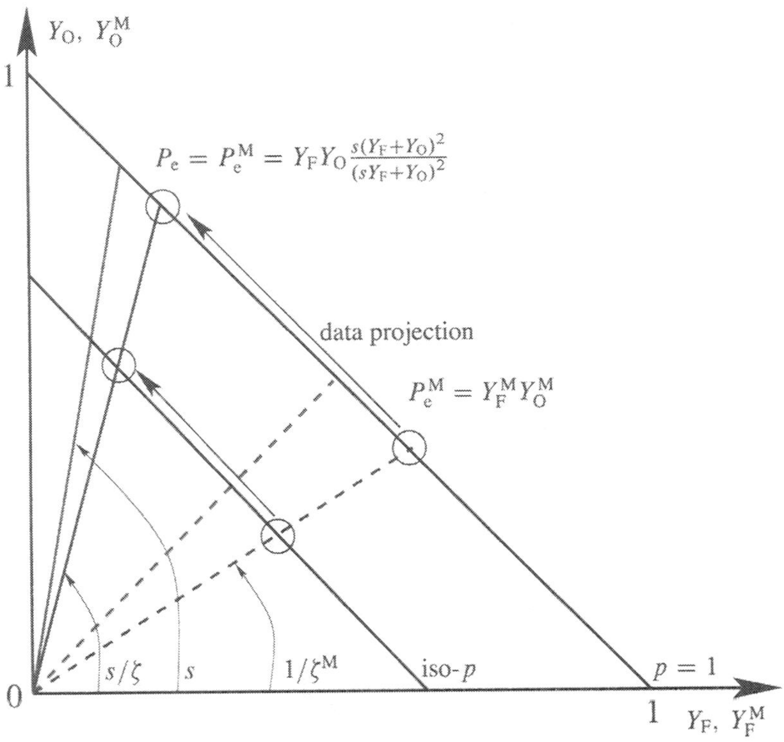

Figure 3. Sketch of the projection parameters for the remapping procedure. ζ is the slope factor and data are sliding along the iso-p line from the $s = 1$ pre-exponential function to be mapped onto any s pre-exponential function.

Remapping As mentioned in the previous section, working with a non-realistic pre-exponential function such as $s = 1$ leads to a ratio $n_O/n_F = 1$ and then

$$P_e(Y_F, Y_O) = Y_F Y_O \,. \tag{5.11}$$

This pre-exponential function is numerically stable and quick. However our objective is to work with higher values of s, which leads to unstable expressions for P_e.

As sketched in Figure 3 the general procedure consists of projecting the values of the stable pre-exponential function (Eq. 5.11) defined for $s = 1$ onto any s pre-exponential function along constant p lines such as $p = Y_F + Y_O$. This procedure allows one to impose the correct shape of the flame velocity curve with respect to the equivalence ratio, while all numerical difficulties linked to the exponents have been avoided. Thus the flame velocity will be at its maximum value when $\Phi = 1$. The mapped pre-exponential function P_e^M is written:

$$P_e^M = Y_F^M Y_O^M \,, \tag{5.12}$$

with Y_F^M and Y_O^M being the 'mapped' mass fractions of fuel and oxidizer, respectively.

To find the transformation function between standard (Y_F, Y_O) and mapped (Y_F^M, Y_O^M) mass fractions, a change of coordinates has to be carried out. First, any line $\mathcal{L}(Y_F, Y_O)$ starting at the origin of the species space $(Y_F = Y_O = 0)$ and including any given (Y_F, Y_O) coordinates may be defined. The slope of this line is defined by the ratio Y_O/Y_F, but it is possible to define the slope ratio ζ such that

$$\frac{Y_O}{Y_F} = \frac{s}{\zeta} . \tag{5.13}$$

The reference stoichiometric line is defined by the slope s. Thus ζ is the slope ratio between the reference line and any $\mathcal{L}(Y_F, Y_O)$ line, and it indicates the 'distance' from stoichiometry of any mixture (Y_F, Y_O).

The remapping procedure must preserve this factor, therefore $\zeta = \zeta^M$ with $\zeta^M = Y_F^M/Y_O^M$. Hence a first relation defining the change of coordinates may be written:

$$\frac{Y_F^M}{Y_O^M} = s\frac{Y_F}{Y_O} . \tag{5.14}$$

A second relation is necessary to define the projection. Indeed, the line describing the relative distance of the mixture from stoichiometry is known, but the position on this line has to be defined. It may be done through the distance from the origin represented by the p parameter, which has to remain constant in both coordinate systems. Then

$$Y_F^M + Y_O^M = Y_F + Y_O \tag{5.15}$$

is the second relation allowing us to define the two mapped mass fractions Y_F^M and Y_O^M as functions of Y_F, Y_O and s:

$$Y_F^M = Y_F\frac{s\,(Y_F + Y_O)}{sY_F + Y_O} , \tag{5.16}$$

$$Y_O^M = Y_O\frac{(Y_F + Y_O)}{sY_F + Y_O} . \tag{5.17}$$

The new pre-exponential function may then be written as

$$P_e^M = Y_F Y_O\frac{s\,(Y_F + Y_O)^2}{(sY_F + Y_O)^2} . \tag{5.18}$$

Now a scaling factor $K(\Phi)$ has to be added to match any experimental or detailed numerical bell-shape curve (BSC) representing the laminar flame velocity with respect to the equivalence ratio.

Scaling The modified Arrhenius law, adapted to any realistic stoichiometry, may be written as

$$\dot{\omega}_r = K(\Phi)\rho Y_F Y_O\frac{s\,(Y_F + Y_O)^2}{(sY_F + Y_O)^2}\exp\left(\frac{\beta}{\alpha} - \frac{\beta}{\alpha(1-\alpha)T}\right) , \tag{5.19}$$

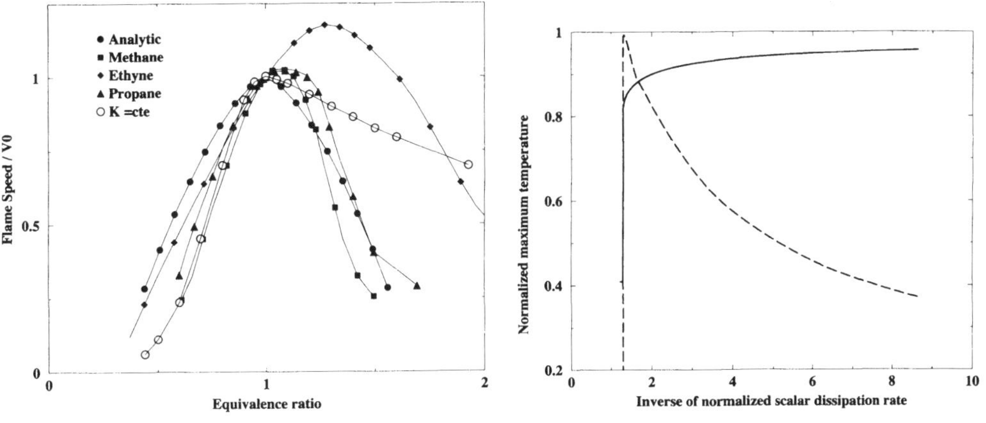

(a) Flame velocity versus equivalence ratio (BSC). Symbols: Experimental results. Lines: GKAS procedure.

(b) Propane extinction limit. Temperature and reaction rates have been normalized by their maximum value along the profile. Scalar dissipation rate is normalized by its asymptotic extinction limit.

Figure 4. Characteristics of flames found from the KGKAS procedure.

where $K(\Phi)$ is a pre-exponential coefficient, which allows one to tune the BSC to a prescribed shape. A first approximation would be to use a constant pre-exponential coefficient $K = K_0$. But even if a maximum value is found when the equivalence ratio Φ equals unity thanks to the remapping procedure, the response velocity of the flame (see Figure 4a, open circles) is not entirely satisfactory. The evolution of the lean side of the BSC is correctly determined with a constant K (as for a classical Arrhenius law). However, the rich-side velocities are overestimated even if there is a distinct improvement compared to the classical Arrhenius law that leads to an unceasingly growing velocity after the unity equivalence ratio. In fact, the reaction rate is decomposed into a pre-exponential and an exponential factor. The imbalance between the lean side and the rich side of the global Arrhenius law is directly due to the imbalance of heat release due to the exponential term.

Another solution is to add a Φ-dependence to the pre-exponential factor K. This can be done according to two procedures. If an extremely accurate determination of the BSC is necessary, it is suggested to use a 1-D code able to compute stabilized premixed flames (Cantera or Chemkin for example), or to use a BSC obtained from experiments. Then, using a root-finding procedure, the coefficient $K(\Phi)$ may be determined according to the correct BSC that gives the velocity reached by the flame for any value of the equivalence ratio. A second procedure, less accurate but more direct, consists of determining $K(\Phi)$ as a generic function with some adjustable parameters. Figure 4a shows various BSC curves obtained with the new Arrhenius law calibrated with experimental BSC for methane, ethane and propane. The analytical result, very close to the first three, has been obtained with an empirical function based on two second-order polynomial expressions: one for the lean side, and one for the rich side. Coefficients of the polynomial have to be adapted to a synthetic fuel or retrieved from a prescribed BSC with a root-finding method

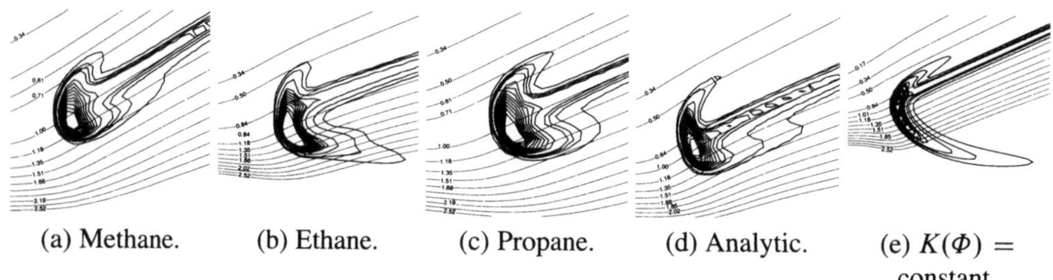

(a) Methane. (b) Ethane. (c) Propane. (d) Analytic. (e) $K(\Phi) =$ constant.

Figure 5. Stabilized edge flames.

and then a least-square estimation.

Both ignition and extinction phenomena are important in order to fully model turbulent combustion. Ignition directly depends on the production of radicals and on the importance of some intermediary kinetics. Peters (2000) and references therein have described the ignition characteristics as a function of the temperature regime. They have shown that three modes may be identified: low-, intermediate- and high-temperature regimes. The GKAS procedure cannot be used to determine ignition delays in the intermediate- and high-temperature regimes. Indeed, for these studies specifically related to the fuel chemistry, an accurate description of the kinetics has to be used. Let us recall that global kinetics are not dedicated to the description of the inner part of the flame, but to its effects on the surrounding flow and vice versa. In particular, the flow can impose on the flame a strong stretch, which can lead to its extinction. So it is of primary importance to capture accurately the extinction limit of the flame. A 1-D counterflow configuration for non-premixed combustion has been used to determine the extinction level of the GKAS procedure. Results have been plotted in Figure 4b. The limit is very close to the theoretical one obtained for a unitary normalized dissipation rate. Thus local extinction due to shear stress, which is intrinsically contained in the untouched exponential term, is captured by the GKAS procedure.

To conclude, various edges flames have been plotted in Figure 5 to demonstrate the impact of an accurate description of the BSC around the stoichiometric level.

6 Dispersion and Evaporation

Although homogeneous turbulence is a straightforward configuration, the addition of an evaporating dispersed phase allows us to be at the center of many poorly understood interactions between turbulence, spray, mixing and combustion.

6.1 Configuration

Using the Eulerian/Lagrangian DNS system previously detailed, the following three-stage procedure has been employed to analyze interactions between the turbulent flow and a dispersed phase (Reveillon and Demoulin, 2006).

Stage 1: Statistically stationary turbulence In this preliminary stage, the turbulent gaseous phase evolves alone in a cubic Cartesian grid until its statistical properties reach a steady

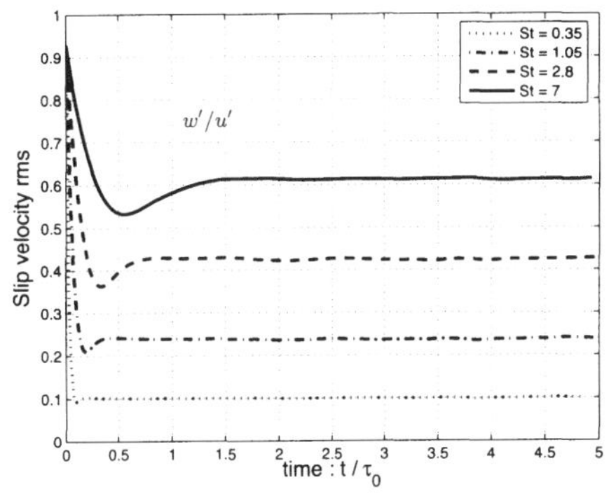

(a) Time evolution of particles slip velocity rms (w'').

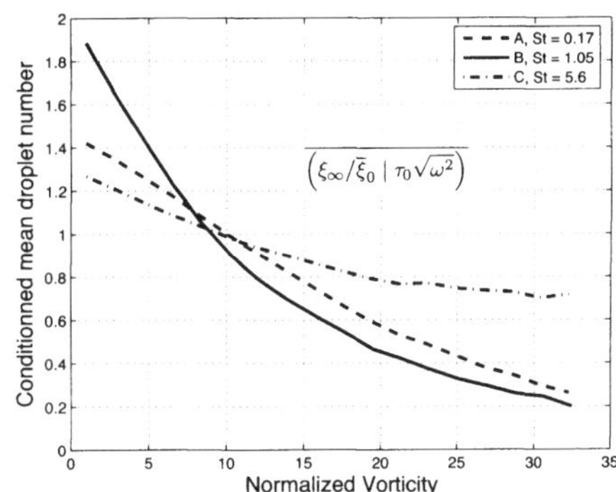

(b) Mean value of stationary droplet number ξ_∞/ξ_0 conditioned by the carrier-phase vorticity level ω.

Figure 6. Spray equilibrium with the turbulent gas phase.

state thanks to a forcing procedure that keeps the mean kinetic energy at a prescribed level (Vervisch-Guichard et al., 2001). The resulting stationary energy spectrum allows us to determine a reference wave number κ_0 corresponding to turbulence scales that contain most of the kinetic energy. The corresponding physical length $l_0 = 2\pi/\kappa_0$ is an integral scale of the flow that has been used as a reference parameter as well as the velocity root mean square u', the eddy turn-over time $\tau_0 = l_0/u'$, and the characteristic time of the velocity fluctuations τ_κ of the smallest structures.

Stage 2: Spray dynamic equilibrium Several eddy turn-over times after the turbulent flow reaches its stationary state, more than two million mono-dispersed, non-evaporating particles are randomly embedded throughout the computational domain with zero initial velocity. The drag force sets particles in motion and the spray reaches a dynamical equilibrium with the turbulence, which corresponds to a stationary value of the slip velocity standard deviation as seen in Figure 6a. Droplet dispersion is usually characterized by the Stokes number $\mathrm{St} = \tau_\mathrm{p}/\tau_\kappa$ (Wang and Maxey, 1993), which indicates the ability of droplets to capture local variations of the carrier-phase velocity. Turbulence properties being fixed, simulations are been carried out by modifying τ_p to achieve a given St.

To characterize droplet dispersion and preferential concentration, a density $\xi(\mathbf{x}, t)$ describing the local mass of liquid per unit volume has been defined. A mean reference mass density is defined by $\overline{\xi}_0 = m_\mathrm{d} N_\mathrm{d}/L^3$ where m_d is the initial mass of each droplet (mono-dispersed spray). Thus, in a non-evaporating mode, $\xi(\mathbf{x}, t)/\overline{\xi}_0$, which may also be considered as a number density, has constant unity mean $\overline{\xi}(t) = L^{-3} \int \xi(\mathbf{x}, t)\, d\mathbf{x} = \overline{\xi}_0$ over the whole domain and its standard deviation $\xi'(t) = L^{-3}(\int (\xi(\mathbf{x}, t) - \overline{\xi}(t))^2\, d\mathbf{x})^{1/2}$, normalized by $\overline{\xi}_0$ in the following, identifies the droplet segregation level. Note that in the following, $\overline{()}$ stands for the mean over the Eulerian grid whereas $\tilde{()}$ is the mean over the Lagrangian particles or droplets.

Stage 3: Liquid-phase evaporation and micro-mixing When droplets are evaporating, $\overline{\xi}/\overline{\xi}_0$ is

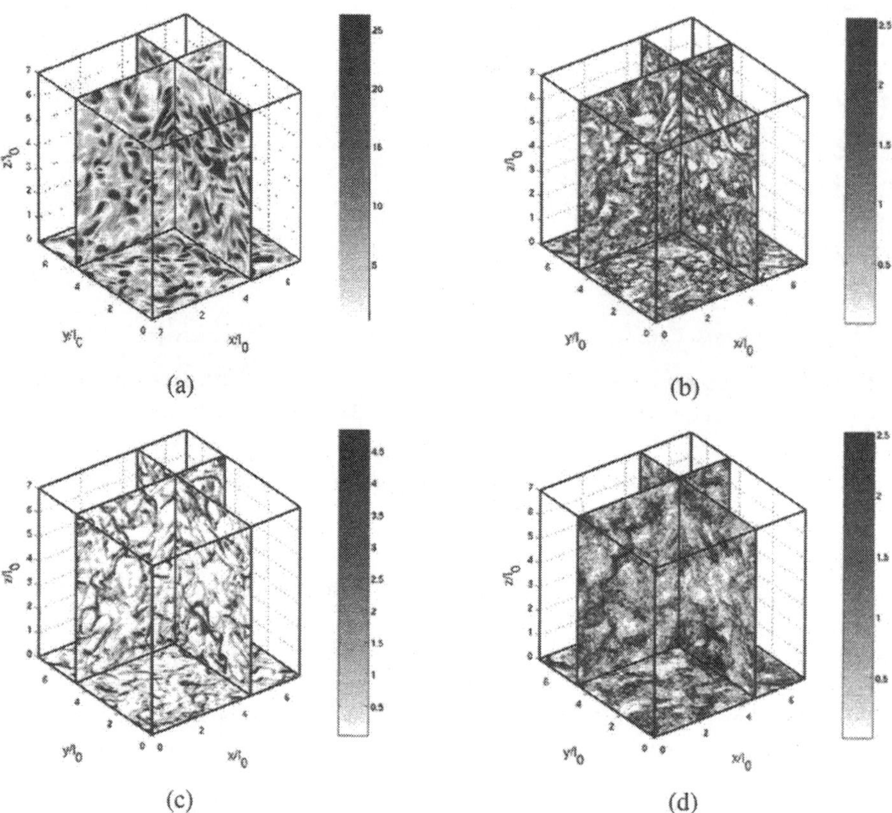

Figure 7. (a) Carrier-phase vorticity $\tau_0\sqrt{\omega^2}$ and spray concentration regions $\xi/\overline{\xi}_0$. (b) St $= 0.17$. (c) St $= 1.05$. (d) St $= 5.6$.

bounded between 1 and 0. $\overline{\xi}/\overline{\xi}_0$ provides information about the evaporation state, whereas $\xi'/\overline{\xi}$ characterizes the droplet segregation. Using the parameter ξ to characterize the spray evolution and preferential segregation is not the usual choice when compared to other works dedicated to the analysis of preferential segregation in sprays, but it allows well-established results to be retrieved (Fessler et al., 1994). Eventually, once the particles have reached a dynamic equilibrium with the surrounding turbulence (stage 2), they are allowed to evaporate according to a specific characteristic evaporation delay τ_v prescribed by the saturation level.

Note that in the case of spray evaporation, the mixture fraction Z cannot reach unity, but rather a local maximum level depending on saturation conditions. Consequently, a normalization of Z is introduced by using the saturation limit Z_s. Z/Z_s is thus bounded between 0 and 1, and it is of practical interest for analyzing correlations between the evaporating spray and turbulent mixing.

6.2 Preferential segregation of non-evaporating droplets

To study the preferential concentration of discrete particles in turbulent flows, several approaches exist (see for instance Squires and Eaton, 1991; Wang and Maxey, 1993; Fessler et al., 1994; Simonin et al., 1993, and Aliseda et al., 2002). If a statistically homogeneously distributed spray is randomly injected, i.e. if there is no preferential segregation, the distribution of the number of particles per control volume (CV) of a given size must follow a binomial distribution, which may be approximated by a Poisson distribution. Hence, the study of preferential concentration is usually based (Fessler et al., 1994) on the difference between the actual segregated distribution and the Poisson distribution. Thus it is characterized by

$$\Sigma = \left(\sigma - \sqrt{\lambda}\right)/\lambda\,, \tag{6.1}$$

where λ is the average number of particles per cell, whereas σ and $\sqrt{\lambda}$ are the standard deviations of the particle distribution and the Poisson distribution, respectively. For a given Lagrangian distribution of the particles, Σ depends strongly on the size of the CV. However, according to Fessler et al. (1994), the length scale corresponding to the characteristic cluster size is equal to $\Delta_{\Sigma\max}$, which is the size of the CV when Σ reaches a maximum value.

In the case considered in this configuration, particles are liquid droplets of fuel that are evaporated to prepare the reactive mixture. Preferential concentration of particles is potentially important when describing flow-induced heterogeneities that could appear in the mixture-fraction field. Following such considerations, another parameter, more representative of the evaporation and turbulent mixing processes, has been considered to describe preferential-segregation effects.

Droplet dispersion and preferential segregation are analyzed from a Eulerian point of view thanks to the local Eulerian liquid density $\xi(\mathbf{x}, t)$. Instantaneous fields of ξ are plotted in Figure 7 for three Stokes numbers (St $= 0.17$, St $= 1.05$ and St $= 5.6$), along with the corresponding vorticity field. These four fields have been captured at exactly the same time after droplet dispersion has reached a stationary value ($t > t_\infty$). Even without any quantitative analysis, it is possible to see the dramatic impact of the particles' inertia on their dispersion properties. Indeed, even with a small Stokes number, particles tend to leave the vortex cores and segregate in weak vorticity areas. This phenomenon may be seen in Figures 7b and 7c where ξ is represented for St $= 0.17$ and St $= 1.05$, respectively. This last case shows a normalized liquid density $\xi/\overline{\xi}_0$ ranging between 0 (no droplets) and 5 (five times the mean density). As will be shown later, density fluctuations reach a maximum when St $= 1$.

When St $= 0.17$, segregation is already clearly visible (maximum deviation: 2.5), although there are more intermediates density areas (see Figure 7b). When the St $= 1$ limit is surpassed, the droplet distribution tends to be totally different than for St ≤ 1 (see Figure 7d). Indeed, kinetic times become large enough for the droplets to cross high vorticity areas, leading to a less segregated spray (maximum deviation: 2.5). This result is confirmed in Figure 6b where the mean liquid density conditioned by the local vorticity level has been plotted for various Stokes numbers. For highly segregated sprays (St $= 1.05$), high vorticity areas are almost empty, with an average value of $\overline{\xi}/\overline{\xi}_0$ equal to 0.2, whereas when vorticity tends to 0, $\overline{\xi}/\overline{\xi}_0$ converges toward 2. Figure 6b is in accordance with classical results such as the work of Squires and Eaton (1991). They found a maximum correlation between the number of particles and the vorticity level for a Stokes number equal to 0.15. However they used a Stokes number based on the integral time scale of the turbulence. By swapping it with the Kolmogorov time scale, the peak of correlation occurs

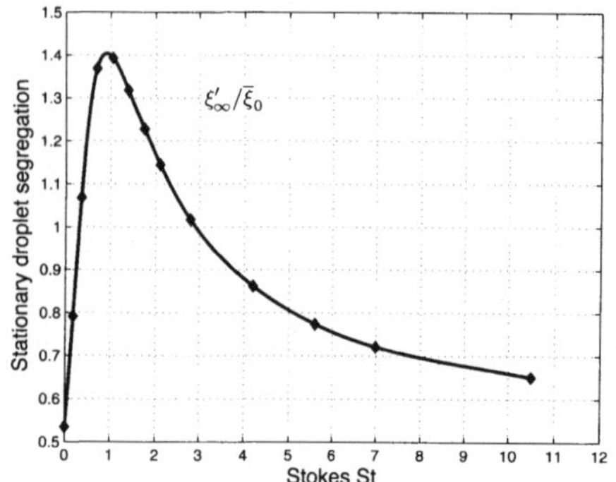

(a) Final stationary values of the droplet segregation parameter ξ'_∞ normalized by the initial mean value of ξ: $\overline{\xi}_0$.

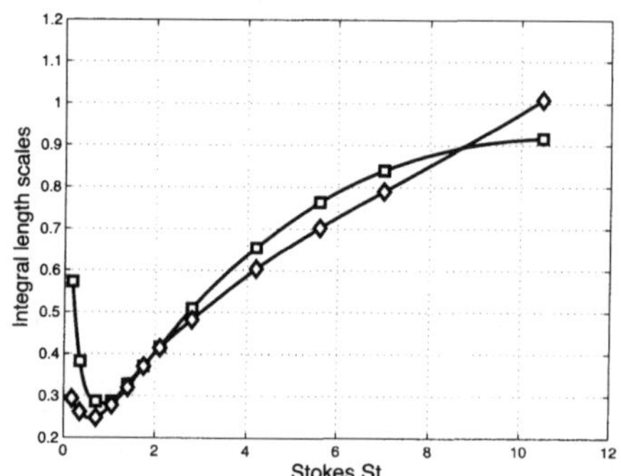

(b) Characteristic cluster size determined by two different methods. Squares: Cell size of the optimal deviation from the Poisson distribution $\Delta_{\Sigma\max}$. Diamonds: Size of the most energetic clusters l_ξ (determined from the energy spectrum of the droplet density).

Figure 8. Preferential segregation properties.

for unit Stokes number. Indeed the Kolmogorov scaling is necessary to characterize segregation effects as outlined by Wang and Maxey (1993). When St $= 0.17$ and 5.6 the correlation between low vorticity areas and high droplet density clusters is less pronounced. However, because of the ballistic nature of the heavy droplets (St $= 5.6$) the high vorticity areas are still densely populated. This confirms the qualitative result of Figure 7d.

To evaluate the preferential segregation of the droplets embedded in the turbulent flow, ξ'_∞ has been plotted in Figure 8a. This parameter is the standard deviation of the field ξ when droplets are in dynamical equilibrium with the carrier phase ($t > t_\infty$). Starting from St $= 0.025$ with $\xi'_\infty/\overline{\xi}_0 = 0.55$, a maximum segregation $\xi'_\infty/\overline{\xi}_0 = 1.4$ is observed for unit Stokes number followed by a progressive decay. This parameter informs us of the characteristic liquid density level in clusters that have been formed by the turbulent structures. The variable ξ is not classically used to capture the properties of dispersed particles, but the results have been successfully compared with the maximum of Σ, which is the deviation from the Poisson distribution as defined in Eq. (6.1). ξ'_∞ reaches a maximum value when St $= 1$. Wang and Maxey (1993) or Fessler et al. (1994) have also shown a similar dependence of the segregation intensity on the Stokes number. This confirms that the parameter ξ is relevant for characterizing segregation.

Similarly, it is possible to determine with various methods the characteristic size of the clouds (or clusters) of particles. In Figure 8b, we use the information obtained from the stationary energy spectrum $E_\xi(\kappa)$ of the variable $\xi(x,t)$, and then we extract $l_\xi = 2\pi/\kappa_\xi$ where κ_ξ is the position, in spectral space, of the most energetic level. Plotting the l_ξ dependence with the Stokes number in Figure 8b demonstrates the large-scale effects of the turbulence on the spray. The cluster sizes computed from E_ξ have been compared (see Figure 8b) with the one obtained thanks to the classical Lagrangian approach of Fessler et al. (1994). In this last approach, the cluster size denoted by $\Delta_{\Sigma\max}$ is the width of the CV when Σ reaches a maximum value. Both l_ξ and $\Delta_{\Sigma\max}$ show a similar evolution with a minimum size for St $= 1$. In their experiment Aliseda et al. (2002) compared the characteristic cluster size $\Delta_{\Sigma\max}$ to the length scale found using a method proposed by Wang and Maxey (1993). In this last approach, the change between the actual distribution to the binomial distribution is measured as the square of the difference of probabilities given by the two distributions summed over all possible values. Aliseda et al. found that both methods lead to the same characteristic length scale for the clusters. Thus using the Eulerian ξ parameter is in accordance with the classical Lagrangian approach for the analysis of preferential segregation. Notice that for the measurement of the characteristic cluster length scale, Aliseda et al. (2002) use all of the droplet class sizes. Therefore, it is not possible to determine the Stokes number dependency of the clusters' characteristic length scale. The two methods they used to find the cluster length scale agree globally when considering all of the droplet sizes. It is still possible that some differences may appear for specific values of Stokes as we found when comparing $\Delta_{\Sigma\max}$ to ξ.

Fessler et al. (1994) conducted experiments with various sets of particles corresponding to Stokes numbers ranging from 1.7 up to 130. Their study does not extend to Stokes numbers smaller than unity. It appears that $\Delta_{\Sigma\max}$ is dependent on the Stokes number, starting with a value of the order of the Kolmogorov length scale, it increases as the Stokes number increases. For Stokes numbers greater than unity, the parameter ξ allows this trend to be recovered. In our DNS, when St < 1 the evolution of the cluster size obtained with both Σ and ξ is similar, although the length scales are different (Figure 8b). However, using ξ as a reference parameter offers a new range of possible analysis.

From a phenomenological point of view, the cluster-size evolution results from the competition between three physical phenomena: the ejection of the droplets from the vortex cores by the turbulence, the turbulent micro-mixing (prevalent when St < 1) and the ballistic effects (prevalent when St > 1). Indeed, droplets tend to be ejected from the turbulent structures to form clusters concentrated in low vorticity areas. However, for droplets with a small Stokes number, turbulent micro-mixing counteracts the segregation process and 'diffuse' clouds are obtained (as seen in Figure 7b): the lighter the droplets, the more effective the mixing and the larger the characteristic size of the clusters. When St $= 1$, an optimal segregation is obtained because micro-mixing's impact is weak and the droplets are not heavy enough to leave low vorticity areas where they are trapped. However, as soon as inertia is prevalent (St > 1) the particles are able to cross turbulent structures no matter what their vorticity is and then the characteristic size of the clusters increases again. In Figure 8b, it is clear that both l_ξ and $\Delta_{\Sigma\max}$ capture this natural dependence on the Stokes number.

To conclude on the non-evaporating dispersion aspect, this first DNS application shows that preferential segregation in sprays cannot be characterized using only the mean segregation parameter ξ'_∞ and the mean density $\overline{\xi}$. Indeed, for two sprays with the same mean density and whose Stokes numbers are 0.17 and 5.6, a similar mean segregation level equal to 0.85 is found in Figure 8a. However, the corresponding topologies of spray density are distinct, as may be seen in Figures 7b and 7d where ξ is plotted for St $= 0.17$ and 5.6, respectively. Consequently, for two fields with identical first moments of ξ, different mixture fraction topologies can be obtained from droplet evaporation and different combustion regimes might be observed if the ignition delay is short compared to the turbulent mixing time. Thus, in addition to $\overline{\xi}$ and ξ'_∞, a third parameter, which has to depend on the droplet's Stokes number, is necessary to describe accurately

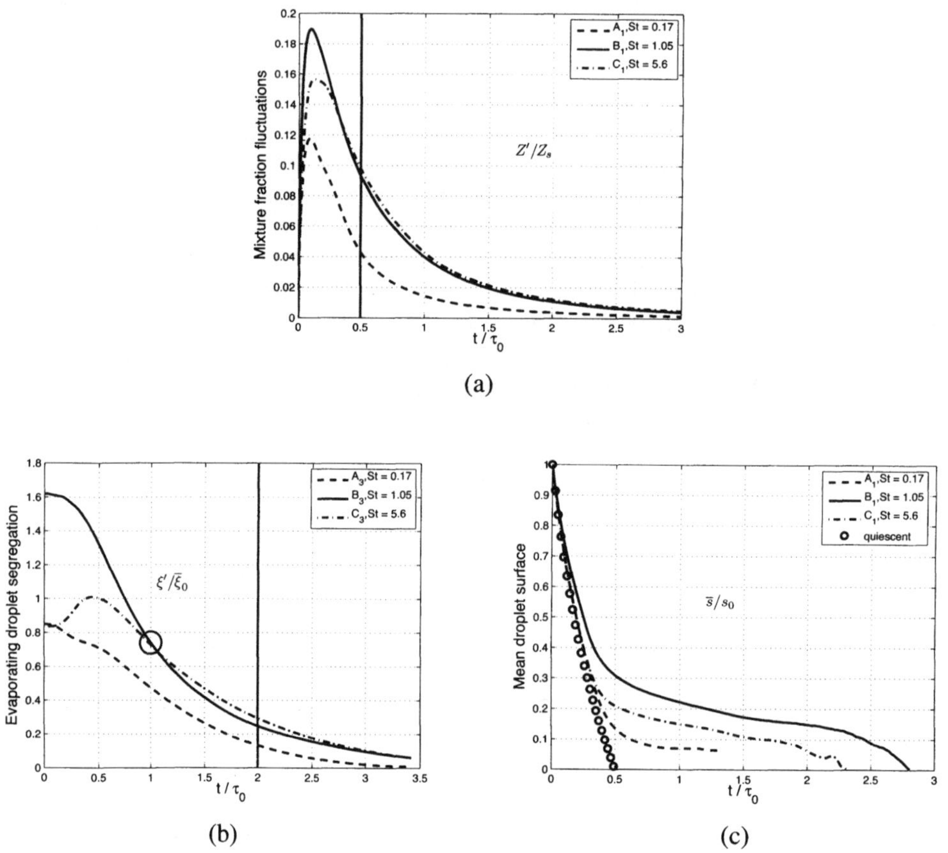

Figure 9. Evaporating spray: (a) Time evolution of spray segregation. (b) Liquid density. (c) Mean droplet surface.

the preferential segregation of the spray and the subsequent mixture-fraction field.

6.3 Evaporation

Once dynamical equilibrium is reached between the turbulent gaseous flow and the dispersed phase, evaporation is activated. A very different behavior may be observed in Figure 9a where the evolution of the standard deviation of the mixture fraction, Z', has been plotted for various Stokes numbers (A_1, B_1 and C_1 correspond to $\tau_v/\tau_0 = 1/2$). From a general point of view, the global shape of the curves is the same: starting from $Z'/Z_s = 0$ where evaporation starts, the curves reach a maximum value long before the characteristic evaporation delay (when $t/\tau_0 \approx 0.2$). At this point, dissipation effects on the mixture-fraction fluctuations become greater than the evaporation effect. Then mixture-fraction fluctuations decrease continuously. Details of the competition between evaporation and dissipation may be found in Reveillon and Vervisch (2000).

A more detailed analysis shows different local evolutions of the mixture fraction. First, the most segregated spray generates the most fluctuating mixture-fraction field (case B1, St $= 1.05$). Indeed, droplets accumulate in small clusters and when evaporation starts, high levels of mixture fraction are obtained in small control volumes leading to a strong deviation Z'. Another important point to notice is the evolution of the deviation Z' of case C1 that initially evolves like case A1 because of the initial similar droplet segregation level. However, it exceeds very quickly (more than 30%) the maximum value of Z'_{A1} before joining, almost exactly, the curve Z'_{B1} corresponding to the highly segregated case. This behavior is due to the decrease of the mean Stokes number of the spray that leads to a very quick segregation of the initially ballistic droplets of case C1. On the other hand, there is no change in segregation for case A1 (St $= 0.17$) as the Stokes number becomes smaller and smaller.

This interpretation is confirmed in Figure 9b where the evolution of the droplet segregation parameter ξ' has been plotted for cases A3, B3 and C3 ($\tau_v = 2\tau_0$), which shows this behavior more clearly. The general trend of cases A and B is a decrease of the segregation because of the diminution of the corresponding Stokes number. The turbulent micro-mixing becomes more and more effective leading to a less segregated spray. When the Stokes number is initially greater than unity (cases C) the segregation first increases before following the general decay. The initial elevation of the segregation level corresponds to an evolution of the droplet dynamics to a level that is more efficient (unit Stokes number) at creating clusters and, therefore, to increase ξ' momentarily. This modification of the evaporating droplet dynamics has a direct impact on the mixture-fraction evolution as seen in Figure 9b.

The evolution of the mean droplet surface divided by its initial value is shown in Figure 9c for various Stokes numbers and various evaporation delays. Additional curves, represented by circles, represent the mean droplet surface evolution that would be obtained if there were no preferential segregation (homogeneous droplet distribution). Figure 9c demonstrates the dramatic effect of the segregation of the dispersed phase on the evaporation process. Two distinct stages may be observed. First, when evaporation starts far from the saturation limit, the decrease in the mean droplet surface area calculated from the DNS is similar to the results obtained with the analytical model that neglects segregation phenomena. Then, depending on the Stokes number (i.e. on the droplet segregation level), a second evaporating stage controlled by saturation may be observed with a slower rate.

7 Laminar Spray Combustion

Depending on the configuration of the studied system, combustion may take place either after the full evaporation of the liquid fuel, or within the evaporating dispersed phase. In the first case, even if classical gaseous combustion models can be utilized, the mixture-fraction topology issued from the spray evaporation is completely different from the one obtained with gaseous fuel injection. As described by the preceding example concerning spray preferential segregation or in Reveillon et al. (1998), partially premixed areas appear and consequently, depending on the dynamics of the droplets, various flame structures may be generated. On the other hand, if flames propagate in the mixture where evaporating droplets are still present, numerous complex interactions appear between combustion, droplet evaporation and turbulence (see Figure 10).

Indeed, the spray modifies local heat transfer, momentum, and dissipation rate and, of course, the mixture-fraction level. All these variables affect deeply not only the flame ignition and propagation, but also the characteristic properties of the turbulence. It is then particularly difficult

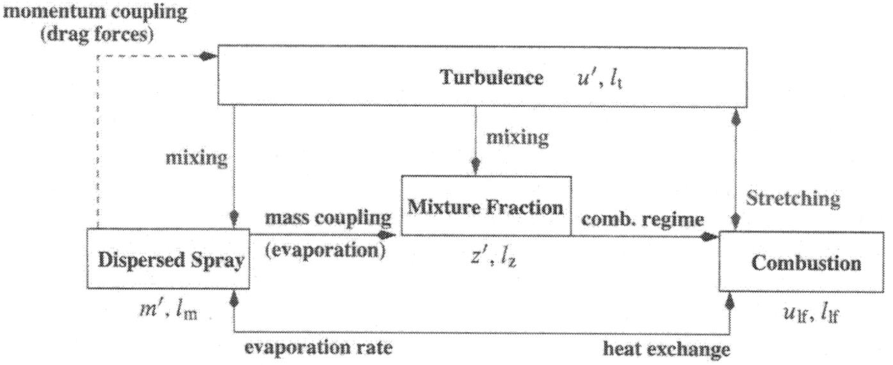

Figure 10. Major interactions between spray, turbulence and combustion.

to isolate physical phenomena and fundamental interactions in order to analyze them. Again, DNS may be of great help in this task. In the following, fundamental questions concerning combustion and two-phase flows are raised: How can one define the local equivalence ratio? How does a laminar 1-D flame propagate in a spray? What are the major spray flame structures and combustion regimes?

7.1 Local equivalence ratio

As far as purely gaseous flows are concerned, the definition of the equivalence ratio is straightforward. It is defined as the ratio of the actual fuel/air ratio to the stoichiometric fuel/air ratio. Following the topology of the mixture fraction, it is thus possible to anticipate with precision combustion regimes and reaction rates with respect to the amount of fuel and oxidizer injected into the combustion chamber. But, in the framework of two-phase combustion, knowing the mass fluxes of liquid fuel and gaseous oxidizer injected in the chamber is not sufficient to determine the local effective equivalence ratio. Indeed, after being atomized, liquid fuel is embedded in the gaseous oxidizer. It may cross various areas whose thermodynamic (mainly temperature and pressure) properties affect (1) the evaporation rate of the droplets, and (2) the local mass fraction of oxidizer and thus the local equivalence ratio. Similar phenomena occur when a flame front is present. Heat transfer may lead to local extinction of the flame. For example, among other phenomena, it appears that a stoichiometric injection of fuel and oxidizer may lead to non-stoichiometric flames. A simple analytical approach may offer some explanation to this phenomenon.

An isolated droplet is considered in a control volume $\mathcal{V}$ that contains a gaseous oxidizer. A similar approach can be employed for a cluster of droplets. The diameter of the droplet determines the initial global equivalence ratio of the problem with respect to the mass of oxidizer. The gas surrounding the droplet undergoes an elevation of temperature by a factor of 5, which correspond to a classical ratio between burnt and fresh gas temperatures of a premixed flame (see Figure 11a). The analysis may be reduced to two characteristic times: the evaporation delay of

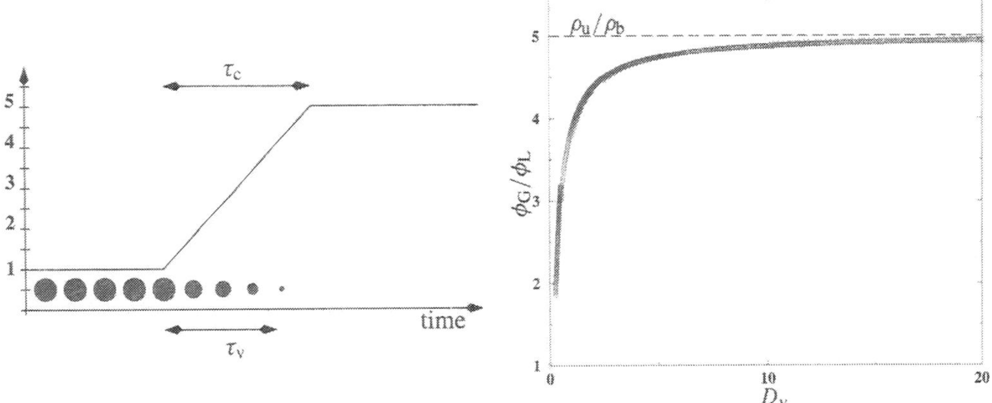

(a) Sketch of the configuration: Droplet is surrounded by a quiescent atmosphere. A sudden temperature growth (1) expands the gas and (2) evaporates the droplet.

(b) Ratio of the final gaseous equivalence ratio to the initial liquid equivalence ratio with respect to D_v.

Figure 11. Analysis of the impact of evaporation dynamics on the gaseous equivalence ratio.

the droplet τ_v, and the heating delay of the gas phase τ_c. An open domain is considered and the pressure remains constant and equal to the atmospheric pressure. The increase of temperature leads to a dilatation of the gas phase. Then the amount of oxidizer in the control volume diminishes. During this process the droplet is evaporating. Consequently, the final equivalence ratio in the volume depends on the ratio of the evaporation delay to the heating delay: $D_v = \tau_v/\tau_c$. Indeed, a very small evaporation delay ($D_v \ll 1$) leads to the quick disappearance of the droplet before the dilatation of the whole gas mixture. The final equivalence ratio is then very close to the initial global equivalence ratio determined from the mass of the liquid droplet. On the contrary, if $D_v \gg 1$ the mass of oxidizer diminishes before the full evaporation of the droplet. This leads to an increase in the final equivalence ratio. These remarks are summarized in Figure 11b, which shows the ratio of the final gas-phase equivalence ratio Φ_g to the initial liquid equivalence ratio Φ_l with respect to the parameter D_v. Theoretically, this ratio may reach the ratio of the burnt gas temperature to the fresh gas temperature ($T_b/T_u = 5$). From a practical point of view, the D_v parameter remains small. However, it concerns the area with the strongest variation of Φ_g/Φ_l, and we know from Figure 4a that only a slight variation of Φ_g may lead to strong variations of the flame structure.

This analysis has been oversimplified and numerous effects have been neglected. Nevertheless, a major point has been put forward: it is necessary to clearly differentiate between the global equivalence ratio of the chamber and the effective local equivalence ratio, which determines the flame structure and the combustion regime. This is one of the reasons that has led us to develop an Arrhenius law capable of determining the local properties of the flame regardless of the equivalence ratio.

7.2 Laminar spray combustion

The second stage, dealing with two-phase flow combustion, is the analysis of the propagation and the structure of 1-D laminar premixed spray flames. Numerous researchers have worked on this subject. One of the first studies was carried out by Williams (1960). He derived an analytical solution for two distinct cases: either all the droplets are evaporated before reaching the flame front, or each droplet burns with a diffusion flame around it. Afterwards, various studies were dedicated to mono- or poly-dispersed spray combustion (Patil and Nicolls, 1978; Richards and Sojka, 1990; Lin et al., 1988; Lin and Sheu, 1991). Most often, in theses studies, the carrier phase contains initially some fuel vapor to stabilize the flame. Then the presence of droplets leads to a modification of the local equivalence ratio. Silverman et al. (1991, 1993) worked on the derivation of an analytical expression for the laminar flame speed in a spray. Nevertheless, in all of these studies, which are mostly numerical, the dilatation of the gas phase is not accounted for. As pointed out above, this phenomenon may lead to some dramatic modifications in the local properties of the mixing and, thus, on the flame structure. However, thanks to DNS, it is possible to carry out a basic 1-D configuration with a compressible formulation for the gas phase. It is a complex study with a lot of different flame behaviors and combustion regimes. In the following, we illustrate with a one simple example the complex behavior of spray flames. Monodisperse droplets are regularly spaced in the 3-D space that may be considered only along the direction of propagation of the planar flame. In this simple configuration, the droplets are anchored in the domain. They may be seen as local sources of fuel that follow the evaporation law, which depends on the local temperature, pressure and vapor mass fraction. The gas phase is initially at rest and the initial liquid equivalence ratio Φ_l is equal to 1.48. Initially, there is no vapor mixed with the oxidizer.

The sketch of the configuration is shown in Figure 12a. Note that the flame propagates from right to left whereas the axis is oriented from left to right. A gaseous premixed flame profile is utilized to initialize the computation (1). The gaseous fuel is removed upstream of the flame front to be replaced by the droplets. The flame evolves within the spray and after a while an equilibrium appears between evaporation and combustion phenomena, and a time-periodic pattern appears (3). The Lewis number of this configuration is equal to 1. In a purely gaseous situation, the flame would be perfectly stationary. But, as far as spray combustion is concerned, oscillations appear. The results (flame thickness, flame speed, maximum reaction rate) have been normalized by the corresponding properties of the stationary gaseous stoichiometric flame obtained with the same fuel.

Flame index Spray combustion involves generally partially premixed and non-premixed combustion regimes. In order to differentiate between heat release due to premixed and diffusion flames, Yamashita et al. (1996) suggested to use a flame index based on the scalar product of the fuel and oxidizer normalized gradients. This flame index has been utilized also by Mizobuchi et al. (2002) to analyze a lifted hydrogen flame. The flame index may be written as

$$G_{FO} = \nabla Y_F \cdot \nabla Y_O \tag{7.1}$$

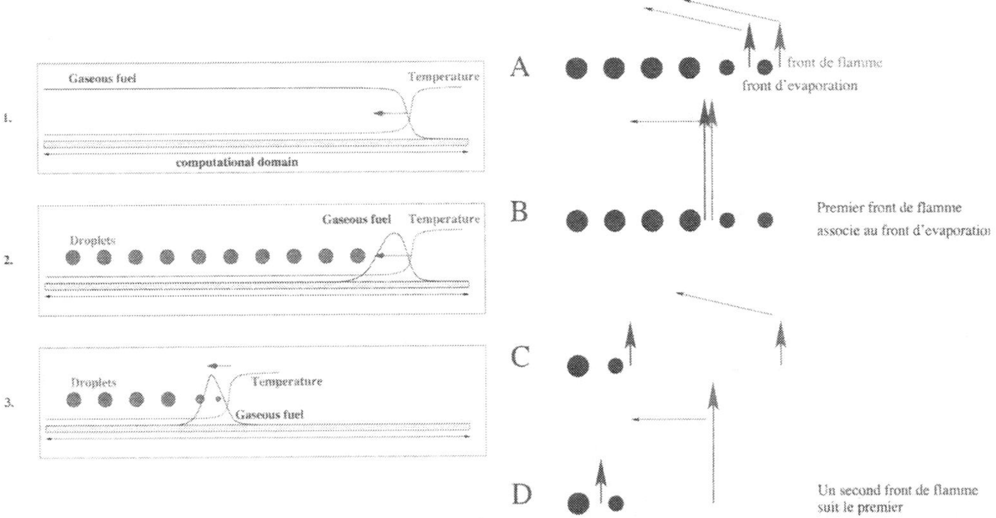

(a) Sketch of the configuration. (1) Purely gaseous flame is generated. (2) Gaseous fuel is replaced by droplets. (3) Spray flame propagates.

(b) Dynamics of the double-flame front.

Figure 12. 1-D spray flame.

It is positive for a premixed flame and negative for a non-premixed one. From this definition, ξ_p, a normalized construction of the flame index, is derived:

$$\xi_p = \frac{1}{2}\left(1 + \frac{\nabla Y_F}{\mid \nabla Y_F \mid} \cdot \frac{\nabla Y_O}{\mid \nabla Y_O \mid}\right) \tag{7.2}$$

When ξ_p vanishes, diffusion flames are observed, while premixed combustion is found when it reaches unity. More or less partially premixed reaction zones develop for values of ξ_p ranging between zero to unity. ξ_p has a sense only where the reaction rate is non-null. To evaluate the amount of burning in premixed and partially premixed regimes and to compare it to the overall heat release rate, a premixed fraction of the burning rate $W_p(x)$ must be introduced. $W_p(x)$ is defined for any control volume V as the average of the amount of fuel burning in premixed modes normalized by the total burning rate:

$$W_p(x) = \frac{\iiint_V \xi_p \dot{\omega}\, \mathrm{d}v}{\iiint_V \dot{\omega}\, \mathrm{d}v} \tag{7.3}$$

This quantity and other flame parameters are now used to seek out combustion regimes observed in the laminar cases.

Flame structure and dynamics Time evolution of the maximum of the reaction rate is plotted in Figure 13a. It appears that the flame follows a periodical pulsating behavior with two maximum reaction peaks over a period of eight flame times. First it is necessary to point out the fact that the interdrop spacing is equal to one-tenth of the reference flame thickness. Thus pulsations do not originate from the possible combustion of isolated droplets (an impossible task to resolve with our DNS formulation), but from the global flame dynamics. Even if initially the mixture is fuel rich, the maximum reaction rate corresponds to the one of the reference stoichiometric flame. To understand what is happening in the core of the flame, five positions: A, B, C, D and E (periodically equivalent to A) have been marked in Figure 13a. The time evolution of the maximum evaporation rate is plotted in Figure 13b. It appears that this data is maximum for position B, which corresponds to the first reaction rate peak. At B the maximum reaction rate and maximum evaporation rate are at the same spatial location (see Figure 13d). (Note that a diminution of the position value corresponds to a propagation of the flame.) The first maximum reaction rate peak corresponds thus to the burning of the vapor that just left the droplet surface. But concerning the second maximum reaction rate peak (position D in Figure 13a), there is no evaporation in the domain (see Figure 13b) and the maximum reaction rate repositions itself downstream (in the burnt gases) the first front (see Figure 13c).

It appears that combustion occurs following two stages summarized in Figures 12b and 13d, which represents the flame trajectory in the (maximum reaction rate/maximum evaporation rate) space. Starting from point A where the maximum reaction and evaporation rates are weak, evaporation begins. As soon as the vapor and oxidizer mixture reach an ideal state, a first flame front propagates strongly (point B) and drives the evaporation front. However, characteristic delays of the flame are shorter than evaporation delays and the first flame front depletes quickly the vapor and extinction occurs (point C). Meanwhile downstream from this flame front, liquid fuel finished to evaporate. A mixture of burnt gases and vapor has been formed because of the originally rich configuration. When the first flame front disappears (point C), oxidizer is able to diffuse backwards into the burnt gases. Then a second flame front appears and follows the path of the first one to reach the cloud of droplets that needed this delay to generate a new evaporation front. This process is periodic and it starts over at point A.

This single example shows the complexity of spray combustion. Indeed, many characteristic parameters are involved: the flame velocity, thickness and heat release, the droplet size, inter-spacing, evaporation delay, etc. By modifying only one of these parameters, the flame dynamics may be dramatically changed. The next section offers a glimpse of the combustion diagrams that have been already developed in the literature.

8 Spray Combustion Diagrams

Two pioneering works have envisioned classifications of spray-flame morphology. The first diagram (see Figure 14), developed by Chiu and coworkers (Suzuki and Chiu, 1971; Chiu and Liu, 1977; Chiu and Croke, 1981; Chiu et al., 1982) consisted of the determination of the structure of flames propagating through a cloud of droplets plunged into a preheated oxidizer. Spray combustion regimes were classified according to a group combustion number G. This dimensionless number may be seen as the ratio between the characteristic evaporation speed and the molecular diffusion speed, or the convective speed of the hot gases inside the cloud. When the Peclet number is large, Candel et al. (1999) showed that in most cases the relation $G \approx 5N^{2/3}/S$ is verified between the group number G, the total number of droplets in the cloud N and the separation parameter S. This last number $S = \delta_s/\delta_{r_f}$ is the ratio between δ_s, the mean droplet inter-spacing, and δ_{r_f}, a characteristic diffusion-flame radius (Kerstein and Law, 1982).

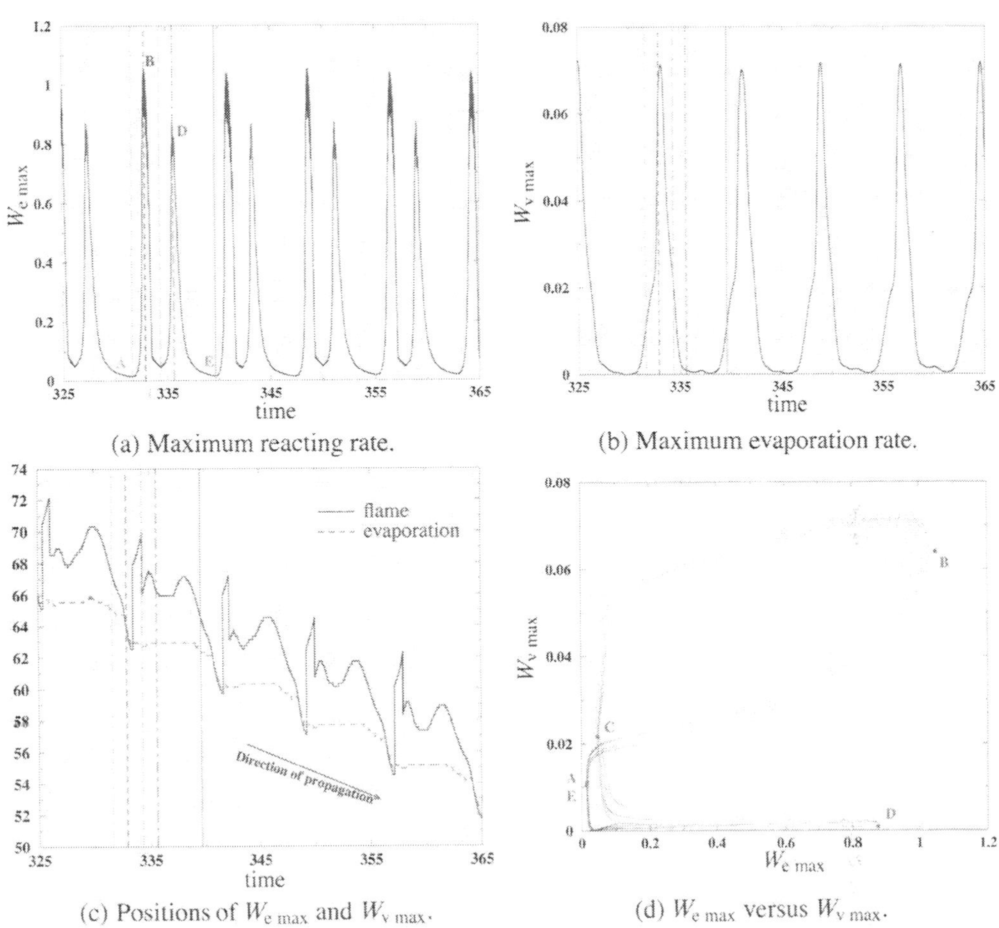

Figure 13. Example of 1-D flame propagation in a rich spray.

In the case of high dilution (volume of the liquid phase much smaller than the volume of the gas phase), the droplet mean inter-spacing, $\delta_s = d^{-1/3}$, is directly determined from the droplet density d that is the number of droplets per unit volume. The length δ_{r_f} denotes the radius of a diffusion flame surrounding a single vaporizing drop in a quiescent oxidizer having the mean properties of the spray (radius and evaporation time). When the separation number S decreases, there is a point where the flame topology evolves from individual droplet combustion to group combustion. For a given value of S, varying N, the number of droplets in the liquid cloud, two major modes (see Figure 14b) of spray combustion may be identified with respect to the group number G. In the first case, $G \gg 1$, the droplets are too close to each other to allow for diffusion of heat inside the cloud. Only an external layer of droplets is evaporated, and the resulting flame remains at a standoff distance from the spray boundary. Under the other limit condition, $G \ll 1$, the droplets are sparse enough so that the hot gases manage to reach the core of the spray.

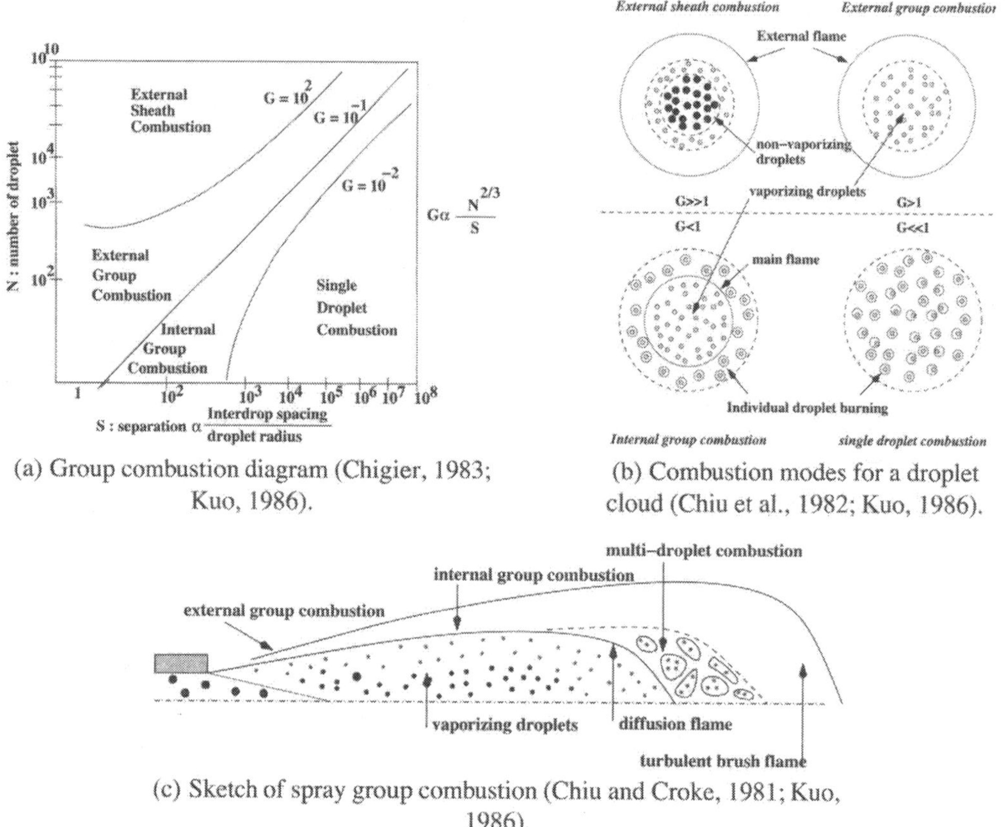

(a) Group combustion diagram (Chigier, 1983; Kuo, 1986).

(b) Combustion modes for a droplet cloud (Chiu et al., 1982; Kuo, 1986).

(c) Sketch of spray group combustion (Chiu and Croke, 1981; Kuo, 1986).

Figure 14. Combustion modes defined by Chiu.

Hence evaporation and combustion processes take place around every individual droplets. Those conditions delineate the so-called "external" combustion regime expected for $G \gg 1$, which is complemented by the "internal" combustion regime, observed for $G \ll 1$ (see Figure 14b). A smooth transition between these limiting regimes was anticipated by Chiu et al. (1982), leading to intermediate submodes depending on the magnitude of G. When G is slightly above unity, the flame stays around the droplet group with a temperature rise of the liquid phase affecting the core of the cloud. For G smaller than unity, a first ring of individual burning droplets is centered on a droplet cloud surrounded by a diffusion flame.

Later on, Chang and also Borghi and coworkers (Chang, 1996; Borghi, 1996b,a; Borghi and Champion, 2000) added to the analysis the control parameters of the reaction zone itself, namely the characteristic flame time τ_f and its thickness δ_f. The corresponding diagram may be seen in Figure 15. In addition, the mean evaporation delay τ_v was introduced. When $\tau_v \ll \tau_f$, the mixture may be locally premixed and a propagating premixed flame develops (see Figure 16a). This regime should be observed for all values of mean droplets inter-spacing δ_s and flame thickness

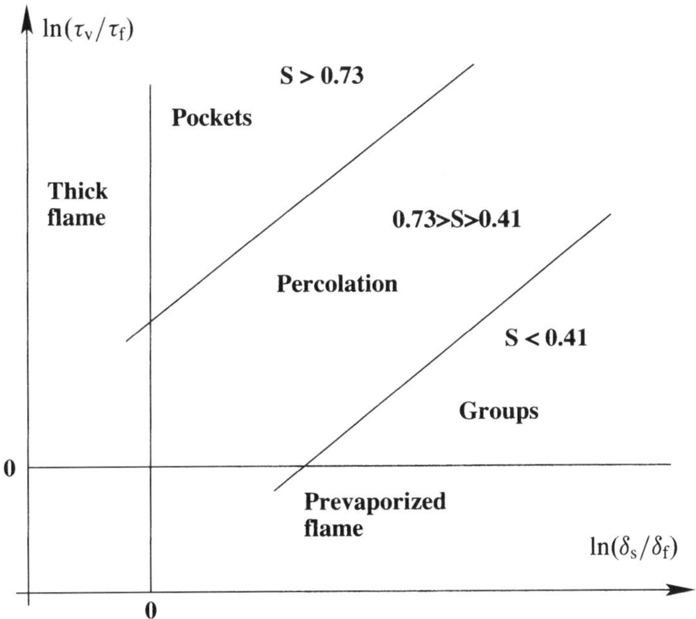

Figure 15. Borghi's diagram for laminar flames propagating in homogeneous and polydispersed droplets (Borghi, 1996a).

δ_f. In practice, the equivalence ratio of the mixture may not be fully uniform, and a weakly varying partially premixed front propagates. If the evaporation time is large enough, for $\delta_f > \delta_s$, the collection of drops penetrates the reacting diffusive layers since the flame is broader than the mean droplet inter-spacing δ_s. This situation should rapidly promote the thickening of the flame (see Figure 16b). Aside from these extreme cases, the separation number (S) should be introduced. After the propagation of a primary partially premixed front, some droplets may remain, leading to a secondary (or back-flame) reaction zone (see Figure 16c). The topology of this secondary combustion zone depends on the magnitude of S. For small values of $S = \delta_s/\delta_{r_f}$ the droplets are burning individually or are clustered in small groups surrounded by a flame. This is called the "group" combustion regime. In complement, Borghi et al. have distinguished a "percolation" combustion regime and a "pocket" combustion regime, subsequently appearing when the separation number S increases (see Figure 16d-e-f). In Figure 17, DNS captured the percolation combustion regime (turbulent V-flame) foreseen by Borghi in a turbulent case.

Chiu et al. and Borghi et al. have defined these flame structures in the case of a quiescent spray without considering the global liquid fuel/air mass ratio. However, within a real spray combustion system, this key ratio is known to modify flame stability along with the overall properties of the combustion chamber. Changing it would affect the distribution of the local equivalence ratio of the gaseous mixture. Specifically, the topology of the primary and secondary reaction zones

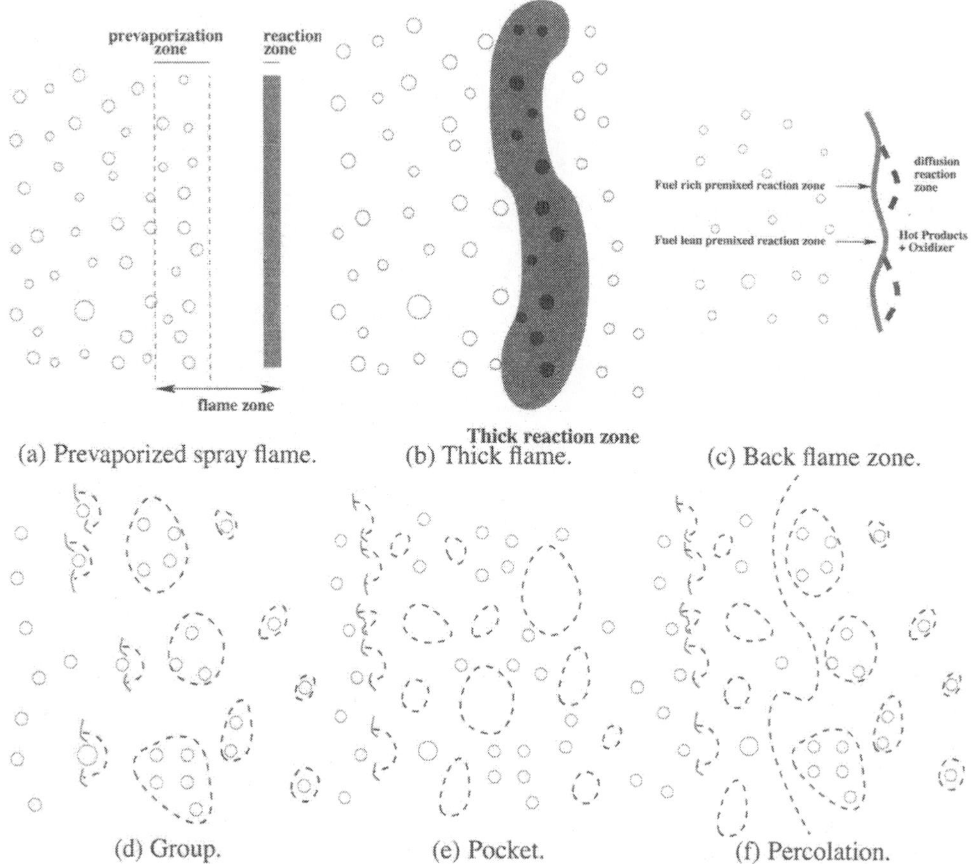

(a) Prevaporized spray flame. (b) Thick flame. (c) Back flame zone.

(d) Group. (e) Pocket. (f) Percolation.

Figure 16. Flames structures developing in a quiescent spray of droplets (Borghi, 1996a). Lines: Premixed fronts. Dashed lines: Diffusion flames.

may vary significantly with this additional parameter. For instance, local extinction may be observed due to local equivalence ratios outside of the flammability limit. Moreover, the droplets and the mixture composition are also sensitive to advection that plays a crucial role in spray combustion. Flame pictures can then hardly be anticipated fully from a quiescent flow analysis, but they constitute a first basis on the top of which additional effects can be included.

Considering the numerous parameters involved in two-phase flow combustion, DNS is of great help to delineate some of the combustion regimes. From DNS results, the flame structures may be organized into three main categories, themselves possibly divided into sub-groups.

1. **External combustion:** This concerns combustion with a continuous flame interface. Two sub-regimes may be observed, depending on the location and topology of the reaction zone.
    - A "closed external" combustion regime when a single flame front, mostly premixed, manages to engulf the droplets and their corresponding amount of evaporated gaseous

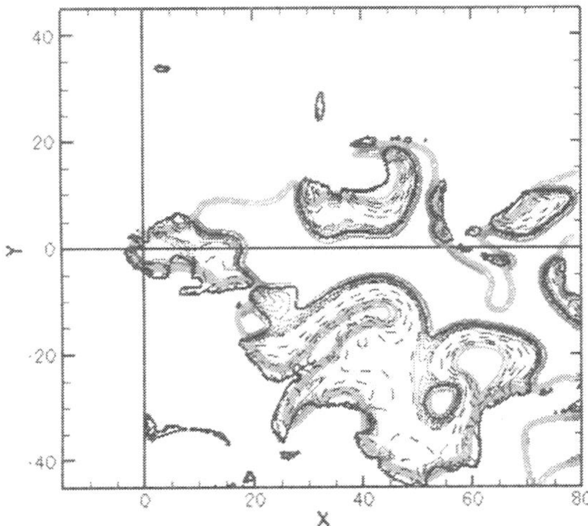

Figure 17. Example of percolating flames obtained from DNS of a turbulent V-shape flame.

fuel to transform it into products (Figure 18a).

- An "open external" combustion regime when two reaction zones develop on each side of the central jet (see Figure 18b).

2. **"Group" combustion:** The droplets are organized in several groups, with flames independently consuming each cluster. Both rich premixed and diffusion flames are observed (see Figure 18c).
3. **"Hybrid" combustion:** This regime is a combination of the two previous ones. The premixed flames are burning in a group combustion mode, whereas the diffusion flames cannot percolate between the clusters of droplets because of the overly rich environment. The leftover fuel is burnt with the co-flowing oxidizer in an additional external diffusion flame (see Figure 18d).

These regimes, and others not detailed here, may be organized into a combustion diagram, and at least three directions are necessary to classify them. The first concerns the ratio τ_v/τ_f between the characteristic evaporation and flame times. Another important ratio, δ_s/δ_f, is built from the mean distance between the droplets and the flame thickness, the droplet interspace δ_s being inversely related to d^I, the spray dilution. Finally, Φ^I_{L0} the equivalence ratio measured within the spray jet, is also needed to account for the amount of oxidizer entrained in the core of the central jet.

To start the classification, the pioneering analysis of Borghi (1996a) is followed. If the evaporation delay of injected droplets is very small compared to the flame characteristic time ($\tau_v/\tau_f \ll 1$), then purely gaseous combustion occurs. Therefore, when $\tau_v/\tau_f \ll 1$ the droplets are vaporized far away from the flame front and combustion develops in a fully gaseous mode.

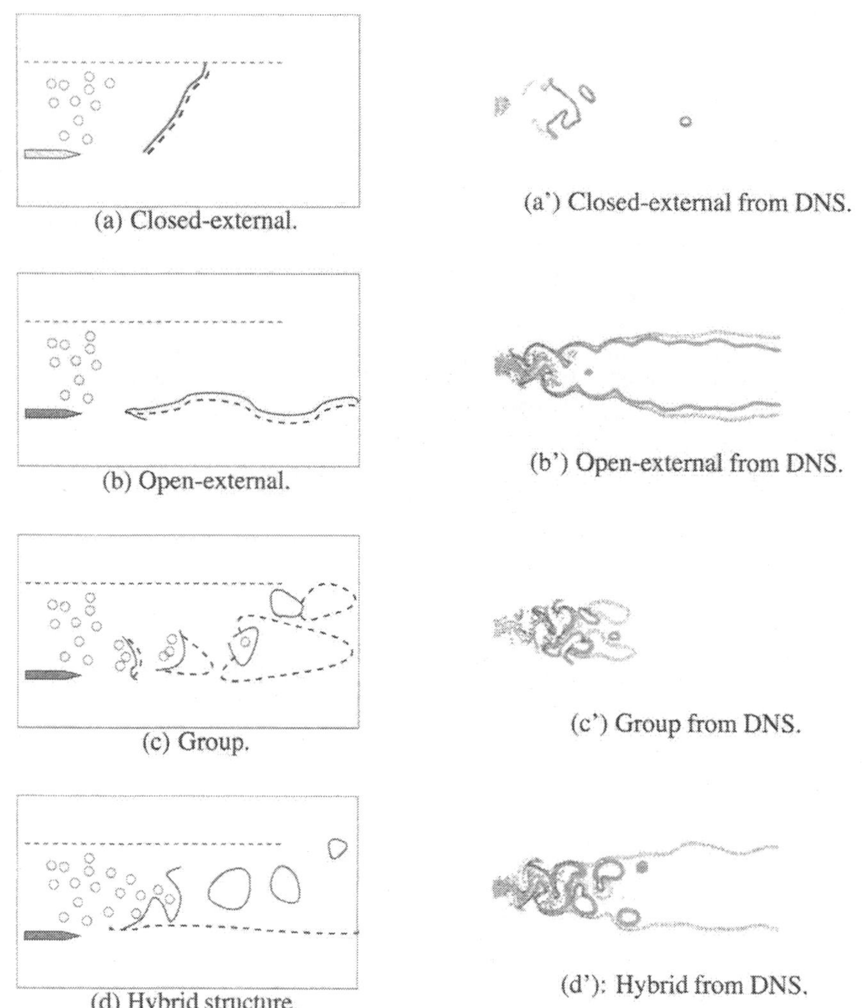

(a) Closed-external.

(a') Closed-external from DNS.

(b) Open-external.

(b') Open-external from DNS.

(c) Group.

(c') Group from DNS.

(d) Hybrid structure.

(d'): Hybrid from DNS.

Figure 18. Summary of generic flames structures. Orange: Premixed burning. Blue: Diffusion burning.

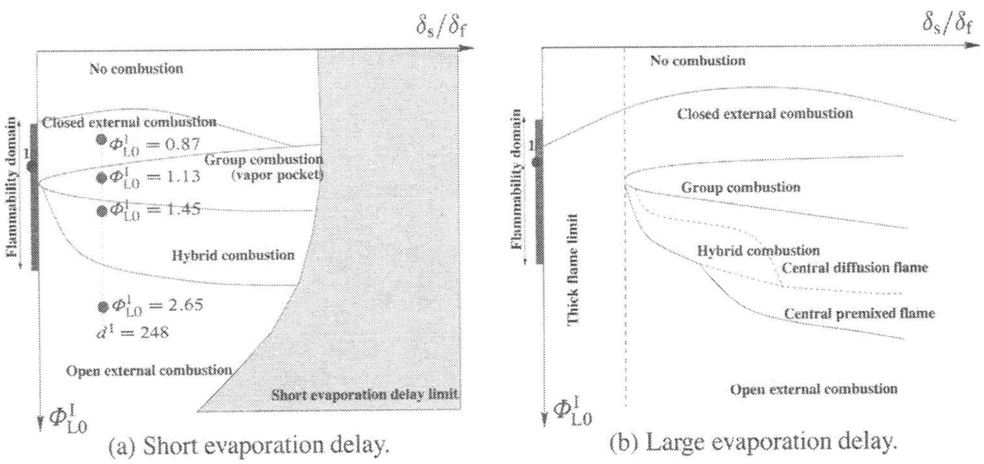

Figure 19. Two-dimensional diagrams for diluted spray combustion. Horizontal planes (τ_v/τ_f = constant) of a three-dimensional diagram.

The jet is thus a mixture of fuel vapor and oxidizer whose equivalence ratio depends on the initial droplet loading. A tentative combustion diagram for this partially premixed jet is presented in Figure 19a, where the six configurations with a short evaporation delay appear. The main combustion modes discussed above are recovered.

To discuss Figure 19a, it is chosen to travel in a direction where Φ^I_{LO} increases. This is done starting from $\Phi^I_{LO} = 0.87$ up to $\Phi^I_{LO} = 2.65$. In the leanest case, a double flame is found. It is wrinkled but remains continuous around the fuel issued from the spray evaporation, corresponding to a "closed external" combustion regime, which is similar to a Bunsen flame. As the injected equivalence ratio increases, local fluctuations of mixture fraction appear because of mixing of the fuel stream with the surrounding oxidizer. For a slightly rich injection $\Phi^I_{LO} = 1.13$, vapor rich areas (or pockets) burn first in a premixed regime that consumes the carrying oxidizer, followed by diffusion flame rings organized between the remaining fuel and the external oxidizer. If the equivalence ratio increases $\Phi^I_{LO} = 1.45$, there is still premixed group combustion in the core of the jet to burn all the carrying oxidizer. However, enough fuel remains to maintain an external diffusion flame, leading to the "hybrid" combustion regime. If the equivalence ratio increases again $\Phi^I_{LO} = 2.65$, the central vapor/oxidizer mixture is too rich to allow for the propagation of a premixed flame and both premixed and non-premixed fronts are pushed away from the core of the spray, to burn in the "open external" combustion regime, which is very similar to a gaseous non-premixed jet flame.

When the evaporation delay increases ($\tau_v/\tau_f \leq 1$), droplets may reach the flame front but it is unlikely that they will cross the burning zone without being fully evaporated. However, because they may interact with the turbulent structures while releasing their vapor, the topology of the gaseous fuel may be highly non-uniform in most cases. The corresponding combustion diagram is proposed in Figure 19b. Within the "hybrid" combustion regime, two specific flame structures, issuing from the droplets that have been flushed away from the carrying oxidizer stream, may be observed with either central diffusion flames or even, for richer mixtures, central premixed flames.

9 Acknowledgments

Numerous reflections, procedures and results presented in this text are the outcome of work carried out in collaboration with other researchers whom I would like to thank deeply and, in particular, F. X. Demoulin, M. Massot and C. Pera for their collaboration and their patience. I also wish to express my full gratitude to R. O. Fox and D. L. Marchisio for their support, encouragement and efficiency.

Bibliography

A. Aliseda, A. Cartellier, F. Hainaux, and J. C. Lasheras (2002). Effect of preferential concentration on the settling velocity of heavy particles in homogeneous isotropic turbulence. *Journal of Fluid Mechanics*, 468:77–105.

M. Baum, T. J. Poinsot, and D. Thévenin (1994). Accurate boundary conditions for multicomponent reactive flows. *Journal of Computational Physics*, 116:247–261.

R. Borghi (1996a). Background on droplets and sprays. In *Combustion and Turbulence in Two-Phase Flows, Lecture Series 1996-02*. Von Karman Institute for Fluid Dynamics.

R. Borghi (1996b). *The links between turbulent combustion and spray combustion and their modelling*, pages 1–18. In , Chang (1996). ISBN 1560324562.

R. Borghi and M. Champion, editors (2000). *Modélisation et théorie des flammes*. Technip. ISBN 2-7108-0758-0.

S. Candel, F. Lacas, N. Darabiha, and C. Rolon (1999). Group combustion in spray flames. *Multiphase Science and Technology*, 11:1–18.

S. H. Chang, editor (1996). *Transport Phenomena in Combustion*. Taylor and Francis. ISBN 1560324562.

N. Chigier (1983). Group combustion models and laser diagnostic methods in sprays: a review. *Combustion and Flame*, 51:127–139.

H. H. Chiu and E. J. Croke (1981). Group combustion of liquid fuel sprays. Technical report, Energy Technology Lab Report 81-2, University of Illinois at Chicago.

H. H. Chiu, H. Y. Kim, and E. J. Croke (1982). Internal group combustion of liquid droplets. In The Combustion Institute, editor, *Proceedings of the Nineteenth Symposium (International) on Combustion*.

H. H. Chiu and T. M. Liu (1977). Group combustion of liquid droplets. *Combustion Science and Technology*, 17:127–131.

C. Crowe, M. Sommerfeld, and Y. Tsuji, editors (1998). *Multiphase Flows with Droplets and Particles*. CRC Press. ISBN 0-8493-9469-4.

S. Elgobashi and G. C. Truesdell (1992). Direct numerical simulation of particle dispersion in a decaying isotropic turbulence. *Journal of Fluid Mechanics*, 242:655–700.

J. R. Fessler, J. D. Kulick, and J. K. Eaton (1994). Preferential concentration of heavy particles in turbulent channel flow. *Physics of Fluids*, (6):3742–3749.

P. Givi (1989). Model free simulations of turbulent reactive flows. *Progress in Energy and Combustion Science*, 15:1–107.

D. G. Goodwin (2003). An open source, extensible software suite for CVD process simulation. In *Proceedings of CVD XVI and EuroCVD Fourteen*. Electrochemical Society, 155–162.

G. Gouesbet and A. Berlemont (1999). Eulerian and Lagrangian approaches for predicting the behaviour of discrete particles in turbulent flows. *Progress in Energy and Combustion Science*, 25:133–159.

C. W. Hirt and B. D. Nichols (1981). Volume of fluid (VOF) - method for the dynamics of free boundaries. *Journal of Computational Physics*, 39:201–225.

A. R. Kerstein and C. K. Law (1982). Percolation in combustion sprays. I: Transition from cluster combustion to percolate combustion in non-premixed sprays. In The Combustion Institute, editor, *Proceedings of the Nineteenth Symposium (International) on Combustion* .

K. K. Kuo, editor (1986). *Principles of Combustion*. John Wiley and Sons. ISBN 0-471-62605-8.

S. Lee, K. Lele, and P. Moin (1991). Numerical simulations of spatially evolving compressible turbulence. *Center for Turbulence Research, Annual Research Briefs, Stanford*, (126).

S. K. Lele (1992). Compact finite difference schemes with spectral like resolution. *Journal of Computational Physics*, (103):16–42.

T. H. Lin, C. K. Law, and S. H. Chung (1988). Theory of laminar flame propagation in off-stoichiometric dilute spray. *International Journal of Heat and Mass Transfer*, 31(5):1023–1034.

T. H. Lin and Y. Y. Sheu (1991). Theory of laminar flame propagation in near stoichiometric dilute spray. *Combustion and Flame*, 84:333–342.

W. Ling, J. N. Chung, T. R. Troutt, and C. T Crowe (1998). Direct numerical simulation of a three-dimensional temporal mixing layer with particle dispersion. *Journal of Fluids Mechanics*, (358):61–85.

F. Mashayek (1998). Direct numerical simulations of evaporating droplet dispersion in forced low Mach number turbulence. *International Journal of Heat and Mass Transfer*, 41(17):2601–2617.

F. Mashayek, F. A. Jaberi, R. S. Miller, and P. Givi (1997). Dispersion and polydispersity of droplets in stationary isotropic turbulence. *International Journal of Multiphase Flow*, 23(2): 337–355.

P. A. McMurtry, W. H. Jou, J. J. Riley, and R. W. Metcalfe (1986). Direct numerical simulation of a reacting mixing layer with chemical heat release. *AIAA Journal*, 24(6):962–970.

R. S. Miller and J. Bellan (1999). Direct numerical simulation of a confined three-dimensional gas mixing layer with one evaporating hydrocarbon-droplet-laden stream. *Journal of Fluids Mechanics*, 384:293–338.

R. S. Miller and J. Bellan (2000). Direct numerical simulation and subgrid analysis of a transitional droplet laden mixing layer. *Physics of Fluids*, 12(3):650–671.

Y. Mizobuchi, S. Tachibana, J. Shinjo, S. Ogawa, and T. Takeno (2002). A numerical analysis on structure of turbulent hydrogen jet lifted flame. *Proceedings of the Combustion Institute*, 29.

S. A. Orszag and G. S. Patterson (1972). Numerical simulation of three-dimensional homogeneous isotropic turbulence. *Physical Review Letters*, 28(2):76–79.

Osher and Sethian (1988). Fronts propagating with curvature-dependent speed: Algorithms based on Hamilton-Jacobi formulations. *Journal of Computational Physics*, 79:12–49.

C. Pantano, S. Sarkar, and F. A. Williams (2003). Mixing of a conserved scalar in a turbulent reacting shear layer. *Journal of Fluid Mechanics*, 481:291–328.

M. Sichel P. B. Patil and J. A. Nicolls (1978). Analysis of spray combustion in a research turbine combustor. *Combustion Science and Technology*, 18:21–31.

N. Peters (2000). *Turbulent Combustion*. Cambridge University Press.

T. Poinsot, S. Candel, and A. Trouvé (1996). Direct numerical simulation of premixed turbulent combustion. *Progress in Energy and Combustion Science*, 12:531–576.

T. Poinsot and S. K. Lele (1992). Boundary conditions for direct simulations of compressible viscous flows. *Journal of Computational Physics*, 1(101):104–129.

T. Poinsot and D. Veynante (2001). *Theoretical and Numerical Combustion*. Edwards. ISBN 1-930217-05-6.

T. Poinsot and D. Veynante (2005). *Theoretical and Numerical Combustion, second edition*. Edwards. ISBN 978-1930217102.

M. W. Reeks (1991). On a kinetic equation for the transport of particles in turbulent flows. *Physics of Fluids*, 3(3):446–456.

J. Reveillon, K. N. C. Bray, and L. Vervisch (1998). Dns study of spray vaporization and turbulent micro-mixing. In *AIAA 98-1028*, 36th Aerospace Sciences Meeting and Exhibit, January 12-15, Reno NV.

J. Reveillon and F. X. Demoulin (2006). Evaporating droplets in turbulent reacting flows. In *Proceedings of the 31th International Symposium on Combustion* , Heidelberg, Germany.

J. Reveillon and L. Vervisch (2000). Accounting for spray vaporization in non-premixed turbulent combustion modeling: A single droplet model (sdm). *Combustion and Flame*, 1(121): 75–90.

J. Reveillon and L. Vervisch (2005). Analysis of weakly turbulent diluted-spray flames and spray combustion regimes. *Journal of Fluid Mechanics*, (537):317–347.

G. A. Richards and P. E. Sojka (1990). A model of H_2-enhanced spray combustion. *Combustion and Flame*, 79:319–332.

J. J. Riley and G. S. Patterson (1974). Diffusion experiments with numerically integrated isotropic turbulence. *Physics of Fluids*, 17:292–297.

R. S. Rogallo (1981). Numerical experiments in homogeneous turbulence. Technical Report, NASA. Technical Memorandum No. 81315.

M. Samimy and S. K. Lele (1991). Motion of particles with inertia in a compressible free shear layer. *Physics of Fluids*, 8(3):1915–1923.

I. Silverman, J. B. Greenberg, and Y. Tambour (1991). Asymptotic analysis of a premixed polydispersed spray flame. *SIAM Journal of Applied Mathematics*, 51(5):1284–1303.

I. Silverman, J. B. Greenberg, and Y. Tambour (1993). Stoichiometry and polydisperse effects in premixed spray flames. *Combustion and Flame*, 118:97–118.

O. Simonin, P. Fevrier, and J. Laviéville (1993). On the spatial distribution of heavy-particle velocities in turbulent flow: from continuous field to particulate chaos. *Journal of Turbulence*, 118:97–118.

K. D. Squires and J. K. Eaton (1991). Preferential concentration of particles by turbulence. *Physics of Fluids*, 3(5):1169–1178.

T. Suzuki and H. H. Chiu (1971). Multi droplet combustion on liquid propellants. In *Proceedings of the Ninth International Symposium on Space Technology Science*, pages 145–154, Tokyo, Japan, AGNE.

S. Tanguy and A. Berlemont (2005). Application of a level set method for simulation of droplet collisions. *International Journal of Multiphase Flow*, 31(9):1015–1035.

L. Vervisch and T. Poinsot (1998). Direct numerical simulation of non-premixed turbulent flame. *Annual Reviews of Fluid Mechanics*, (30):655–692.

L. Vervisch-Guichard, J. Reveillon, and R. Hauguel (2001). Stabilization of a turbulent premixed flame. In TSFP, editor, *Proceedings of the Second Symposium on Turbulent Shear Flow Phenomena, Stockholm.*

L. P. Wang and M. R. Maxey (1993). Settling velocity and concentration distribution of heavy particles in homogeneous isotropic turbulence. *Journal of Fluid Mechanics*, 256:27–68.

Y. Wang and C. J. Rutland (2005). Effects of temperature and equivalence ratio on the ignition of n-heptane fuel droplets in turbulent flows. *Proceedings of the Combustion Institute*, 30.

F. A. Williams (1960). Progress in spray combustion analysis. In The Combustion Institute, editor, *Proceedings of the 8th Symposium (International) on Combustion*.

F. A. Williams (1985). *Combustion Theory*. Addison Wesley Publishing Company; 2nd edition.

A. A. Wray (1990). Minimal storage time-advancement schemes for spectral methods. Technical report, Center for Turbulence Research Report, Stanford University.

H. Yamashita, M. Shimada, and T. Takeno (1996). A numerical study on flame stability at the transition point of jet diffusion flames. *Proceedings of the Combustion Institute*, 26:27–34.

Zeitfracht Medien GmbH
Ferdinand-Jühlke-Straße 7
99095 Erfurt, Deutschland
produktsicherheit@kolibri360.de